SYMPOSIA OF THE
SOCIETY FOR EXPERIMENTAL BIOLOGY

NUMBER XLV

PROCEEDINGS OF A MEETING
HELD AT THE UNIVERSITY OF GLASGOW,
SCOTLAND
28–31 AUGUST 1990

SYMPOSIA OF THE
SOCIETY FOR EXPERIMENTAL BIOLOGY

I	Nucleic Acid
II	Growth, Differentiation and Morphogenesis
III	Selective Toxicity and Antibiotics
IV	Physiological Mechanisms in Animal Behaviour
V	Fixation of Carbon Dioxide
VI	Structural Aspects of Cell Physiology
VII	Evolution
VIII	Active Transport and Secretion
IX	Fibrous Proteins and their Biological Significance
X	Mitochondria and other Cytoplasmic Inclusions
XI	Biological Action of Growth Substances
XII	The Biological Replication of Macromolecules
XIII	Utilization of Nitrogen and its Compounds by Plants
XIV	Models and Analogues in Biology
XV	Mechanisms in Biological Competition
XVI	Biological Receptor Mechanisms
XVII	Cell Differentiation
XVIII	Homeostasis and Feedback Mechanisms
XIX	The State and Movement of Water in Living Organisms
XX	Nervous and Hormonal Mechanisms of Integration
XXI	Aspects of the Biology of Ageing
XXII	Aspects of Cell Motility
XXIII	Dormancy and Survival
XXIV	Control of Organelle Development
XXV	Control Mechanisms of Growth and Differentiation
XXVI	The Effects of Pressure on Organisms
XXVII	Rate Control of Biological Processes
XXVIII	Transport at the Cellular Level
XXIX	Symbiosis
XXX	Calcium in Biological Systems
XXXI	Integration of Activity in the Higher Plant
XXXII	Cell–Cell Recognition
XXXIII	Secretory Mechanisms
XXXIV	The Mechanical Properties of Biological Materials
XXXV	Prokaryotic and Eukaryotic Flagella
XXXVI	The Biology of Photoreceptors
XXXVII	Neural Origin of Rhythmic Movements
XXXVIII	Controlling Events in Meiosis
XXXIX	Physiological Adaptations of Marine Animals
XL	Plasticity in Plants
XLI	Temperature and Animal Cells
XLII	Plants and Temperature
XLIII	Mucus and Related Topics
XLIV	Hormone Perception and Signal Transduction in Animals and Plants

SYMPOSIA OF THE
SOCIETY FOR EXPERIMENTAL BIOLOGY

NUMBER XLV

MOLECULAR BIOLOGY OF PLANT DEVELOPMENT

EDITED BY

GARETH I. JENKINS

Department of Biochemistry, University of Glasgow, Glasgow, Scotland

AND WOLFGANG SCHUCH

ICI Seeds, Jealott's Hill Research Station, Bracknell, UK

Published for the Society for Experimental Biology
by The Company of Biologists Limited
Department of Zoology, University of Cambridge
Downing Street, Cambridge CB2 3EJ

Typeset, Printed and Published by The Company of Biologists Limited,
Department of Zoology, University of Cambridge
Downing Street, Cambridge CB2 3EJ

ISBN 0 948601 31 0
ISSN 0081–1386

A CIP Catalogue record for this book is available from The British Library

SOCIETY FOR EXPERIMENTAL BIOLOGY SYMPOSIA

SEB Symposia form a long-standing series of volumes first published in 1947. The series is annual and each publication is a collection of authoritative articles on an aspect of modern experimental Biology. The contributors are all invited and speak on specific topics within the chosen field. Meetings are held annually, in September, over a period of two or three days.

The aims of the Symposium series are to stimulate discussion and communication between scientists of all nationalities, to foster the development of, and research on, modern aspects of plant, animal, and cell biology.

The cover photograph is an *apetala1* mutant of *Arabidopsis thaliana*. Photograph courtesy of Norman Tait, University of Glasgow.

CONTENTS

PAGE

Preface   i

KOORNNEEF, M. Isolation of higher plant developmental mutants   1

CHORY, J., AGUILAR, N. AND PETO, C. A. The phenotype of *Arabidopsis thaliana det*1 mutants suggests a role for cytokinins in greening   21

COVE, D. J., KAMMERER, W., KNIGHT, C. D., LEECH, M. J., MARTIN, C. R. AND WANG, T. L. Developmental genetic studies of the moss, *Physcomitrella patens*   31

HAUGE, B. M., HANLEY, S., GIRAUDAT, J. AND GOODMAN, H. M. Mapping the *Arabidopsis* genome.   45

GRILL, E. AND SOMERVILLE, C. Development of a system for efficient chromosome walking in *Arabidopsis*   57

DEAN, C., SJODIN, C., BANCROFT, I., LAWSON, E., LISTER, C., SCOFIELD, S. AND JONES, J. Development of an efficient transposon tagging system in *Arabidopsis thaliana* . .   63

MARKS, M. D., ESCH, J., HERMAN, P., SIVAKUMARAN, S. AND OPPENHEIMER, D. A model for cell-type determination and differentiation in plants   77

BOWMAN, J. L. AND MEYEROWITZ, E. M. Genetic control of pattern formation during flower development in *Arabidopsis*   89

SCHUCH, W. Using antisense RNA to study gene function   117

LAZARUS, C. M., NAPIER, R. M., YU, L.-X., LYNAS, C. AND VENIS, M. A. Auxin-binding protein – antibodies and genes   129

BLEEKER, A. B. Genetic analysis of ethylene responses in *Arabidopsis thaliana*   149

CHEN, Q., BRGLEZ, I. AND BOSS, W. F. Inositol phospholipids as plant second messengers   159

FRICKER, M. D., GILROY, S., READ, N. D. AND TREWAVAS, A. J. Visualisation and measurement of the calcium message in guard cells   177

WEIßHAAR, B., BLOCK, A., ARMSTRONG, G. A., HERRMANN, A., SCHULZE-LEFERT, P. AND HAHLBROCK, K. Regulatory elements required for light-mediated expression of the *Petroselinum crispum* chalcone synthase gene.   191

SCHINDLER, U., ECKER, J. R. AND CASHMORE, A. R. An *Arabidopsis thaliana* G-box-binding protein similar to the wheat leucine zipper protein identified as HBP-1 . .   211

BARNES, S. R. RFLP analysis of complex traits in crop plants   219

McCORMICK, S., TWELL, D., VANCANNEYT, G. AND YAMAGUCHI, J. Molecular analysis of gene regulation and function during male gametophyte development   229

MAU, S.-L., ANDERSON, M. A., HEISLER, M., HARING, V., McCLURE, B. A. AND CLARKE, A. E. Molecular and evolutionary aspects of self-incompatibility in flowering plants.   245

MARIANI, C., GOLDBERG, R. B. AND LEEMANS, J. Engineered male sterility in plants. .   271

Index of Subjects   281

Molecular Biology of Plant Development

PREFACE

Aspects of development have fascinated plant scientists for centuries. The early plant anatomists were intrigued by developmental changes in the structure of plant tissues and organs and their work effectively laid the foundations of plant developmental biology. Later, as the ideas of inheritance and evolution gained acceptance, the significance of the genetic makeup of the individual in specifying the developmental programme was realised. Plant physiologists became interested in the regulation of developmental transitions, such as germination and flowering, and they demonstrated the importance of interactions between the genetic programme and environmental variables in the control of growth and differentiation. These early studies revealed the complexity of plant development. One only has to study, for example, the pattern of organ formation during flower morphogenesis to appreciate the remarkable temporal and spatial organisation of developmental changes. Moreover, a consideration of the number of factors that can induce flowering in different plant species illustrates the complex ways plant growth regulators and environmental variables can interact in different genotypes to control developmental processes.

Despite the evident difficulty of some of the problems facing plant developmental biologists, many advances have been made in recent years in understanding the molecular and cellular basis of differentiation, morphogenesis and the regulation of developmental transitions. Undoubtedly, much of this progress has resulted from the application of recombinant DNA technology coupled with the ability to produce transgenic plants. A significant number of plant genes have been cloned and changes in their expression analysed during the differentiation of various tissues and organs. Increasingly, attempts are being made to correlate changes in gene expression with morphogenetic and physiological processes in the plant. However, whereas many of the genes studied to date are expressed as a consequence of developmental changes, we have only recently entered the era where genes controlling development are being isolated. Further progress promises to be rapid, as a powerful combination of genetic and molecular approaches is being applied to problems in plant development enabling important genes to be cloned knowing only their mutant phenotypes.

It was therefore timely for the Society for Experimental Biology to choose *The Molecular Biology of Plant Development* as the subject of its 1990 Symposium. The meeting was held in Glasgow from 28th to 31st August, and was attended by over 230 participants. The delegates were addressed by 28 speakers of international standing, and they were able to peruse 70 poster contributions. This book is based on the proceedings of that meeting and we hope that readers will find its chapters reveal both how existing techniques have provided considerable insights into plant development and how new approaches promise dramatic findings in the next few years.

The molecular biology of plant development is a broad topic and indeed our intention, in organising the Symposium, was to show how current research is providing insights into a wide range of developmental problems. Hence the Symposium discussions ranged from the identification of genes controlling aspects of morphogenesis to the investigation of cellular components concerned with the detection and transduction of signals that initiate developmental transitions. Lectures were included on appropriate aspects of animal development because some of the ground now being covered by plant developmental biologists has already been trodden by our animal counterparts. It should be stated, however, that the rapid progress that has been made in several areas of plant development means that we are no longer the poor relations of the animal developmental biologists. The availability of excellent plant developmental systems has made a substantial contribution to progress and throughout the meeting one plant received more attention than its insignificant form would seem to justify: *Arabidopsis thaliana*. This small plant has become a model system for developmental research because it possesses features that facilitate genetic and molecular studies. It has a small genome with little interspersed repetitive DNA, is easy to grow, has a relatively short life-cycle and is amenable to transformation by *Agrobacterium tumefaciens*. Furthermore, *Arabidopsis* is excellent for mutant production and numerous mutants have now been described, many of which have been located on a genetic map. Research on *Arabidopsis* is therefore well represented in this book. The final part of the meeting focussed on aspects of research relevant to crop improvement, since it is evident that some of the fundamental discoveries being made will have applications in this domain. The chapters concerned with these different aspects are outlined briefly below.

In recent years plant scientists have started to tackle the problems of development with approaches that have proved successful in studies of animal development. In particular, two strategies are currently being developed to identify and clone plant genes: chromosome walking based on the known genetic map locations of mutant genes and gene tagging by insertional mutagenesis. The basis of the chromosome walking strategy is the availability of known mutants altered in specific developmental processes. At the Symposium Dr Michael Akam (Cambridge, UK) discussed how the production of mutants has facilitated the analysis of development in the fruit fly, *Drosophila*. In this volume Koornneef discusses the approaches available to isolate plant developmental mutants and the problems associated with mutant characterisation in higher plants. The ability to generate fascinating mutants for genetic analysis is demonstrated by Chory *et al.* who describe *Arabidopsis* mutants altered in the light-mediated regulation of leaf and chloroplast development. Further interesting mutants altered in flower morphogenesis (Bowman and Meyerowitz), trichome formation (Marks *et al.*) and ethylene response (Bleecker) are described in later chapters. Certain lower plants, of course, are well suited to the production of mutants because they are haploid for much of their life cycle, and Cove and co-workers have developed a genetic approach to analyse development of the moss *Physcomitrella patens*. Their chapter indicates that it is now possible to transform *Physcomitrella*, raising the prospect that gene tagging and the complementation of mutants by cloned DNA molecules may now be possible.

The strategy for cloning a gene based on its mutant phenotype requires the genetic mapping of the mutant allele and the ability to integrate the genetic map with a physical map of the genome. Restriction fragment length polymorphisms (RFLPs; see the chapter by Barnes) provide markers that can link the genetic map with a physical map of the genome represented by a genomic library of large, ordered DNA fragments. To clone a gene of interest, closely linked RFLP markers are used to select cloned genomic DNA fragments which represent the starting point for a walk towards the gene. The completion of the walk is determined by successful complementation of the mutant by transformation with a defined DNA fragment cloned from the wild-type. This approach is being used successfully in the nematode *Caenorhabditis*, and Dr John Sulston (Cambridge, UK) was able to give the Symposium delegates an indication of the progress that might now be expected in plant systems. In this volume Hauge *et al.* discuss the application of this approach in *Arabidopsis* and describe progress in producing an RFLP map integrated with an ordered physical map of the *Arabidopsis* genome. Grill and Somerville describe the generation of a library of *Arabidopsis* genomic DNA cloned into yeast artificial chromosome vectors (YACs), which are able to carry particularly large DNA fragments (at least 500 kb). The success of the chromosome walking approach in cloning genes important in development is described in a later chapter by Bleecker.

The strategy of gene tagging has also resulted in the successful cloning of genes controlling aspects of plant development. Dean *et al.* discuss the basis of this technique and describe the development of a tagging system based on transposable elements. This approach is impressively sophisticated and promises to generate numerous important mutants in the future. However, the successes to date have all resulted from insertional mutagenesis by T-DNA, introduced into *Arabidopsis* by *Agrobacterium*. Marks *et al.* describe the cloning of a gene responsible for the *glabra1* mutation which results in a lack of trichomes, and Bowman and Meyerowitz report the cloning of *AGAMOUS*, a gene involved in controlling flower morphogenesis in *Arabidopsis*. Both of these genes encode putative transcription factors. Bowman and Meyerowitz further discuss the known mutants altered in flower development in *Arabidopsis* and *Antirrhinum* and present a model to account for the genetic control of this process.

A further approach that is being used increasingly to investigate the function of cloned genes is expression of antisense copies of genes in transgenic plants. The presence of antisense transcripts has been shown to impair the accumulation of normal, sense transcripts of the corresponding gene resulting in a decrease in synthesis of its protein product and a consequent loss of function. This approach has now been used to ascribe functions to previously unidentified, cloned genes. Schuch discusses the application of the antisense approach and describes some of the successes resulting from it. Since the isolation of genes from cDNA libraries prepared from tissues at particular stages of development remains an important way of cloning plant genes, often of unknown function, antisense technology is an important part of the armoury of techniques now available to the plant molecular biologist. The use of the more traditional cDNA cloning techniques is represented in this volume in the work of McCormick *et al.*, Mau *et al.* and others (see below) and was much in evidence in the posters presented at the Symposium. Recent advances in cDNA cloning techniques, employing subtractive hybridisation and the polymerase chain reaction, will undoubtedly lead to the isolation of many important genes previously inaccessible because of their low abundance or restricted localisation in small amounts of tissue.

As mentioned above, the second major theme of the Symposium was the discussion of cellular processes concerned with the detection of external stimuli and the transduction of these signals to initiate changes in gene expression and other cellular responses. Such stimuli include environmental factors, such as light, but also the plant growth regulators auxins, cytokinins, ethylene, abscisic acid and gibberellins. In animal cells the receptors for several external 'agonists' are well characterised and much is known about the cellular signalling pathways that serve to amplify and transduce the signals. This topic was introduced at the Symposium by Dr Michaël Wakelam (Glasgow, UK). However, much less is known about plant receptors and signal transduction processes. The only receptor which has been studied in great detail is the photoreceptor phytochrome. In this volume, Lazarus *et al.* describe progress in the biochemical characterisation of putative auxin receptors and report the cloning of a possible receptor. Bleecker has cloned a gene, by chromosome walking, that is likely to be the site of the defect in the *etr* mutant of *Arabidopsis*, which has much reduced sensitivity to ethylene and may be deficient in the ethylene receptor. The cloned gene resembles a trans-membrane protein kinase. This impressive piece of work shows that the new approaches promise real progress in one of the most intractable aspects of plant developmental physiology: the mechanism of action of plant growth regulators. Aspects of signal transduction are discussed by Chen *et al.* and Fricker *et al.* Chen *et al.* show that the inositol phospholipid second messenger pathway, which is responsible for many aspects of signal transduction in animal cells, is unlikely to be so ubiquitously important in plants, and discuss alternative mechanisms of signal transduction. Fricker *et al.* emphasise the importance of changes in intracellular calcium ion concentration in plant signal transduction.

The result of the detection and transduction of external signals is often a change in the rate of transcription of specific genes, which can be of dramatic importance in development. The perception of changes in the light environment, for example, triggers a wide range of developmental responses in plants, such as leaf and chloroplast development, stem extension and, in many species, flowering. It is therefore important to identify the DNA sequence elements of individual genes that are required for the regulation of transcription in response to specific signals. Moreover, it is important to identify the sequence-specific DNA-binding proteins that interact with the sequence elements to control transcription. Animal molecular biologists have been working on these questions for some time and Dr Richard Treisman (London, UK) presented a paper on the human serum response factor, which controls transcription of several genes. In plants there has been much progress in this area in recent years and several plant transcription factors have now been cloned and sequenced. The chapter by Weisshaar *et al.* shows how complex the transcriptional control elements of particular genes may be, in this case the parsley chalcone synthase gene. The transcription of this gene is regulated by several factors, notably UV light and fungal elicitors. Cashmore, who has studied the regulatory sequence elements of photoregulated genes in detail, reports the cloning of a putative transcription factor.

The final chapters of the book are concerned with aspects of development relevant to crop improvement. The first of these chapters, by Barnes, discusses the principles of RFLP mapping and the application of this technology to the analysis of complex traits in crop plants. McCormick *et al.* describe the functional analysis of a series of genes that are expressed specifically in pollen, using expression in transgenic plants to analyse the regulatory regions of the genes. The chapter by Mau *et al.* is also concerned with plant reproduction, in this case with genes that are expressed in parts of the female reproductive organs and which are involved in self-incompatibility. The identification of such gene products is clearly of interest in the understanding of plant reproduction and the development of the reproductive organs, but it is likely to have applications in the manipulation of reproductive systems in plant breeding. The chapter by Mariani *et al.* demonstrates that this is not a far-fetched notion in that male-sterile plants have been produced by genetic manipulation.

In conclusion, we would like to thank those organisations and individuals who made the Symposium possible. The Society for Experimental Biology provided funding for the meeting, but additional financial contributions enabling us to invite speakers from leading laboratories throughout the world were generously made by the following organisations: ICI Seeds, The Agricultural and Food Research Council, The Royal Society, The Gatsby Charitable Foundation, The Scottish Development Agency, Unilever Research, Schering Agrochemicals Ltd, Advanced Technologies (Cambridge) Ltd, Agricultural Genetics Company Ltd and Shell Research Ltd. In addition, several organisations contributed to other costs of organising the meeting, to whom we are very grateful: The City of Glasgow District Council, Strathclyde Regional Council, Town and Country Refrigeration Ltd, John Wiley and Sons Ltd and Boehringer Mannheim UK. We would also like to thank all those individuals in the University of Glasgow and elsewhere who helped in the organisation of the meeting, in particular Dr Richard Firn, the SEB Plant Biology Secretary, for his invaluable advice, encouragement and hard work, Dr Len Evans, the SEB treasurer, for financial advice, and Vicky Wragg of the SEB for her administrative skills. The publication of this book was facilitated by Dr Richard Skaer of the Company of Biologists Ltd, to whom we are also grateful.

Gareth I. Jenkins
Wolfgang Schuch
December 1990

Printed in Great Britain © *Society for Experimental Biology 1991*

ISOLATION OF HIGHER PLANT DEVELOPMENTAL MUTANTS

MAARTEN KOORNNEEF

Department of Genetics, Wageningen Agricultural University, Dreijenlaan 2, NL-6703 HA Wageningen, The Netherlands

Summary

A variety of biological restraints limit the number of plant species suitable for genetic and molecular analysis of development. A number of approaches to isolate developmental and physiological mutants are described as well as complications that arise in mutant selection procedures. Examples of mutant types especially in *Arabidopsis* are presented for a number of processes.

Introduction

Plant development is a broad term, described by Steeves and Sussex (1989) as the sum total of events that contribute to the progressive elaboration of the plant body. According to this description the emphasis in studies of development is on irreversible changes in the morphology of the plant during its life cycle. These morphogenetic changes are assumed to be induced by both internal and external signals that change the fate of cells in the meristems, including modulation of gene expression patterns. In contrast to animals, many morphogenetic changes in plants occcur post-embryonically from the continuous differentiation of shoot and root meristematic cells. This aspect makes the distinction between plant physiology and plant developmental biology unclear. 'Typical' topics of plant physiology and plant biochemistry such as basic metabolism, photosynthesis, respiration, water relations, mineral uptake, tropisms etc., in general, are not considered to be subjects of plant developmental biology. However, it should be realized that an increase in size (growth), almost always coincides with morphogenesis, which in many ways depends on cell metabolism and is regulated by internal and external environmental signals. The observation that many of the known regulatory factors in plants, such as hormones and light, affect both growth and development as well as physiological processes indicates the complex interaction between these processes. The role of the plant hormone auxin illustrates this complexity since it affects cell division, a growth phenomenon; induces root primordia, a typical developmental change, where differentiation into specific tissues and cell types take place, and induces tropism: a physiological response associated with differential growth. Light is another example of a factor controlling many aspects of plant life, e.g. photosynthesis, phototropism and the evocation of flowering.

Key words: *Arabidopsis*, mutant induction, plant hormones

These considerations make the definition of developmental mutants for higher plants an arbitrary one.

The present review will emphasize those mutations that result in a phenotype that deviates morphologically from 'normal'. Mutations that affect primary metabolism (housekeeping genes) and secondary metabolism are not discussed, despite the fact that often these also affect the appearance of the plant. Mutations in secondary metabolism e.g. anthocyanin synthesis and the formation of epicuticlar waxes are restricted to specific tissues and organs and, therefore, have a developmental component. Physiological mutants, especially those affecting the interaction of the plant with its environment are included since the underlying regulatory factors in many cases also control development.

Developmental mutants can be classified according to the plant part or organ that they influence (Marx, 1983; Sheridan, 1988) or according to the role of the gene product that they affect (Meinke, 1990). This latter type of mutant may either directly affect the pattern of development or the realization of this pattern. Since it is assumed that the regulatory signals are diffusable, it is expected that many of the genes primarily controlling developmental changes are non-cell autonomous, whereas those affecting pattern expression are cell-autonomous. The difficulty of the latter classification is that it can only be used after an in depth analysis of a mutant.

Although developmental mutants have been known for many years, their analysis has not contributed very much to the understanding of plant morphogenesis. Probably the main reason for this is the black box between a describable phenotype and its underlying biochemical and physiological basis. Plant hormones and one photoreceptor, phytochrome, were, in fact, the only regulatory molecules that were known for several decades, resulting in plant hormone and phytochrome mutants being the best characterized.

Recently, the interest in plant developmental mutants has increased considerably, as procedures for their analysis at the molecular level have been developed. Mutants are one of the major tools that allow the cloning and subsequent analysis of the genes controlling plant development. Although there are effective procedures such as the differential screening of developmentally specific mRNAs, and the use of heterologous probes, even from other organisms, these are not easily applicable in the case of genes with low mRNA levels and genes that are specific for (certain) plants. The molecular procedures that 'require' a mutant phenotype controlled by a single gene are more widely applicable and have been developed in 'more advanced' multicellular organisms such as *Drosophila* and mammals. These procedures can also be effectively applied in plant species that have certain specific features such as a low DNA content per haploid genome and a low level of interspersed repetitive DNA (Meyerowitz, 1987). However, more work is necessary to build up genetic and molecular knowledge, such as the construction of genetic maps and ordered genomic libraries. Molecular techniques that use mutants for cloning genes are: (a) chromosome walking; (b) tagging; (c) deletion cloning; (d) differential screening of mutant and wild-type mRNAs. The applicability of the different procedures depends on the way the mutant was generated and the genetic knowledge about it.

An important advance in cloning a gene *via* a mutant is that one can directly observe what the effect of the gene is on plant development by comparing the mutant and wild-type phenotype. Apart from their use in cloning genes, mutants have proved helpful for the analysis of the site of gene-action by studying genetic mosaics and genotypes where the embryo genotype differs from the maternal genotype. In addition, mutants can be very useful in dissection of complicated developmental pathways. For the latter two aspects the 'classical' genetic techniques in combination with morphological and physiological descriptions are essential.

Genetic Model Systems

A number of biological restraints: amenability to cell and tissue culture, generation time, size of the plant at the seedling and adult stage, size of the genome and the amount of pre-existing genetic knowledge and material, limit the number of plant species suitable for genetic and molecular analysis of development.

Plant biologists traditionally used a wide variety of plant species, mainly selected on the basis of their suitability for a specific type of research and often (Marx, 1983) because of the ecomomic importance of a particular species. However, since the current challenge is the integrated study of complex and broad phenomena, such as development, the number of plant species used will probably be reduced, because of the restraints mentioned above. The power of focusing on one species as a model system of one group of organisms has been shown by scientists working with *Drosophila melanogaster*, *Caenorhabditis elegans*, mouse and man. *Arabidopsis thaliana* has recently emerged as a model system for basic plant research that should prove beneficial for plant science as a whole. However, a number of considerations should warn us not to throw away all other higher plant model systems: (a) Apart from many principles of plant development common to all higher plants the variation among morphogenetic mechanisms is an interesting topic of research. A maize plant obviously differs in many aspects from *Arabidopsis*. *Petunia* is much better suited for studying flower pigmentation than *Arabidopsis*. Plant microbe interactions are often very specific. (b) There are genetic peculiarities in some species, such as duplicated genes and the presence of endogenous transposons that will make them less or more suitable to search for specific mutants. The ease of detecting a certain gene from one plant species in another with molecular probes will enable an efficient switching between species. (c) Practical considerations, such as the size of the plant and its organs will remain important. Some characteristics of the most frequently used genetic model systems are listed in Table 1.

The main advantage of *Arabidopsis* is its small genome, facilitating chromosome walking. Its short generation time and low chromosome number allow efficient

Table 1. *Some characteristics of genetic model species*

Plant species	Haploid chromosome number	Haploid genome (kbp)	Genetic maps C*	Genetic maps R†	Generation time‡	Routine transformation§	Tagging applied¶
Antirrhinum	8	1.5×10^6	+	±	4 m	−	+(e)
Arabidopsis	5	7×10^4	+	+	2 m	+	+(T)
Barley	7	5.3×10^6	+	±	6 m	−	−
Maize	10	2.3×10^6	+	+	6 m	−/±	+(e)
Pea	7	4.7×10^6	+	±	5 m	±	−
Petunia	7	2×10^6	+	±	4 m	+	±(e)
Rice	12	5.8×10^5	+	+	6 m	±	−
Tomato	12	7.1×10^5	+	+	6 m	+	±(T)

* 'Classical' genetic map is available.
† RFLP map has been published (+) or is under construction.
‡ 'Practical' generation time for genetic experiments in months, in case of 'early' genotypes and depending on favourable growing conditions.
§ Transgenic plants can be obtained routinely for many genotypes (+) or is limited to specific genotypes (±) or not yet reported (−).
¶ Either tagging by T-DNA (T) or endogenous transposons (e) has been reported (+) or is being developed (±).

genetic analysis. Several efficient transformation procedures have been developed (Feldmann and Marks, 1987; Valvekens *et al.* 1988) allowing the tagging of genes by insertion of 'foreign' DNA and the complementation of mutant phenotypes by cloned DNA. A disadvangtage for some types of research is the small size of *Arabidopsis*, particularly of its seeds and seedlings. So, for seed development studies, maize (*Zea mays*), a species that probably represents the best genetically characterized plant system, might be more attractive (Sheridan, 1988). However, many of the efficient seedling selection systems (Estelle and Somerville, 1986), studies on intragenic recombination (Koornneef *et al.* 1983) and seed transformation (Feldmann and Marks, 1987) in *Arabidopsis* are only possible because of its small size and short generation time. The disadvantage of the large genome of maize is somewhat balanced by the ability to clone genes by endogenous transposons (Wessler and Hake, 1990). The limitation of a suitable transformation technique may be only temporary, since technology is developing rapidly in this field (Klein *et al.* 1990). Because transposons from species such as maize and *Antirrhinum* also appear to function in other species, the lack of endogenous transposons may be overcome.

If *Arabidopsis* and maize are considered as the 'leading' plant species for genetic studies, then tomato (*Lycopersicon esculentum*) is, in many respects an interesting alternative model for dicot species (Hille *et al.* 1989) and rice (*Oryza sativa*) for monocots. Wheat (*Triticum aestivum*) might have been included in Table 1, however, its hexaploid nature complicates mutational analysis. Some *Brassica* genotypes also have many attractive genetic features, especially the so-called 'rapid cycling' types. The use of *Petunia* and *Antirrhinum* at the moment is focused mainly on floral development and pigmentation. Apparently, pea (*Pisum sativum*) remains recalcitrant for many molecular genetic techniques because of its large genome and problematic transformability. Soybean (*Glycine max*) is another representative of a legiminous species together with the diploid *Medicago* species, although some of its problems are similar to those in pea and genetics of the species is less developed. *Nicotiana tabacum* and even more so the true diploid *N. plumbaginifolia*, together with carrot (*Daucus carota*), have as a specific advantage their amenability to cell culture, and in carrot this has led to the isolation and use of an interesting set of embryo developmental mutants (Terzi and Lo Schiavo, 1990). It was difficult to explain how recessive mutations could be isolated from diploid cell cultures of this species, however, this is now less of a problem since it was recently found that meiotic events take place during somatic embryogenesis in carrot (Ronchi *et al.* 1990) resulting in haploid cells. Starting with haploid cells also allows the isolation of recessive mutants and has resulted in the isolation of many mutants affecting nitrate reductase (e.g. Gabard *et al.* 1988).

However, the relatively poorly developed genetics of these species, the amphidiploid nature of *N. tabacum* and the long generation time of carrot severely limit the use of mutants in whole-plant studies.

Mutagenesis

Mutant induction

Mutations in higher plants can be induced by different types of mutagens, including transposable elements and the insertion of 'foreign' DNA by transformation. Genetic variation of 'spontaneous' origin can be found among cultivated and wild genotypes of a species and may appear to be due to transposon like insertions, as in the case of Mendel's *r* mutant (Bhattacharyya *et al.* 1990).

Tissue culture is also a mutagenic procedure which seems to be more popular for applied purposes (Evans, 1989) than for basic reseach. Probably the reason for this is that mutagenic procedures using chemicals and irradiation are very efficient, especially in genetic model species. The generation of this 'somaclonal variation' by tissue culture, which is hard to manipulate quantitatively, is much more laborious, and selfing is also necessary to recover recessive mutations in cases of homozygous species.

Chemical mutagens and ionizing irradiation are most commonly applied to seeds, pollen grains or tissue culture cells. The plants growing from such treated seeds, or plants pollinated with treated pollen grains are the M_1 generation from which, after selfing, M_2 seeds are harvested.

Individual seeds contain a small number of cells that give rise to the primary reproductive stem. If a mutation is induced in one of these cells the sector of the plant developing from this cell will be heterozygous for the induced mutation. When flowers in this sector self-fertilize, 25 % of the seeds that develop will be homozygous for this (recessive) mutation. When individual M_1 plants are harvested as a whole, recessive mutants are usually found at lower frequencies because other sectors, not mutated in the same gene(s) are also harvested. If seeds are harvested only from the top of the main inflorescence a normal 3:1 segregation is usually observed because chimerism is progressively lost on the main stem. Since in maize the male and female gametes are products of seperate cell lineages that become established in different regions of the embryonic shoot apex, seed treatment followed by selfing will not result in the segregation of recessive mutants, in contrast to bisexual plants. For this reason and since pollen can be collected easily in maize, a mutagenic treatment of pollen is preferred. This also results in non-chimeric M_1 plants.

Different chemical and physical mutagens may result in different mutant spectra as shown by the extensive analysis of *eceriferum* mutants in barley (Lundqvist and von Wettstein, 1962). Irradiation generally gives more chromosome damage than chemical mutagens, resulting in a lower frequency of monogenic mutants at a given level of M_1 sterility. The larger proportion of deletion mutants among the monogenic mutants than with chemical mutagens, such as ethylmethane sulphonate (EMS), allows the application of deletion cloning procedures as developed by Straus and Ausubel (1990). However, several 'leaky' mutant alleles have been isolated after irradiation (Koornneef *et al.* 1982a), showing that it does not only induce true null (deletion type) mutants.

Insertional mutagenesis, either by endogenous transposable elements, foreign transposable elements introduced by transformation (Jones *et al.* 1990) or random DNA, mostly introduced *via* the T-DNA of *Agrobacterium tumefaciens*, deals with similar considerations as 'classical' mutagenesis such as chimerism. An additional problem is the preferential insertion of transposable elements near the original integration site. The use of genotypes with elements at defined locations close to the gene to be targeted and crossing such unstable genotypes with already existing mutants of that gene, allows an efficient search for a specific insertional mutant.

One of the advances of mutant induction in *Arabidopsis* is the relatively easy growth of tens of thousands of M_1 plants, which are often bulk harvested to give a single M_2 population. This has the disadvantage that if several mutants with the same phenotype are found, they may be derived from the same mutation event. However, when the M_1 plants are subdivided into a number of groups, if the same mutant phenotype is found in different sublots they definitely derive from different mutation events and, therefore, tests for allelism can be limited to those mutants. Individually harvesting (if possible of non-chimeric sectors) of M_1 plants is very laborious but has the advantage that in the M_2, heterozygous sisterplants of a particular mutant can be saved. This allows the maintenance of sterile and lethal mutants and makes a study of the genetics of the mutant possible without having to cross it with the wild-type.

Genetic analysis of mutants

If a mutant phenotype has been recognized in an M_2 population the first step in the analysis is to harvest seeds and to confirm the mutant phenotype in the M_3 generation. In cases of male sterility the mutant will have to be crossed with the parental wild-type and the mutant phenotype should reappear in the subsequent

F_2 generation. Heterozygous M_2 sisterplants have to be saved for lethal or completely sterile mutants.

A genetic analysis of a new mutant is necessary to establish the monogenic inheritance of different (pleiotropic) characters associated with a mutant phenotype and to establish the recessiveness/dominance of the mutant. When several independent mutants with a similar phenotype have been isolated, a complementation analysis has to be performed by intercrossing the different mutants. Two recessive mutations of the same gene will, in general, give a mutant phenotype when combined in a hybrid of the two mutants, whereas recessive mutations in different genes give a wild-type phenotype as indicated in the following scheme:

Mutant 1		Mutant 2		Hybrid		Conclusion
				Genotype	Phenotype	
a^1a^1	$\times$	a^2a^2	$\rightarrow$	a^1a^2	mutant	allelic
aa,BB	$\times$	AA,bb	$\rightarrow$	Aa,Bb	wild-type	non-allelic

In the case of both mutants being dominant, the segregation of wild-type phenotypes, in the subsequent F_2 generation or the backcross with the recessive wild-type, demonstrates non-allelism. Very close linkage is hard to distinguish from allelism when both mutants are dominant.

The map position of a new locus can be obtained from linkage analysis with previously mapped morphological or restriction fragment length polymorphism (RFLP), markers. For several species, genotypes homozygous for a number of markers with well defined map positions, are available for this purpose.

Double or digenic mutants, combining in one individual two mutations in different genes that both affect a similar developmental pathway, can be important to study the interaction between the two genes. If the two genes are acting in the same pathway, the double mutant should show either the same phenotype as the mutant with the most severe phenotype or have a slightly more severe phenotype due to the combination of two leaky mutations. On the other hand, if the genes are acting through different pathways the double mutant should have a much more severe phenotype. An example of the first situation is the *aurea/high pigment* (*au/hp*) double mutant in tomato that predominantly resembles the phytochrome deficient *au* mutant, indicating that phytochrome action is a prerequisite for expression of the *hp* phenotype (Peters *et al.* 1989), which is characterized by enhanced sensitivity to light. Selection for segregants of the second mutation in F_3 lines, obtained after selfing F_2 plants that are homozygous for the first mutation, allowed the isolation of this double mutant without an *a priori* prediction of its phenotype (Adamse *et al.* 1989). Examples of double mutants with a strongly additive effect in *Arabidopsis* are the combination of different *hy* mutants (Koornneef *et al.* 1980) and the combination of some but not all ABA insensitive (*abi*) mutants (Finkelstein and Somerville, 1990). The usefulness of double mutants to study genes affecting floral development has been shown by Bowman *et al.* (1989).

Mutant selection

Knowledge of relevant biochemical or physiological processes is important in mutant selection, since this is essential for setting up the right screening procedure. Morphogenetic mutants are of course mostly isolated on the basis of their abberant phenotype and different researchers have focused on alterations in specific organs and structures. For a long time, changes in root development were not taken into consideration. However, by growing M_2 plants on translucent media, such as agar, several root-hair mutants have now been isolated in *Arabidopsis* (Schiefelbein and Somerville, 1990). Selection schemes, designed for the isolation of specific physiological mutants, may also result in mutants with an altered morphogenesis, which give an immediate indication of the physiological factor involved. An example of this is the auxin resistant *axr2* mutant that has no root hairs (Wilson *et al.* 1990).

The isolation of abscisic acid (ABA) deficient mutants in *Arabidopsis* (Koornneef *et al.* 1982*b*) provides an example of a selection scheme using a particular mutant to isolate mutations in other genes acting on the same process. Some gibberellin (GA) deficient mutants in this species have an absolute GA requirement for germination (Koornneef and van der Veen, 1980). By selecting for germinating seeds amongst the progeny of mutagen-treated non-germinating (*ga1* mutant) seeds, mutants were isolated that were reverted at least for the non-germination trait. Genetic analysis showed this reversion to be due to a mutation in a suppressor gene, segregating independently from the *ga1* gene. Physiological and biochemical characterization indicated that this suppressor gene controls ABA biosynthesis. The absence of seed-dormancy, which is induced by ABA during seed development, indicates that no GA or only very low levels of GA are required for germination (Karssen *et al.* 1987).

To 'saturate' a developmental pathway with mutations one should not only try to get mutations at all target loci, but it is also important to isolate a number of different mutant alleles at each locus, which might differ quantitatively in their effect.

Even with a well designed selection scheme there are a number of reasons why specific mutants are not found or cannot be maintained without specific measures.

(a) A mutant allele disturbs the normal development of the gametophyte so that no zygote will be formed. For example, when only the male gametophyte is affected no mutant embryo will be formed if the mutation is recessive.

(b) The mutation is lethal at some stage during embryo development. Embryonic lethals are a frequent and heterogeneous class of mutants that will be described in detail later.

(c) The mutation leads to non-germinating seeds, either because the dormancy breaking mechanism is affected (e.g. GA deficient mutants) or because mutant seeds are not desiccation resistant.

(d) Mutations are lethal after the seedling stage e.g. many chlorophyll deficient mutants and the more extreme nitrate reductase-deficient mutants.

(e) Mutants are either male, female or completely sterile.

(f) Mutants have (in a specific environment) no distinguishable phenotype. This may be because a duplicated gene provides sufficient gene function or that the gene product is not essential for normal growth and development.

(g) Large differences between mutation frequencies per locus have also been described for mutants with similar phenotypes (Koornneef *et al.* 1982*a*).

The factors described will result in no mutants being found in the screening procedure employed or will make their maintenance impossible. The latter problem can be solved by saving seeds from the heterozygous sisterplants and in some cases (auxotrophs) by supplementing the substance lacking or by applying tissue culture. Specific selection for conditional lethals (e.g. temperature sensitive mutants) is another alternative.

Mutants without a distinguishable phenotype require a sophisticated selection procedure or methods to assay individual M_2 plants for the absence of a specific metabolite or enzyme activity (Estelle and Somerville, 1986).

Duplicated loci can frustrate the search for mutants that could be isolated in other plant species where these genes are not duplicated (e.g. nitrate reductase in *Arabidopsis*; Cheng *et al.* 1988). However, since in many cases similar genes are not expressed at the same level and/or in the same tissue, it may be possible to find mutants with a 'leaky' phenotype due to a mutation in the gene that is preferentially expressed.

Examples of developmental and physiological mutants

The study of developmental mutants is just starting to move from its 'juvenile' phase of descriptive genetics to the 'adult' phase of finding out the molecular mechanism of plant development with all its interactions between signal molecules and the expression of different genes. This is indicated by the limited number of typical morphogenetic genes that have been cloned so far. In *Antirrhinum* (Sommer *et al.* 1990) and *Arabidopsis* (Yanofsky *et al.* 1990) two genes affecting floral development have been characterized at the molecular level. Genes affecting kernel development in maize, such as *vp1* and *o2* and regulatory genes of flavonoid expression (*C1*, *R*) were cloned by transposon tagging. Cloned genes affecting vegetative development are represented by *GL1* in *Arabidopsis* (Herman and Marks *et al.* 1989) and *Kn* in maize (recent progress is summarized by Wessler and Hake, 1990). It appears from this limited number of examples that several of these genes code for proteins that control transcriptional activators. Different combinations of transcription factors might regulate specific processes. If one factor is involved in several processes a complicated phenotype or even lethality might be the result of a mutation in a gene coding for such a protein. Although the genes that switch on a whole set of genes changing the differentiation pattern of cells within a meristem are of considerable interest, the nature of the activated genes should also be known, in order to understand plant development.

Plant hormones and phytochrome are also known to affect transcription of specific genes. Fig. 1 shows a scheme for hormones which predicts different classes

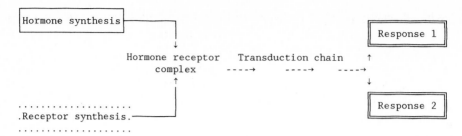

Fig. 1. A simplified signal transduction chain involving hormones. Since each step is controlled by (several) genes, mutations in all steps might be isolated.

of mutants that might be isolated. In *Arabidopsis*, mutants defective in light perception, some of them lacking the photoreceptor phytochrome (Parks *et al.* 1989) were isolated and described by Koornneef *et al.* (1980) and Chory *et al.* (1989*a*). Finding a mutant that develops as a light-grown plant (de-etiolated=*det* mutants) suggests that there is negative control of seedling development in the absence of light (Chory *et al.* 1989*b*). Mutants affecting the biosynthesis of plant hormones ABA (Koornneef *et al.* 1982*b*), GA (Koornneef and van der Veen, 1980) and ethylene (Guzman and Ecker, 1990) have been isolated as well as mutants insensitive for ABA (Koornneef *et al.* 1984; Finkelstein and Somerville, 1990), auxin (Maher and Martindale, 1980; Estelle and Somerville, 1987; Wilson *et al.* 1990), ethylene (Bleecker *et al.* 1988; Guzman and Ecker, 1990) and GA (Koornneef *et al.* 1985*b*). Mutants representing similar classes and even others resulting in hypersensitivity to light perception and hormones have also been found in other species such as tomato (Peters *et al.* 1990), and pea (Reid, 1986).

Since light and hormones both affect several developmental and physiological processes, mutants altered with respect to these factors are expected to change many aspects of plant development as shown in Table 2. Complications arise because different plant hormones influence the level and sensitivity of other hormones as was shown for secondary auxin and ethylene effects in ABA deficient tomato mutants (Tal *et al.* 1979) and for the auxin insensitive *diageotropica* mutant showing ethylene deficiency (Kelly and Bradford, 1986) as a secondary effect. The possibility that there are common steps in the signal transduction chains of the different hormones may explain mutations with insensitivity to several plant hormones (Wilson *et al.* 1990; Bitoun *et al.* 1990).

Mutants that are lethal because the compounds involved are essential for plant growth and development are obviously not present among the mutant collections mentioned. It has been suggested that this explains the lack of viable auxin and cytokinin auxotrophs (Blonstein *et al.* 1988).

Most genes affecting development in higher plants are indentified only by a mutant phenotype and a genetic characterization that in some, but not all cases, includes a map position. Lists of mapped genes can be found for example in The Handbook of Genetics vol.2 (ed. King, 1974) and in Genetic Maps (1990, ed. S.J. O'Brien). Monographs and reviews describe mutants for several plant species, e.g.

Table 2. *Developmental and physiological processes affected by different classes of mutants in* Arabidopsis

	Specific*	Photoreceptor	Plant hormones†			
			ABA	Auxin	GA	Ethylene
Early embryo	+					
Late embryo	+		+			
Seed dormancy			+		+	+
Hypocotyl growth	+	+		+	+	+
Phototropism	+			+		
Root development	+			+		
Leaf shape	+			+		
Trichome development	+					
Stem elongation	+	+		+	+	
Floral evocation	+	+				+
Floral development	+			+		

* Mutants were selected only on the basis of one specific defect.
† Including both biosynthesis and hormone action mutants.

Antirrhinum (Stubbe, 1966), maize (Sheridan, 1988), *Petunia* (Sink, 1984), tomato (Atherton and Rudich, 1986).

In the present review mutants controlling specific developmental stages in *Arabidopsis* are presented as an example of the variation in mutant types that can be found in a higher plant species (Haughn and Somerville, 1988 and Finkelstein *et al.* 1988).

Embryo and seed development

Embryonic morphogenesis in *Arabidopsis* occurs over a 7-day period according to the pattern described by Müller (1963) and Meinke (1986). Since the seed coat stays transparent until just before seed ripening, mutations, viz. embryonic lethals and chlorophyll deficient mutants that are expressed at the embryo stage, can be screened by opening the siliquas on M_1 plants. Mutations expressed at these stages are relatively frequent. Müller (1963) and Meinke and co-workers (Meinke, 1986) isolated many mutants of this class that differ with respect to the lethal phase, pattern of abnormal development, response in culture, ultrastructure, extent of cellular differentiation, accumulation of seed-storage proteins and the expression of mutant genes throughout the life cycle. Some of the genes affected the control of housekeeping functions, one embryonic lethal being a biotin auxotroph (Meinke, 1990).

Embryonic pattern mutants that specifically affect the fate of the meristem were identified by Mayer *et al.* (1990) as those mutants that alter the embryo/seedling organization without arresting embryonic development. By selection of such mutants among germinated M_2 seeds, Jürgens and co-workers (Mayer *et al.* 1990) isolated at least 200 mutants of this class.

Some mutations resulting in reduced seed dormancy have been associated with ABA deficiency (*aba*) or ABA insensitivity (*abi* mutants) (Koornneef et al. 1982; Koornneef et al. 1984). Despite the fact that the monogenic mutants form normal seeds, the genotype recessive for both *aba* and *abi3* was severely affected in the latter stages of seed development (Koornneef et al. 1989). Many mutant alleles at the *ga1*–*ga3* loci controlling early steps in GA biosynthesis do not germinate without exogenously applied GA (Koornneef and van der Veen, 1980).

Vegetative development

Mutants have been isolated with defects in almost every aspect of vegetative development (McKelvie, 1962). Monogenic mutations resulting in round (e.g. *cp3*) or asymmetric (*as*) leaf blades, twisted or narrow leaves have been identified. These mutants are also often affected in other aspects of development, and are sometimes associated with auxin defects (e.g. *dwf*, *axr1*, *axr2*). GA deficient and GA non-responsive mutants are characterized by reduced plant height and reduced apical dominance (Koornneef et al. 1985b).

At least 8 different genes control trichome development in *Arabidopsis* (Haughn and Somerville, 1988). Mutations at the *ttg* and *gl2* loci also affect the surface structure of the seed coat (Koornneef, 1981) resulting in the absence of mucilage excretion.

Root-hair (*rhd*) mutants have been isolated by Schiefelbein and Somerville (1990). However, screens for auxin resistant mutants also resulted in mutants with abberant roots. Root geotropism was disturbed in all auxin resistant mutants (Maher and Martindale, 1980; Estelle and Somerville, 1987; Wilson et al. 1990). Direct screens for geotropic mutants, affected either in hypocotyl, root, or both hypocotyl and root geotropism were developed by Bullen et al. (1990). In addition to mutants with altered geotropic responses, Poff and co-workers (Khurana and Poff, 1989) also isolated phototropic mutants. In some of these mutants, but not all, geotropic curvature was also altered.

The evocation of flowering

Flowering occurs in *Arabidopsis* after bolting when a number of leaves have been formed in the rosette and on the elongated stem. Inflorescences, which are open racemes, can also develop from the axils of the upper rosette and leaves present on the main stem below the first flower. If flowering is postponed, either by adverse physiological factors, such as short days and the absence of a vernalization treatment or by mutations in specific genes, more leaves are formed before the shoot meristem starts producing flowers. Since flowering time also depends on factors determining plant growth, the number of leaves formed prior to flowering seems a more reliable parameter for the developmental stage at which flowering is expressed. Genetic differences in flowering time are observed both within and between natural populations of *Arabidopsis* (Napp-Zinn, 1985).

Mutants delayed in flowering can be obtained from M_2 populations of early flowering ecotypes such as Lansberg *erecta* or Columbia. These mutants, representing at least 12 different loci, show no other morphogenetic defect than that the timing of the conversion of the apical meristem from vegetative to reproductive growth is changed.

Observations that mutants at some of these loci (e.g. *co* and *gi*) do not respond to differences in day length (Rédei, 1962; Koornneef *et al.* unpublished data) indicate that some of these genes control the perception of the induction signal. However, mutants at the loci *fca*, *fve*, *fy*, and *fpa* (Martinez-Zapater and Somerville, 1990; Koornneef *et al.* unpublished data) show a more pronounced response to vernalization both under long and short days. Such a phenotype is not in agreement with a transduction chain mutant, but may be explained by the assumption that flowering in *Arabidopsis* is controlled by at least two parallel pathways, only one of which is environmentally controlled. Mutations in the constitutive pathway would result in more pronounced effects of the environmental conditions. Such a model is in agreement with the observation that most ecotypes do not have an absolute requirement for either vernalization and/or long days and also with the fact that non-flowering mutants have not yet been found in *Arabidopsis*. Mutants that flower earlier than the original (early) wild-types have less dramatic phenotypes and are mostly associated with other physiological changes e.g. the phytochrome mutants *hy1*, *hy2*, *hy3* and *hy6* (Koornneef *et al.* 1980; Chory *et al.* 1989a) However, a mutant (*hls1*) selected by its absence of an apical hook, flowered and senesced much earlier than its corresponding Columbia wild-type (Guzman and Ecker, 1990).

Floral development

Flowers are specialized structures growing out from shoot meristems and composed of a species-specific number of floral organs arranged in a number of concentric whorls. In *Arabidopsis* a number of loci have been identified in which this very regular pattern is disturbed in such a way that some organs seem to be replaced by others or organs are absent (listed by Drews and Goldberg, 1989). All these mutants seem to have no obvious changes in other aspects of the plant's morphology or physiology and for this reason represent a class of specifically developmental mutants. The morphology and genetics (including the interaction between different mutants) has been described by several authors (Haughn and Somerville, 1988; Komaki *et al.* 1988; Bowman *et al.* 1989; Kunst *et al.* 1989). Recently, one of these genes (*agamous*) has been cloned by Yanofsky *et al.* 1990) and models explaining the interrelation and determination of the different floral organs have been presented (Bowman *et al.* 1989; Kunst *et al.* 1989).

The use of mutants to study cell and tissue autonomy

Morphogenetic changes are often assumed to be induced by diffusable signal molecules, whereas the genes responding to these factors are likely to be expressed only in the target-cell and are therefore cell-autonomous. A mutant cell, defective in a non-diffusable gene product should not be influenced by adjacent wild-type cells. Diffusable gene products should result in a non-cell phenotype. This aspect can be studied if one can obtain individuals composed of genetically different cells. Such genetic mosaics or chimeras have also been used to study cell lineages in plants and can be constructed in several ways (Poethig, 1989).

(a) By irradiation of genotypes heterozygous for two genes, where the recessive allele of the mutant to be studied is closely linked to a recessive allele of a gene with a known cell-autonomous expression. The latter marker is often a chlorophyll or anthocyanin deficiency. Irradiation of seeds or developing young plants will result in some cells lacking the corresponding wild-type alleles by chromosome breakage or mitotic recombination. If the phenotype of both markers can be observed in the sectors developing from such cells it is concluded that both mutations are cell-autonomous. In cases where only the phenotype of the known cell-autonomous marker (e.g. the chlorophyll deficiency) is observed the other mutation is non-cell-autonomous. Harberd and Freeling (1989) showed, with this procedure, that the GA insensitivity due to the *D1* mutation in maize is cell-autonomous, as might be expected for a gene product that receives the diffusable hormone signal.

(b) Grafting. Plants in which organs are genetically different can be obtained by grafting. This technique can be easily applied in *Solanaceae* species and has also been applied to pea. This procedure allows e.g. the root system to be made genetically different from the shoot. Since a wild-type scion resulted in normal roots on a *diageotropica* (*dgt*) root stock, it could be shown that the abberant root phenotype due to the *dgt* mutation in tomato is caused by a diffusable factor from the roots (Zobel, 1974; Hicks *et al.* 1989).

Non cell-autonomy was demonstrated for gibberellin deficiency in pea (Reid *et al.* 1983) and tomato (J.A.D. Zeevaart personal communication).

(c) Reciprocal crosses of mutants and wild-type. To investigate the autonomy of the embryo in relation to its surrounding maternal tissue, using a recessive mutant as the mother plant in a cross with a wild-type, will result in a mutant plant bearing a wild-type embryo (Fig. 2). Using such genotypes it was shown by Karssen *et al.* (1983) that only ABA produced by the embryo induces dormancy in the embryo, where the large amount of maternal ABA did not seem to be effective. However, the abberant seed development in the *aba/abi3* double mutant was influenced by maternal ABA (Koornneef *et al.* 1989).

By using a similar approach it has been shown that GA produced by the embryo is important for fruit growth. However, a significant maternal effect was also observed (Barendse *et al.* 1986). The fact that embryonic lethals are observed on wild-type plants indicates embryo autonomy for these factors.

Conclusion

Mutants can be very useful for a proper understanding of plant development.

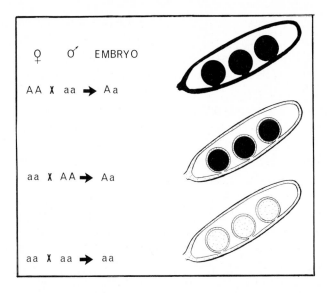

Fig. 2. Crosses between a recessive mutant (*aa*) and wild-type (*AA*) plant that result in different combinations of mutant and wild-type embryos and maternal tissues.

However, the ultimate benefit from mutants depends very much on how they will be utilized. Producing endless lists of mutant descriptions and even the cloning of many of the genes will, as such, not give the answer to all questions about plant development. Experimental studies, with the mutants, on the physiology, the study of interactions, tissue and cell autonomy and the analysis and location of gene expression will all be necessary to unravel a complicated network of interactions that results in a functional plant.

The sessile life habit, which requires a strong but also flexible adaptation to its environment, has also resulted in a number of different physiological and morphological adaptations in plants. For this reason it is important to focus research in plant development on more than one species. However, having one or two reference species about which a lot is known will be extremely useful and requires a large research investment in the coming years.

I thank Chris and Dick Kendrick, David Meinke and Scot Poethig for their help in preparing this article.

References

ADAMSE, P., PETERS, J. L., JASPERS, P. A. P. M., VAN TUINEN, A., KOORNNEEF, M. AND KENDRICK, R. E. (1989). Photocontrol of anthocyanin synthesis in tomato seedlings: a genetic approach. *Photochemistry and photobiology* **50**, 107–111.

ATHERTON, J. G. AND RUDICH, J. (eds.) (1986). *The Tomato Crop*. Chapman and Hall, London, New York. 661 pp.

BARENDSE, G. W. M., KEPCZYNSKI, J., KARSSEN, C. M. AND KOORNNEEF, M. (1986). The role of endogenous gibberellins during fruit and seed development. Studies on gibberellin-deficient genotypes of Arabidopsis thaliana. *Physiol. Plantarium* **67**, 315–319.

BHATTACHARYYA, M. K., SMITH, A. M., ELLIS, T. H. N., HEDLEY, C. AND MARTIN, C. (1990). The wrinkled-seed character of pea described by Mendel is caused by a transposon-like insertion in a gene encoding starch-branching enzyme. *Cell* **60**, 115–122.

BITOUN, R., ROUSSELIN, P. AND CABOCHE, M. (1990). A pleiotropic mutation results in cross-resistance to auxin, abscisic acid and paclobutrazol. *Molec. gen. Genet.* **220**, 234–239.

BLEECKER, A. B., ESTELLE, M. A., SOMERVILLE, C. R. AND KENDE, H. (1988). Insensitivity to ethylene conferred by a dominant mutation in *Arabidopsis*. *Science* **241**, 1086–1089.

BLONSTEIN, A. D., VAHALA, T., KOORNNEEF, M. AND KING, P. J. (1988). Plants regenerated from auxin-auxotrophic variants are inviable. *Molec. gen. Genet.* **215**, 58–64.

BOWMAN, J. L., SMYTH, D. R. AND MEYEROWITZ, E. M. (1989). Genes directing flower development in *Arabidopsis*. *Plant Cell* **1**, 37–52.

BULLEN, B. L., BEST, T. R., GREGG, M. M., BARSEL, S. E. AND POFF, K. L. (1990). A direct screening procedure for gravitropism mutants in *Arabidopsis thaliana* (L.) Heynh. *Pl. Physiol.* **93**, 525–531.

CHENG, C., DEWDNEY, J., NAM, H., DEN BOER, B. AND GOODMAN, H. M. (1988). A new locus (*nia1*) in *Arabidopsis thaliana* encoding nitrate reductase. *EMBO J.* **7**, 3309–3314.

CHORY, C., PETO, C. A., ASHBAUGH, M., SAGANICH, R., PRATT, L. H. AND AUSUBEL, F. (1989*a*). Different roles for phytochrome in etiolated and green plants reduced from characterization of *Arabidopsis thaliana* mutants. *Pl. Cell* **1**, 867–880.

CHORY, J., PETO, C. A., FEINBAUM, R., PRATT, L. H. AND AUSUBEL, F. (1989*b*). *Arabidopsis thaliana* mutant that develops as a light-grown plant in the absence of light. *Cell* **58**, 991–999.

DREWS, G. N. AND GOLDBERG, R. B. (1989). Genetic control of flower development. *Trends Genet.* **5**, 256–261.

ESTELLE, M. A. AND SOMERVILLE, C. R. (1986). The mutants of *Arabidopsis*. *Trends Genet.* **2**, 89–92.

ESTELLE, M. A. AND SOMERVILLE, C. R. (1987). Auxin resistance mutants of *Arabidopsis thaliana* with altered morphology. *Molec. gen. Genet.* **206**, 200–206.

EVANS, D. A. (1989). Somaclonal variation – Genetic basis and breeding applications. *Trends Genet.* **5**, 46–50.

FELDMANN, K. A. AND MARKS, M. D. (1987). *Agrobacterium*-mediated transformation of germinating seeds of *Arabidopsis thaliana*: A non-tissue culture approach. *Molec. gen. Genet.* **208**, 1–9.

FINKELSTEIN, R., ESTELLE, M., MARTINEZ-ZAPATER, J. AND SOMERVILLE, C. (1988). *Arabidopsis* as a tool for the identification of genes involved in plant development. In *Temporal and spatial regulation of plant genes* (eds. D. P. S. Verma and R. B. Goldberg), pp. 1–25. Springer Verlag, New York.

FINKELSTEIN, R. R. AND SOMERVILLE, C. R. (1990). Three classes of abscisic acid (ABA)-insensitive mutations of *Arabidopsis* define genes which control overlapping subsets of ABA responses. *Pl. Physiol.* **94**, 1172–1179.

GABARD, J., PELSY, F., MARION-POLL, A., CABOCHE, M., SAALBACH, I., GRAFE, R. AND MÜLLER, A. J. (1988). Genetic analysis of nitrate reductase deficient mutants of *Nicotiana plumbaginifolia*: Evidence for six complementation groups among 70 classified molybdenum cofactor deficient mutants. *Molec. gen. Genet.* **213**, 206–213.

GUZMAN, P. AND ECKER, J. R. (1990). Exploiting the triple response of *Arabidopsis* to identify ethylene-related mutants. *Plant Cell* **2**, 513–523.

HARBERD, N. P. AND FREELING, M. (1989). Genetics of dominant gibberellin-insensitive dwarfism in maize. *Genetics* **121**, 827–838.

HAUGHN, G. W. AND SOMERVILLE, C. R. (1988). Genetic control of morphogenesis in *Arabidopsis*. *Devl Genet.* **9**, 73–89.

HERMAN, P. L. AND MARKS, M. D. (1989). Trichome development in *Arabidopsis thaliana*. II. Isolation and complementation of the *GLABROUS1* gene. *Pl. Cell* **1**, 1051–1055.

HICKS, G. R., RAYLE, D. L. AND LOMAX, T. L. (1989). The *diageotropica* mutant of tomato lacks high specific activity auxin binding sites. *Science* **245**, 52–54.

HILLE, J., KOORNNEEF, M., RAMANNA, M. S. AND ZABEL, P. (1989). Tomato: a crop species amenable to improvement by cellular and molecular methods. *Euphytica* **42**, 1–24.

JONES, J. D. G., BISHOP, G., HARRISON, K., JONES, D., SCOFIELD, S. AND THOMAS, C. M. (1990). Establishing a tomato transposon tagging system with heterologous transposons. In *Progress *in Plant Cellular and Molecular Biology* (eds. H. J. J. Nijkamp, L. H. W. van der Plas and J. van Aartrijk) pp. 175–182. Kluwer Academic Publ. Dordrecht/Boston/London.

KARSSEN, C. M., BRINKHORST-VAN DER SWAN, D. L. C., BREEKLAND, A. E. AND KOORNNEEF, M. (1983). Induction of dormancy during seed development by endogeneous abscisic acid: Studies on abscisic acid deficient genotypes of *Arabidopsis thaliana* (L.) Heynh. *Planta* **157**, 158–165.

KARSSEN, C. M., GROOT, S. P. C. AND KOORNNEEF, M. (1987). Hormone mutants and seed dormancy in Arabidopsis and tomato. *Developmental Mutants in Higher Plants.* SEB seminar Series 32, Cambridge University Press, Cambridge (1987) 119–133.

KELLY, M. O. AND BRADFORD, K. J. (1986). Insensitivity of the *diageotropica* tomato mutant to auxin. *Pl. Physiol.* **82**, 713–717.

KHURANA, J. P. AND POFF, K. L. (1989). Mutants of *Arabidopsis thaliana* with altered phototropism. *Planta* **178**, 400–406.

KING, R. C. (ed.) (1974). *Handbook of Genetics*, vol. 2. New York, Plenum. pp. 3–377.

KLEIN, T. M., GOFF, S. A., ROTH, B. A. AND FROMM, M. E. (1990). Applications of the particle gun in plant biology. In *Progress in Plant Cellular and Molecular Biology* (eds. H. J. J. Nijkamp, L. H. W. van der Plas and J. van Aartrijk) pp. 56–66. Kluwer Academic Publ. Dordrecht/Boston/London.

KOMAKI, M. K., OKADA, K., NISHINO, E. AND SHIMURA, Y. (1988). Isolation and characterization of novel mutants of *Arabidopsis thaliana* defective in flower development. *Development* **104**, 194–203.

KOORNNEEF, M. (1981). The complex syndrome of *ttg* mutants. *Arabidopsis Information Service* **18**, 45–51.

KOORNNEEF, M., CONE, J. W., DEKENS, R. G., O'HERNE-ROBERS, E. G., SPRUIT, C. J. P. AND KENDRICK, R. E. (1985a). Photomorphogenic responses of long hypocotyl mutants of tomato. *J. Plant Physiol.* **120**, 153–165.

KOORNNEEF, M., DELLAERT, L. M. W. AND VAN DER VEEN, J. H. (1982a). EMS- and radiation-induced mutation frequencies at individual loci in *Arabidopsis thaliana* (L.) Heynh. *Mutation Research* **93**, 109–123.

KOORNNEEF, M., ELGERSMA, A., HANHART, C. J., VAN LOENEN MARTINET, E. P., VAN RIJN, L. AND ZEEVAART, J. A. D. (1985b). A gibberellin insensitive mutant of *Arabidopsis thaliana*. *Physiologia Pl.* **65**, 33–39.

KOORNNEEF, M., HANHART, C. J., HILHORST, H. W. M. AND KARSSEN, C. M. (1989). *In vivo* inhibition of seed development and reserve protein accumulation in recombinants of abscisic acid biosynthesis and responsiveness mutants in *Arabidopsis thaliana*. *Pl. Physiol.* **90**, 463–469.

KOORNNEEF, M., JORNA, M. L., BRINKHORST-VAN DER SWAN, D. L. C. AND KARSSEN, C. M. (1982b). The isolation of abscisic acid (ABA) deficient mutants by selection of induced revertants in non-germinating gibberellin sensitive lines of *Arabidopsis thaliana* (L.) Heynh. *Theor. appl. Genet.* **61**, 385–393.

KOORNNEEF, M., REULING, G. AND KARSSEN, C. M. (1984). The isolation and characterization of abscisic acid-insensitive mutants of *Arabidopsis thaliana*. *Physiologia Pl.* **61**, 377–383.

KOORNNEEF, M., ROLFF, E. AND SPRUIT, C. J. P. (1980). Genetic control of light-inhibited hypocotyl elongation in *Arabidopsis thaliana* (L.) Heynh. *Z. Pflphysiol.* **100**, 147–160.

KOORNNEEF, M. AND VAN DER VEEN, J. H. (1980). Induction and analysis of gibberellin sensitive mutants in *Arabidopsis thaliana* (L.) Heynh. *Theor. Appl. Genet.* **58**, 257–263.

KOORNNEEF, M., VAN EDEN, J., HANHART, C. J. AND DE JONGH, J. A. M. (1983). Genetic fine structure of the *ga-1* locus in the higher plant *Arabidopsis thaliana*. *Genet. Res.* **41**, 57–68.

KUNST, L., KLENZ, J. E., MARTINEZ-ZAPATER, J. AND HAUGHN, G. W. (1989). *Ap2* gene determines the identity of perianth organs in flowers of *Arabidopsis thaliana*, *Plant Cell* **1**, 1195–1208.

LUNDQVIST, U. AND VON WETTSTEIN, D. (1962). Induction of *eceriferum* mutants in barley by ionizing radiations and chemical mutagens. *Hereditas* **48**, 342–362.

MAHER, E. P. AND MARTINDALE, S. J. B. (1980). Mutants in *Arabidopsis* with altered responses to auxins and gravity. *Biochem. Genet.* **18**, 1041–1053.

MARTINEZ-ZAPATER, J. M. AND SOMERVILLE, C. R. (1990). Effect of light quality and vernalization on late-flowering mutants of *Arabidopsis thaliana*. *Pl. Physiol.* **92**, 770–776.

MARX, G. A. (1983). Developmental mutants in some annual seed plants. *A. Rev. Pl. Physiol.* **34**, 389–417.

MAYER, U., TORRES RUIZ, R. A., BERLETH, T., MISERA, S. AND JÜRGENS, G. (1990). Towards a genetic analysis of embryonic pattern formation in *Arabidopsis*. Fourth Int. Conf. on *Arabidopsis* Research, Vienna 1990 (Abstract S7/O) pp. 113.

MCKELVIE, A. D. (1962). A list of mutant genes in *Arabidopsis thaliana* (L.) Heynh. *Radiat. Bot.* **1**, 233–241.

MEINKE, D. W. (1986). Embryo-lethal mutants and the study of plant embryo development. *Oxford Surv. Plant molec. cell. Biol.* **3**, 122–165.

MEINKE, D. W. (1990). Genetic analysis of plant development. In *Plant Physiology*, vol. X (ed. F. C. Steward and R. G. S. Bidwell), Academic Press (in press).

MEYEROWITZ, E. M. (1987). *Arabidopsis thaliana*. *A. Rev. Genet.* **21**, 93–111.

MÜLLER, A. J. (1963). Embryonentest zum Nachweis rezessiver Lethalfaktoren bei *Arabidopsis thaliana*. *Biol. Zentralbl.* **82**, 133–163.

NAPP-ZINN, K. (1985). *Arabidopsis thaliana*. In *Handbook of flowering*, vol. 1 (ed. A. H. Halevy) pp. 492–503. CRC Press. Boca. Raton, Fla.

O'BRIEN, S. J (ed). (1990). Genetic Maps, Cold Spring Harbor Laboratory Press.

PARKS, B. M., SHANKLIN, J., KOORNNEEF, M., KENDRICK, R. E. AND QUAIL, P. H. (1989). Immunochemically detectable phytochrome is present at normal levels but is photochemically nonfunctional in the *hy-1* and *hy-2* long hypocotyl mutants of *Arabidopsis*. *Pl. mol. Biol.* **12**, 425–437.

PETERS, J. L., VAN TUINEN, A., ADAMSE, P., KENDRICK, R. E. AND KOORNNEEF, M. (1989). High pigment mutants of tomato exhibit high sensitivity for phytochrome action. *J. Pl. Physiol.* **134**, 661–666.

PETERS, J. L., WESSELIUS, J. C., GEORGHIOU, K. C., KENDRICK, R. E., VAN TUINEN, A. AND KOORNNEEF, M. (1990). The physiology of photomorphogenetic tomato mutants. In *Phytochrome properties and biological action* (ed. B. Thomas) (in press).

POETHIG, S. (1989). Genetic mosaics and cell lineage analysis in plants. *Trends Genet.* **5**, 273–276.

RÉDEI, G. P. (1962). Supervital mutants of *Arabidopsis*. *Genetics* **47**, 443–460.

REID, J. B. (1986). Gibberellin mutants. In *A Genetic Approach to Plant Biochemistry* (eds. A. D. Blonstein and P. J. King), pp. 1–34. Springer Verlag, Wien.

REID, R. B., MURFET, I. C. AND POTTS, W. C. (1983). Internode length in *Pisum* II. Additional information on the relationship and action of loci *Le*, *La*, *Cry*, *Na* and *Lm*. *J. exp. Bot.* **34**, 349–364.

RONCHI, V. N., GIORGETTI, L. AND TONELLI, M. G. (1990). The commitment to embryogenesis: a cytological approach. In *Progress in Plant Cellular and Molecular Biology* (eds. H. J. J. Nijkamp, L. H. W. van der Plas and J. van Aartrijk) pp. 437–442. Kluwer Academic Publ. Dordrecht/Boston/London.

SCHIEFELBEIN, J. W. AND SOMERVILLE, C. (1990). Genetic control of root hair development in *Arabidopsis thaliana*. *Plant Cell* **2**, 235–243.

SHERIDAN, W. F. (1988). Maize developmental genetics: Genes of morphogenesis. *A. Rev. Genet.* **22**, 353–385.

SINK, K. C. (1984). 'Petunia'. Springer Verlag, New York.

SOMMER, H., BELTRAN, J. P., HUIJSER, P., PAPE, H., LONIG, W. E., SAEDLER, H. AND SCHWARZ-SOMMER, S. (1990). *Deficiens*, a homeotic gene involved in the control of flower morphogenesis in *Antirrhinum majus*: the protein shows homology to transcription factors. *EMBO J.* **9**, 605–613.

STEEVES, T. A. AND SUSSEX, I. M. (1989). Patterns in plant development. Cambridge Univ. Press, Cambridge 388 pp.

STRAUS, D. AND AUSUBEL, F. M. (1990). Genomic substraction for cloning DNA corresponding to deletion mutations. *Proc. natn. Acad. Sci. U.S.A.* **87**, 1889–1893.

STUBBE, H. (1966). Genetik und Zytologie von *Antirrhinum* L. sect. *Antirrhinum*. VEB Gustav Fisher Verlag, Jena. 421 pp.

TAL, M., IMBER, D., EREZ, A. AND EPSTEIN, E. (1979). Abnormal stomatal behaviour and hormonal imbalance in *flacca*, a wilty mutant of tomato. *Pl. Physiol.* **63**, 1044–1048.

TERZI, M. AND LO SCHIAVO, F. (1990). Developmental mutants in carrot. In *Progress in Plant*

Cellular and Molecular Biology (eds. H. J. J. Nijkamp, L. H. W. van der Plas and J. van Aartrijk) pp. 391–397. Kluwer Academic Publ. Dordrecht/Boston/London.

VALVEKENS, D., VAN MONTAGU, M. AND VAN LIJSEBETENS, M. (1988). *Agrobacterium tumefaciens*-mediated transformation of *Arabdopsis thaliana* root explants by using kanamycin selection. *Proc. natn. Acad. Sci. U.S.A.*. **85**, 5536–5540.

WESSLER, S. AND HAKE, S. (1990). Maize harvest. *Plant Cell* **2**, 495–499.

WILSON, A. K., PICKETT, F. B., TURNER, J. C. AND ESTELLE, M. (1990). A dominant mutation in *Arabidopsis* confers resistance to auxin, ethylene and abscisic acid. *Mol. gen. Genet.* **222**, 377–383.

YANOFSKY, M. F., MA, H., BOWMAN, J. L., DREWS, G. N., FELDMANN, K. A. AND MEYEROWITZ, E. M. (1990). The protein encoded by the *Arabidopsis* homeotic gene *agamous* resembles transcription factors. *Nature* **346**, 35–39.

ZOBEL, R. W. (1974). Control of morphogenesis in the ethylene-requiring tomato mutant *diageotropica*. *Can. J. Bot.* **52**, 735–741.

Printed in Great Britain © Society for Experimental Biology 1991   21

THE PHENOTYPE OF *ARABIDOPSIS THALIANA DET*1 MUTANTS SUGGESTS A ROLE FOR CYTOKININS IN GREENING

JOANNE CHORY[1], *NATHAN AGUILAR*[1], *and CHARLES A. PETO*[2]

[1] Plant Biology Laboratory and [2] Laboratory of Neuronal Structure and Function, The Salk Institute for Biological Studies, San Diego, CA 92138, USA

Summary

When grown in the absence of light, the *det*1 mutants of *Arabidopsis thaliana* (L.) Heynh. develop characteristics of light-grown plants as determined by morphological, cellular, and molecular criteria. Further, in light-grown plants, mutations in the *DET*1 gene affect cell-type-specific expression of light-regulated genes and the chloroplast developmental program. Here we show that the addition of exogenously added cytokinins (either 2-isopentenyl adenine, kinetin, or benzyladenine) to the growth medium of dark-germinated wild-type seedlings results in seedlings that resemble *det*1 mutants, instead of having the normal etiolated morphology. Like *det*1 mutants, these dark-grown seedlings now contain chloroplasts and have high levels of expression of genes that are normally 'light'-regulated. These results suggest an important role for cytokinins during greening of *Arabidopsis*, and may implicate abnormal cytokinin levels or an increased sensitivity to cytokinins as explanations for some of the observed phenotypes of *det*1 mutants.

Introduction

The mechanisms underlying developmental processses in multicellular eukaryotic organisms are largely unknown. In plants, the environmental stimulus of light triggers a profound change in the developmental program of the organism, leading to alterations in gene expression, tissue differentiation, and gross morphology (Dale, 1988; Mullet, 1988). The development of photosynthetically active chloroplasts from the small, undifferentiated proplastids present in meristematic cells (a process called greening) is one dramatic transition that occurs in response to light signals (Leech, 1976; Mullet, 1988; Gruissem, 1989). In addition to light, the development of chloroplasts is also regulated by intrinsic developmental signals that control leaf differentiation. Only particular leaf-cell types, namely the mesophyll cells, house mature chloroplasts, indicating that cell-specific signals are also important determinants of chloroplast biogenesis. Several of the growth

Key words: gene expression, cytokinins, light regulation.

regulators have been implicated in greening, including cytokinins and gibberellins (Stetler and Laetsch, 1965; Flores and Tobin, 1986; Mathis *et al.* 1989). How light might interact with these hormone signal transduction pathways is not understood.

Aside from the red-light photoreceptor, phytochrome (Colbert, 1988), the molecular biology of light-regulated developmental pathways in higher plants is unknown. In order to understand better the molecular mechanisms that control this process, we isolated mutations that mimic the light signal and induce the light developmental program in darkness. From among a population of mutagenized dark-grown seedlings, we obtained mutants that displayed many phenotypic characteristics of light-grown wild-type plants (Chory *et al.* 1989*a*, Chory and Peto, 1990). We have currently assigned these mutants to three complementation groups, designated *det1*, *det2*, and *det3*, and have studied alleles of *det1* in the most detail. Alleles *det1-1* and *det1-2* were shown to be recessive single gene mutations. When grown in the dark, these mutants have the gross morphology of light-grown plants, including the development of chloroplasts and leaf mesophyll tissue. The mRNA levels for several nuclear and chloroplast photoregulated genes are similar in dark-grown *det1* mutants to those found in light grown wild-type plants, and are 20–100-fold higher than those found in dark-grown wild-type seedlings (Chory *et al.* 1989*a*). These results suggest that DET1 may be a master regulatory molecule exerting negative control over the light response.

det1 mutants are small, pale-green, and lack apical dominance when grown in the light, implying that DET1 has a function in light-grown plants, as well as in dark-grown plants. We have recently shown, by histology and RNA analysis, that the role of *DET1* in light-grown plants is likely to be in regulating the cell-type-specific expression of light-regulated genes and chloroplast development (Chory and Peto, 1990). Using several light-regulated promoters fused to screenable marker genes that were introduced into *det1* and wild-type plants, we showed that in light-grown *det1* plants, these promoters are active in cell-types where they are normally silent or expressed at very low levels in wild-type plants. Taken together with the dark-grown *det1* phenotypes, these results suggest that *DET1* is involved in the integration of the light and tissue-specific signal transduction pathways that regulate greening in *Arabidopsis*.

When *det1* tissue was put into culture on synthetic medium containing only auxin, we observed that the calli were green, instead of achlorophyllous like the wild-type, indicating that the undifferentiated calli derived from *det1* leaves contained chloroplasts. Since the growth regulator, cytokinin, is normally required for greening in calli, this observation implied that the *det1* mutants had become cytokinin attenuated. Here, we test the possibility that aberrant cytokinin physiology is related to the numerous phenotypes observed in *det1* seedlings. On the basis of our results with exogenously added cytokinins, we suggest that cytokinins play an important role during greening of *Arabidopsis*, and that abnormally high cytokinin levels or an increased sensitivity to cytokinins may be explanations for some of the observed phenotypes of *det1* mutants.

Results

Review of the phenotype of det1 *mutants*

Dark-grown dicotyledonous seedlings are developmentally arrested. They have extended hypocotyls, no cotyledon expansion, no true leaves, and are white. A set of well-characterized genes, the so-called light-regulated genes (e.g. *cab*, the chlorophyll a/b binding protein gene; *rbc*S, the small subunit of the ribulose bisphosphate carboxylase gene; *chs*, the chalcone synthase gene, and a number of chloroplast genes) are not expressed or are expressed at very low levels. In contrast, recessive mutations in the *DET1* gene affect a wide variety of light-regulated traits, including leaf and chloroplast development, gene expression, anthocyanin accumulation, and germination (Table 1) (Chory *et al.* 1989*a*). *det1* mutants also have an aberrant phenotype when grown in the light (Table 1, Chory and Peto, 1990). Light-regulated traits that are affected in light-grown *det1* mutants include: ectopic development of chloroplasts in root cortex cells, inappropriate expression of nuclear photosynthesis genes in roots, and ectopic expression of the *chs* promoter in leaf mesophyll cells and flowers (Chory and Peto, 1990). In addition, *det1* mutants lack apical dominance, have green roots, and green in tissue-culture in the absence of exogenously added cytokinins. These characteristics indicated that *det1* mutations affect cytokinin physiology, either directly or indirectly. To test this hypothesis, we undertook the experiments described below.

Exogenously applied cytokinins cause morphological changes in wild-type plants

We tested the effects of exogenously supplied cytokinins on dark- and light-

Table 1. *Summary of greening phenotypes in wild-type and* det1 *plants*

	Dark Grown		Light-Grown	
	DET1[+]	*det1*[−]	*DET1*[+]	*det1*[−]
Leaves	Unexpanded cotyledons	Expanded cotyledons, leaves		
Hypocotyl	Long	Short	Short	Short
Pigments	Absent	Anthocyanins	Green	Pale-green
Germination	20 % of light levels	100 % of light levels	100 %	100 %
Plastid type (leaf)	Etioplast	Young chloroplast	Chloroplast	Chloroplast
(root)	n.d.	n.d.	Amyloplast	Chloroplast
Gene expression*:				
Leaf *chs* (nuclear)	Undetected	100	100	200
cab (nuclear)	1–2	25	100	50
*rbc*S (nuclear)	2–5	80		
Chloroplast genes	1–2	100	100	70
Root *cab*	n.d.	n.d.	0	10
chloroplast genes			1–2	10

*Gene expression values are expressed as a percentage of the total RNA accumulated in wild-type plants grown in the light.

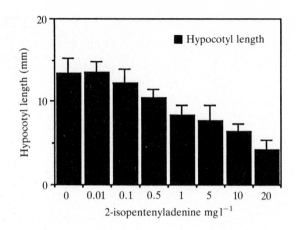

Fig. 1. Effect of 2-isopentenyl adenine on hypocotyl length of dark-grown wild-type *Arabidopsis* seedlings. Seedlings were grown for 7 days in the dark on MS medium, 2% sucrose, supplemented with increasing concentrations of 2-isopentenyl adenine as indicated. At least 100 hypocotyls per sample were measured after growth in the dark.

grown wild-type plants. Seeds were surface-sterilized, and plated on MS medium containing various concentrations of cytokinin, either in the light or in the dark. The cytokinins tested included: 2-isopentenyl adenine, benzyladenine, and kinetin; the range of concentrations tested were from 0 to $20\,\mathrm{mg\,l^{-1}}$. The effects of each cytokinin on dark-grown seedlings was examined at 7 days and 20 days post-germination. Each cytokinin had a similar effect on wild-type morphology, and only the results from the 2-isopentenyl adenine experiments are presented. 2-isopentenyl adenine had a dramatic effect on hypocotyl elongation of wild-type seedlings (Fig. 1). The hypocotyl length was increasingly shorter with increasing concentrations of cytokinin, up to $20\,\mathrm{mg\,l^{-1}}$. At concentrations higher than this (e.g. $30\,\mathrm{mg\,l^{-1}}$), the seedlings did not germinate well. Further, at 7 days, the cotyledons were expanded and open, unlike wild-type seedlings grown in the absence of cytokinin (Fig. 2a). By 20 days, the cytokinin-treated seedlings had developed true leaves (Fig. 2b). All of these characteristics, i.e. inhibition of hypocotyl elongation, cotyledon expansion, and leaf development, are similar to those of the *det1* mutant. Thus, we were able to make an almost perfect phenocopy of *det1* by the addition of cytokinin to our growth media. Interestingly, when *det1* seedlings were subjected to the same cytokinin treatments, they were much more sensitive to the concentration of cytokinin in the medium than the wild-type. At concentrations of $10\,\mathrm{mg\,l^{-1}}$, the *det1* seeds did not germinate well (data not shown).

Exogenously supplied cytokinin also had an effect on the morphology of light-grown wild-type plants. These plants were smaller than wild-type without cytokinin, paler, had increased anthocyanins, and decreased apical dominance. As for dark-grown seedlings, we observed phenotypes that were very similar to *det1* seedlings. The one noticeable difference is that wild-type seedlings treated with

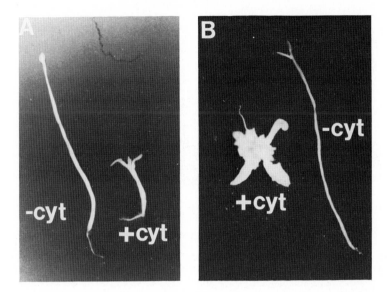

Fig. 2. Morphology of wild-type seedlings after growth in the dark on 2-isopentenyl adenine. (A) 7-day old seedlings after growth in the presence or absence of cytokinins. (B) 20-day old seedlings after growth in the presence or absence of cytokinins. Seedlings were grown in the dark on MS medium supplemented with 2 % sucrose and $20 \, \text{mg} \, l^{-1}$ 2-isopentenyl adenine.

cytokinin did not develop extensive roots, while *det*1 mutants had fairly normal root growth.

Cytokinins allow chloroplast development in the absence of light

Since the morphological changes observed after cytokinin treatment were reminescent of *det*1 mutants, we wanted to test if other light-regulated traits were also affected in the cytokinin-treated wild-type seedlings. We used electron microscopy to examine the plastid types of wild-type seedlings either grown in the absence or presence of 2-isopentenyl adenine ($20 \, \text{mg} \, l^{-1}$). Fig. 3A shows the typical etioplast structure of dark-grown wild-type *A. thaliana* seedlings. These etioplasts were small, irregularly shaped, and contained a central paracrystalline assembly of tubules, the prolamellar body. In contrast, plastids in the cytokinin-treated wild-type plants (Fig. 3B) showed clear signs of chloroplast development, as evidenced by the lack of prolamellar bodies in the plastids, the somewhat larger size and more regular lens shape of the plastid, and the formation of some thylakoid membrane structures. These developing chloroplasts appeared similar to the chloroplasts that we previously observed in dark-grown *det*1 seedlings (Chory *et al.* 1989*a*).

'Light'-regulated genes are expressed in the dark in the presence of cytokinins

A remarkable trait of *det*1 mutants is that light-regulated gene expression is uncoupled from the presence of a light signal (Table 1). Since cytokinin-treated

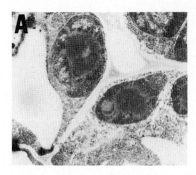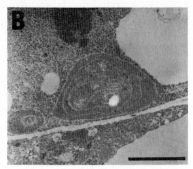

Fig. 3. Electron micrographs of representative plastids from dark-grown wild-type seedlings grown in the absence (A) or the presence (B) of cytokinins. Note the large prolamellar body in (A), while in (B) the plastid is enlarged, has a lens-shape, and contains thylakoid membranes. Bar=1 μm.

wild-type seedlings had many of the morphological phenotypes of *det1* mutants, we tested to see if light-regulated genes were now expressed in the dark in cytokinin-treated wild-type seedlings. To do this experiment, we used previously constructed transgenic lines that contained either the *chs* or *cab3* promoter fused to the *Escherichia coli* β-glucuronidase (GUS) gene (Chory and Peto, 1990). We chose these two promoters because work in our lab (Chory *et al.* 1989*b*), and by others (Feinbaum and Ausubel, 1988; Karlin-Neumann *et al.* 1988) has shown that the *chs* gene is regulated primarily by blue-light signals, while *cab3* is red-light controlled. Thus, these transgenic lines would allow us to assess two divergently light-regulated genes. Measurements of GUS activity indicated that the *chs* and *cab3* promoters were at least 25-fold more active in cytokinin-treated dark-grown seedlings than in control seedlings (Table 2). The levels observed were similar to those that we observed when transgenic *det1* mutants were plated and grown in the dark in the absence of cytokinins (Table 2). Thus, by a third criterion, gene expression, cytokinin-treated wild-type seedlings had similar characteristics to *det1* seedlings.

Table 2. *Light-dark expression of* cab *and* chs *promoters in seedlings treated with cytokinins*

Construct	Wild-type (no cytokinins)		Wild-type (+20 mg l⁻¹, 2-IP)		*det1*	
	Light	Dark	Light	Dark	Light	Dark
p*cab3*–GUS	6500	60	5100	2100	5700	3200
p*chs*–GUS	31 840	1990	60 000	49 000	71 000	57 000

GUS units are pmol 4-MU min⁻¹ mg⁻¹ protein (where 4-MU is 4-methyl umbelliferone).
  Plants were grown for either 10 days in the light (light) or were germinated and grown in the dark for 7 days (dark). The wild-type (no cytokinins) and *det1* data are from Chory and Peto (1990) and are shown here for comparison.

Putative cytokinin-insensitive mutants have greening phenotypes

Since dark-grown wild-type seedlings had such a striking phenotype when grown on cytokinins, we decided to screen for mutants that looked etiolated in the presence of cytokinins. These cytokinin-insensitive mutants might be opposite to *det* mutants, and may aid in the further analysis of greening in *Arabidopsis*. Cytokinin-insensitive mutants were identified in a population of mutagenized M_2 generation seeds that were sown at a density of approximately 1000 seeds per 100 mm Petri dish and allowed to grow in complete darkness in the presence of $20 \, mg \, l^{-1}$ 2-isopentenyl adenine. Following 7 days of growth in the dark, the plates were scored for seedlings that appeared etiolated. Out of 60 000 M_2 seeds tested, 16 putative cytokinin-insensitive mutants were identified. Of these, 5 never developed leaves and died; 11 grew to maturity and were re-screened for the cytokinin-insensitive phenotype in the M_3 generation. Three of the 11 tested positive for the cytokinin-insensitive phenotype in the M_3 generation. Interestingly, 2 of these are pale yellow-green, i.e. have a greening phenotype. Further molecular and genetic analyses are being performed on these mutants. It will clearly be of interest to cross these mutants to the *det*1 mutants, and examine the phenotype of the double mutants.

Conclusions

We have shown that many of the phenotypes of *det*1 mutants can be mimicked by the addition of 2-isopentenyl adenine to the growth medium. The traits we examined included leaf and chloroplast development and gene expression. To our knowledge, this is the first report of de-etiolating a dicotyledonous seedling in the dark in the presence of cytokinins.

The similarity of red-light and cytokinin effects was first noted in 1956 by Miller (Miller, 1956). Since then, work has concentrated on studying the effects of adding cytokinins to undifferentiated tissue culture cells or to excised cotyledons (e.g. Stetler and Laetsch, 1965; Teyssendier de la Serve *et al.* 1985). These studies have indicated a role for cytokinins in chloroplast development and expression of genes for chloroplast-destined proteins. More recently, *Lemna gibba* has been used to study cytokinin effects on light-regulated gene expression (Flores and Tobin, 1986). From these latter studies, it was concluded that kinetin regulation of *cab* and *rbc*S mRNA levels was primarily post-transcriptional. In *Arabidopsis*, the increased expression of *cab* and *chs* in the presence of cytokinins is at the transcriptional level, since we show that the promoter is all that is required for increased GUS activity. Indeed, our studies show that greening in *Arabidopsis* is particularly sensitive to the levels of cytokinin.

The primary mode of action of phytochrome, the blue-light photoreceptor, or cytokinins is currently unknown. Based on their data with red-light and cytokinin treatments, Flores and Tobin (1986) speculated that phytochrome and cytokinins independently change the pool size of a common intermediate, and this intermediate more directly regulates gene expression. It is interesting to speculate that this intermediate is specified by the *DET*1 gene. Since high cytokinin to auxin ratios are known to promote organ (shoots and leaves) formation from callus tissue, it is further tempting to speculate that the cytokinin effects are related to tissue-specificity. In favor of this last argument, it was recently shown that tobacco seedlings, stressed by transformation with *Agrobacterium tumefaciens* over-expressing the T-cytokinin gene, had changes in the tissue-specific control of the levels of several defense-related mRNA species (Memelink *et al.* 1987). Alternative explanations for the *det*1 phenotype that relate to the experiments described here are that the cytokinin levels in *det*1 seedlings are abnormally high or *det*1 mutants have an increased sensitivity to cytokinins.

Clearly, many questions need to be addressed to define the role of cytokinin in the regulation of chloroplast development and gene expression. Quantitative measurements of endogenous cytokinins need to be made in light-grown and etiolated seedlings, and compared with levels in the *det*1 mutant. Molecular clones of the *DET*1 gene will aid in the analysis of its function. Finally, isolation and characterization of additional cytokinin mutants might aid in the elucidation of the mode of action of cytokinin and how cytokinin interacts with the developmental pathways involving light in plants. All of these experiments are in progress.

This research was funded by grants from the Department of Energy and the National Science Foundation to J.C, and by The Samuel Roberts Noble Foundation. N.A. was funded through a summer research program at The Salk Institute in conjunction with The Elementary Institute of Science, San Diego. We thank Dr Paul Sawchenko for use of E. M. facilities and Dr A. Binns, University of Pennsylvania, for helpful suggestions.

References

CHORY, J. AND PETO, C. (1990). Mutations in the *DET*1 gene affect cell-type-specific expression of light-regulated genes and chloroplast development in *Arabidopsis*. *Proc. natn. Acad. Sci. U.S.A.* **87**, 8776–8780.

CHORY, J., PETO, C., ASHBAUGH, M., SAGANICH, R., PRATT, L. AND AUSUBEL, F. (1989*b*). Different roles for phytochrome in etiolated and green plants deduced from characterization of *Arabidopsis thaliana* mutants. *The Plant Cell.* **1**, 867–880.

CHORY, J., PETO, C., FEINBAUM, R., PRATT, L. AND AUSUBEL, F. (1989*a*). *Arabidopsis thaliana* mutant that develops as a light-grown plant in the absence of light. *Cell.* **58**, 991–999.

COLBERT, J. T. (1988). Molecular biology of phytochrome. *Pl. Cell Environ.* **11**, 305–318.

DALE, J. E. (1988). The control of leaf expansion. *A. Rev. Pl. Physiol. Pl. molec. Biol.* **39**, 267–295.

FEINBAUM, R. AND AUSUBEL, F. (1988). Transcriptional regulation of the *Arabidopsis thaliana* chalcone synthase gene. *Molec. cell. Biol.* **8**, 1985–1992.

FLORES, S. AND TOBIN, E. M. (1986). Benzyladenine modulation of the expression of two genes for nuclear-encoded chloroplast protein in *Lemna gibba*: Apparent post-transcriptional regulation. *Planta.* **168**, 340–349.

GRUISSEM, W. (1989). Chloroplast gene expression: How plants turn their plastids on. *Cell* **56**, 161–170.

KARLIN-NEUMANN, G. A., SUN, L. AND TOBIN, E. M. (1988). Expression of light-harvesting chlorophyll a/b protein genes is phytochrome-regulated in etiolated *Arabidopsis thaliana* seedlings. *Pl. Physiol.* **88**, 1323–1331.

LEECH, R. M. (1976). Plastid development in isolated etiochloroplasts and isolated etioplasts. In *Perspective in Experimental Biology* (ed. N. Sunderland). New York: Pergamon Press. **2**, 145–162.

MATTHIS, J. N., BRADBURNE, J. A. AND DUPREE, M. A. (1989). Gibberellic acid effects on greening in pea seedlings. *Pl. Physiol.* **91**, 19–22.

MEMELINK, J., HOGE, J. AND SCHILPEROORT, R. (1987). Cytokinin stress changes the developmental regulation of several defense-related genes in tobacco. *EMBO J.* **6**, 3579–3583.

MILLER, C. O. (1956). Similarity of some kinetin and red light effects. *Pl. Physiol.* **31**, 318–319.

MULLET, J. E. (1988). Chloroplast development and gene expression. *A. Rev. Pl. Physiol. Pl. molec. Biol.* **39**, 475–502.

STETLER, D. A. AND LAETSCH, W. M. (1965). Kinetin induced chloroplast maturation in cultures of tobacco tissue. *Science.* **149**, 1387–1389.

TEYSSENDIER DE LA SERVE, B., AXELOS, M. AND PEAUD-LENOEL, C. (1985). Cytokinins modulate the expression of genes encoding the protein of the light-harvesting chlorophyll a/b complex. *Pl. molec. Biol.* **5**, 155–163.

Printed in Great Britain © *Society for Experimental Biology 1991*

DEVELOPMENTAL GENETIC STUDIES OF THE MOSS, *PHYSCOMITRELLA PATENS*

D. J. COVE[1], *W. KAMMERER*[1], *C. D. KNIGHT*[1], *M. J. LEECH*[2], *C. R. MARTIN*[2] and *T. L. WANG*[2]

[1] Department of Genetics, University of Leeds, Leeds, LS2 9JT, UK and [2] John Innes Institute, Colney Lane, Norwich NR4 7UH, UK

Summary

The development of the haploid gametophyte stage of *Physcomitrella patens* presents excellent opportunities for the detailed study of plant morphogenesis at the cellular level. The filamentous protonema undergoes a number of developmental transitions that can be observed directly in living material using time-lapse video microscopy and that can be manipulated both by treatment with phytohormones and by environmental stimuli. Mutants affecting these processes can be isolated and can be analysed using conventional as well as parasexual methods. Molecular biological techniques are now being established since these will be essential if the mechanisms that bring about morphogenetic processes are to be understood at the molecular level. A technique for genetic transformation has been devised and is being exploited to attempt to tag developmentally-relevant genes using maize transposons. Changes in gene activity associated with developmental transitions and in response to treatment with phytohormones and environmental stimuli are being studied using cDNA library subtraction techniques. Heterologous genes involved in the control of transcription and of the cell cycle are being used to probe the *P. patens* genome to identify possible homologues involved in developmental regulation.

Introduction

The study of development, and in particular the cellular basis of morphogenesis, requires the choice of an organism that is amenable to analysis by the methods of both cell and molecular biology. We believe that mosses are very suitable for such studies, but the development of molecular genetic procedures for this group of organisms is crucial if they are to be fully exploited for the study of plant development. The potential of mosses for the study of development was recognised over 60 years ago (see for example: von Wettstein, 1924) and physiological studies of development were well advanced by 1960 (see for example: Mitra and Allsopp, 1959*a*,*b*; Mitra *et al.* 1959). Developmental genetic studies, principally of *Physcomitrella patens*, have since confirmed many of the conclusions made from physiological studies and have added to our detailed

Key words: bryophytes, cellular morphogenesis, mosses, *Physcomitrella patens*, transformation, transposon-tagging.

knowledge of development. Our understanding of developmental events at a molecular level is however, still very poor and much of our research is now directed towards this. This paper gives an audit of our progress to date.

Reviews are available both of the techniques for the study of *P. patens* (Ashton *et al.* 1988; Knight *et al.* 1988) and of the developmental studies that have been carried out (Cove and Ashton, 1984, 1988). These will not, therefore, be covered in detail here.

Gametophyte development in *Physcomitrella patens*

The gametophyte generation of mosses has two advantages for the developmental genetic study of morphogenesis. Firstly, being haploid, mutant isolation is straightforward with no necessity to breed from mutagenised material to detect mutants. This may not necessarily be an advantage, since mutants with developmental blocks are lethal in organisms where sexual reproduction is essential for propagation. However, the second advantage of moss gametophytes is that they show considerable powers of regeneration; cells from almost any tissue behave like spores, dividing to produce at least the earliest cell-type characteristic of gametophyte development that can be cultured indefinitely. Thus even mutants blocked early in development can be propogated vegetatively. Prolonged storage of such mutants can be achieved by cryogenic treatment of gametophyte tissue (Grimsley and Withers, 1983).

P. patens is usually cultured in the laboratory axenically in continuous white light at 25°C on a simple solid medium containing inorganic salts. Under these conditions, spore germination or cell regeneration leads to the production of filamentous tissue, the protonemata. Protonemal filaments elongate by the repeated division of the apical cell and branch by division of sub-apical cells. The first filament type to be produced contains chloronemal cells (Fig. 1, 2A), characterised by being densely packed with large near-spherical chloroplasts. The apical cells of chloronemal filaments divide at about 24h intervals, and subapical cells usually divide once or twice to produce lateral chloronemal filaments. A few days after spore germination, some chloronemal apical cells develop into the second protonemal cell type, caulonemata. Caulonemal cells contain fewer, spindle-shaped chloroplasts and also differ from chloronemata in having oblique rather than perpendicular cross walls (see Fig. 2B). The apical cells of caulonemal filaments divide more frequently than those of chloronemal filaments. When first formed their cell-cycle time is about 12h but the apical cells of more mature caulonemal filaments divide every 6h. The transition of a chloronemal to a caulonemal apical cell usually takes only one or two cell divisions.

In high light intensities (e.g. $>2\ \mathrm{W\,m^{-2}}$ white light), caulonemal sub-apical cells divide several times to produce single-celled side branch initials that can adopt a number of alternative developmental fates. In these conditions, most side branch initials develop into chloronemal filaments. A small proportion (1 to 5 %) develop into caulonemal filaments and a similar proportion develop first into buds and then into gametophores, the leafy shoots which are most usually associated with moss plants (see Fig. 3).

Lower light intensities lead to an altered balance between these alternative fates and under some conditions the majority of side branch initials develop no further (see Fig. 4B).

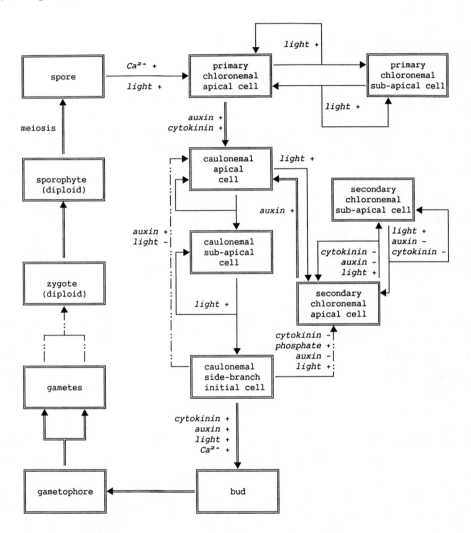

Fig. 1. Developmental cycle of *Physcomitrella patens*. Unless stated to be diploid, stages are haploid. Transitions between stages that are connected by a broken-lined arrow do not require cell division. Transitions between stages that are connected by a single-lined arrow require a single cell division. Transitions between stages that are connected by a double-lined arrow require more than one cell division. The (+) sign beside a phytohormone or environmental signal indicates that it is required for or enhances the frequency of the transition, the (−) sign indicates that it decreases the frequency of the transition.

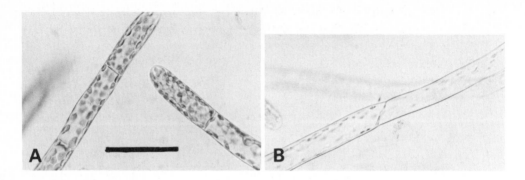

Fig. 2. Chloronemal (A) and caulonemal (B) cells from protonema of *P. patens*. Scale bars=100 μm.

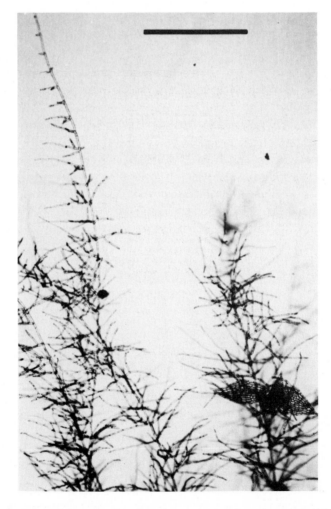

Fig. 3. Caulonemal filaments showing the pattern of caulonemal side branches characteristic of growth in high intensity white light (10 W m^{-1}). Scale bar=1 mm.

Studies of developmental mutants have identified roles for auxins and cytokinins in the development of the *P. patens* gametophyte

A large variety of mutants of *P. patens*, altered in the development of their protonema, can be isolated following mutagenesis of spores or of somatic tissue (Boyd *et al.* 1988; Knight *et al.* 1988). Mutants altered in their responses to auxin and cytokinin are particularly easy to isolate (Ashton *et al.* 1979). The treatment of wild-type cultures with auxin leads to increased production of caulonemal filaments while treatment with cytokinin leads to a massive increase in bud production (Cove and Ashton, 1984; see Fig. 4). Two classes of mutants can be isolated that are resistant to the developmental effects of these phytohormones. The first produces only chloronemal filaments and the second produces both chloronemal and caulonemal filaments but no buds. Analyses of these mutants in parallel with experiments involving the replacement of endogenous phytohormones by a continuous external supply of medium (Cove, 1984), have identified two stages of protonemal development that involve auxin and cytokinin. Low levels of auxin and cytokinin are required in combination for the transition of chloronemal apical cells to caulonemal apical cells but higher levels of both together are required for the production of buds and for the modulation of secondary chloronemal growth (Cove and Ashton, 1984 and 1988 – see Fig. 1). Many of the mutants isolated appear to be affected in their ability to synthesise either auxins or cytokinins, since they respond to their exogenous supply. Other mutants, however, are insensitive to the supplied phytohormones and therefore have phenotypes that suggest their ability to respond to these phytohormones is blocked. It is this class of mutants that we are particularly interested, since these should have defects in the transduction pathway for the hormone signal. The isolation of genes that can mutate to this phenotype is one of our highest priorities.

Light and gravity affect protonemal development

The intensity of light, its wavelength and its direction all effect protonemal

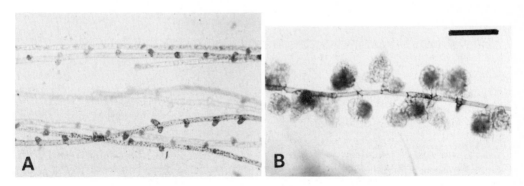

Fig. 4. Caulonemal filaments from *P. patens* culture grown in low intensity white light $(100\,\mathrm{mW\,m^{-1}})$ in the absence (A) and presence (B) of cytokinin ($1\,\mu\mathrm{M}$ benzylaminopurine). Scale bar$=200\,\mu\mathrm{m}$.

development (Cove *et al.* 1978; Cove and Knight, 1987). The development of caulonemal filaments is particularly sensitive to light intensity (see Fig. 1); bud production, for example, requires a higher level of light than the formation of side branches by caulonemal subapical cells. Mutants altered in these responses have not yet been specifically identified but it is possible that some mutants, which are unable to produce buds, even when treated with cytokinin or auxin, may be blocked in the detection of light. Gametophores and the apical cells of both chloronemal and caulonemal filaments show phototropism (Jenkins and Cove, 1983*a*; Cove and Knight, 1987). The phototropic response of protonemal apical cells does not result in the bending of the filament but in a change of the direction of extension of the tip of the apical cell. Mutants with abnormal phototropic responses have been isolated (Cove *et al.* 1978; Jenkins and Cove, 1983*b*) and these show pleiotropically abnormal responses from both protonemal apical cells and gametophores, indicating that there must be similar mechanisms involved in the multi-cell response of the gametophore and the single-cell response of the protonema.

Gametophores and caulonemal apical cells also respond to gravity, growing negatively gravitropically. These gravitropic responses are however, only shown in darkness (Cove and Knight, 1987) or at very low light intensities ($<10\,\mathrm{mW\,m^{-2}}$ white light). Mutants have been isolated that show abnormal gravitropic responses by their caulonemal apical cells but these mutants retain the normal response to gravity by their gametophores. For gravitropism therefore, the response of the two tissues must be, at least to some extent, independent. Mutants with abnormal phototropism show no abnormalities in their response to gravity and *vice versa* (Jenkins *et al.* 1986; Cove and Knight, 1987) and so we have not so far obtained evidence that two tropic responses involve common mechanisms. The tropic responses of protonemal apical cells are particularly suitable for the study of cellular aspects of morphogenesis since perception and response take place within the same cell. The isolation of the genes that can mutate to affect these tropisms is, therefore, also a high priority.

The development of techniques for the molecular analysis of development in *P.*
patens

Transformation

Considerable effort has been put into developing a transformation system for *P. patens*, and this has only recently been successful (Schaefer *et al.* 1991). Although a number of techniques have been attempted, including *Agrobacterium tumefaciens* T-DNA-mediated transfer (C. D. Knight and B. Hohn, unpublished data), success has only been achieved using a technique based on one devised for higher plants (Saul *et al.* 1988), which utilises the direct uptake of DNA by protoplasts, mediated by polyethylene glycol. We have used plasmids containing genes coding for antibiotic-resistance, coupled to promoters known to be active constitutively in flowering plants. Selection for antibiotic-resistant regenerants following treatment with plasmid-DNA-containing genes conferring resistance either to the synthetic aminoglycoside antibiotic, G418, or to hygromycin, yields initially from about 1 to 25 regenerants per 1000 viable protoplasts (1–150 regenerants per μg of plasmid DNA). No such regenerants are found when DNA is used from control plasmids containing no genes for antibiotic resistance or an antibiotic resistance gene that lacks a promoter (W. Kammerer and D. Cove, unpublished data).

The behaviour of these initial regenerants varies. Approximately half do not grow beyond a plant containing about 100 cells and it is likely that this class of regenerant represents plants in which antibiotic resistance is only expressed transiently. Almost all the remaining regenerants do continue to grow in the presence of selection although growth is considerably slower than a control wild-type grown in the absence of antibiotic. We describe this class of transformant as unstable since antibiotic resistance is lost rapidly when selection is relaxed. We have not yet established the status of the transforming DNA in these unstable transformants. The third class of transformant retains antibiotic resistance after relaxation of selection and we therefore describe this class of transformant as stable. Stable transformants grow faster than unstable transformants on media containing antibiotic. The most resistant transformant so far obtained, which was selected after treatment with plasmid DNA coding for neomycin phosphotransferase II (*npt*II), grows as well on medium containing $50 \, \text{mg} \, l^{-1}$ G418, as it or the wild-type grows on medium containing no antibiotic. Stable transformants are not common (about 1 per 10^6 viable protoplasts or 1 per 100 μg of plasmid DNA). The rate of recovery of regenerants is lower when linear DNA is used instead of covalently closed circular DNA, however, there is some evidence that the rate of occurrence of stable transformants is higher for linear than for circular DNA (D. Schaefer and J.-P. Zryd, unpublished data).

Analysis of the DNA from stable transformants shows that from 10 to 30 copies of the plasmid have become integrated at a single site. The antibiotic-resistant phenotype is not only stable upon vegetative subculture and in the absence of selection but is also transmitted meiotically. In a cross of a stable transformant strain to a sensitive strain, resistance to sensitivity segregates 1:1 as is expected for Mendelian inheritance of a character observed in haploid tissue. Analysis of the DNA from the progeny of such a cross shows that antibiotic-sensitive progeny contain no plasmid DNA whereas resistant progeny resemble the stable transformant parent and contain a similar number of copies of the plasmid (See Fig. 5). The resistant progeny retain their antibiotic resistance in the absence of selection.

Work is now in hand to increase the frequency of stable transformants although the unstable transformant phenotype can be exploited for certain procedures (see below). We are testing whether exposure of protoplasts prior to transformation to recombinogenic treatments such as UV irradiation is effective. We are also examining whether the inclusion of *P. patens* DNA in the construct used for transformation has any effect.

The behaviour of the maize transposon Ac in the moss Physcomitrella patens

To establish a method to isolate genes from *P. patens*, we are developing a gene-tagging system using maize transposons. The pKU plasmid family (Baker *et al.* 1987) has been introduced into the moss by transformation. The system comprises three plasmids; pKU2 is a positive control and contains the *npt*II gene under the control of the 1' promoter from *Agrobacterium* T-DNA. The second plasmid, pKU3 contains a complete Ac transposon (4.6 kb), inserted into the untranslated leader region of the *npt*II gene, thus abolishing any expression of the gene as long as the transposon remains inserted. A third plasmid, pKU4 provides a negative control being similar to pKU3 except that a 1.6 kb internal *Hind*III fragment of the transposon, which includes part of the transposase coding region, is deleted. pKU4 thus contains a Ds rather than an Ac element. After transformation of *P. patens*, we have only been able to recover G418 resistant colonies with the pKU2 plasmid.

We therefore conclude that Ac is unable to excise in *P. patens*, probably because

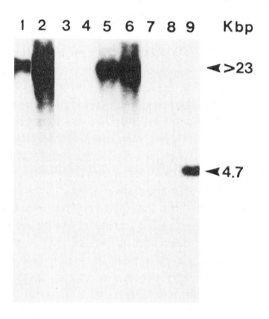

Fig. 5. Southern hybridisation of restricted DNA from the hygromycin-resistant transformant 15.03 and progeny from a cross between 15.03 and the hygromycin-sensitive strain, nicA4. The 1 kbp *Bam*H1 fragment of plasmid pGL2 was used as a probe. Lane 1 was loaded with 0.35 μg of uncut DNA from transformant 15.03. Lanes 2 to 8 were loaded with 3.5 μg of moss DNA digested with *Hind*III, from strains as follows: lane 2, 15.03; 3 and 4, hygromycin-sensitive progeny; 5 and 6, hygromycin-resistant progeny; 7 and 8, pabA3. pGL2 DNA was linearised by *Hind*III digestion and loaded in lanes 8 (24 pg) and 9 (240 pg), equivalent to 1 and 10 copies per haploid moss genome respectively. A longer exposure was required to reveal a 4.7 kb band in lane 8. *Hind*III should normally cut the plasmid once to generate a band of 4.7 kb but because this enzyme was used to linearise the plasmid prior to transformation, this site has not been reconstituted and, as with the undigested DNA from 15.03 (lane 1), the hybridisation signal remains with the high relative molecular mass DNA.

the Ac transposase is not functionally expressed in the moss. This may be because the transposase promoter is not active or because splicing of the pre-mRNA is not accurate. To address this problem, we have used a plasmid containing a cDNA sequence for the Ac transposase coupled to the 2′ promoter from *Agrobacterium* T-DNA and containing a hygromycin-resistance marker for selection (Coupland *et al.* 1988). Co-transformation experiments were first carried out using this transposase expression vector and pKU3. Transformants were selected on hygromycin and after one to two weeks growth, selected colonies were tested for G418 resistance. About 3–5 % of the hygromycin resistant colonies showed G418 resistant sectors. We conclude that the Ac cDNA-containing construct expresses the Ac transposase thus allowing the excision of the element from the plasmid DNA. Similar results were obtained with the pKU4 plasmid containing the Ds element, as well as with some other constructs where additional DNA was inserted in the Ds element.

As described above, transformation in *P. patens* usually leads to unstable expression of antibiotic resistance in that resistance is only retained providing selection is maintained. However, following transformation with the Ac cDNA-containing plasmid, moss lines have been established which stably maintain the hygromycin resistance marker and these have been shown to contain several copies of the structurally-intact Ac cDNA sequence. Southern blot analysis indicates that the plasmid has integrated in the transposase region of one of the copies. The *P. patens* genome contains sequences homologous to Ac and we do not know whether integration took place in one of the homologous sequences of the moss DNA.

When the transformed line that stably expresses hygromycin resistance is retransformed with pKU plasmids containing either Ac or Ds, G418 resistant colonies are recovered. The transformed strain containing the Ac cDNA therefore functionally expresses the Ac transposase. We have not so far been able to establish a selection system using a third marker and so we have been unable to select for transformation independently of excision and cannot, as yet, estimate the rate of excision. Transformation using other Ds-containing plasmids supports the data from earlier experiments. We can only recover G418 resistant colonies in the transformed line expressing the Ac transposase; no resistant colonies are recovered in control experiments employing an isogenic moss strain. We therefore conclude that the selection system is tight and we are not observing some uncontrolled events. However, in this series of experiments we have so far been unable to detect an integrated transposon in the moss genome.

In a further series of experiments, we have attempted to exploit the phenomenon of unstable transformation shown by our system. As described above, most transformants lose resistance after two weeks growth in the absence of selection. An *npt*II gene has been integrated into a Ds element to give a plasmid, which although unsuitable for the direct selection of excision events, can be used to select for transformation. After initial selection on G418, selection was relaxed for two weeks by growing the regenerants on antibiotic-free medium.

Selection was then re-established by transfer to medium containing G418. This scheme will select for stable G418-resistant transformants that could arise either by integration of the whole plasmid or by transposition of the Ds element containing the *npt*II gene into the *P. patens* genome. About 3 % of the colonies initially selected retained resistance after selection had been relaxed. One colony was completely resistant and others showed resistant sectors. Surprisingly, not only the completely resistant colony but also one that showed a sector, seems to have integrated the complete plasmid. Further colonies of this type are being analysed.

The identification of developmentally-regulated genes

cDNA subtractive hybridisation techniques allow the identification of mRNAs that are produced at specific developmental stages or in response to particular treatments. These techniques have considerable potential in the study of *P. patens* development and should allow the identification, for example, of genes expressed only in one cell type. To develop these techniques, we are seeking first to identify changes in gene expression associated with cytokinin treatment. Specific and rapid transcriptional changes have been found in flowering plants following treatment with several classes of plant hormone, particularly auxins (Hagen, 1988). However, no such changes have yet been identified in bryophytes following cytokinin treatment and none of the changes which have been identified, are closely associated with morphogenetic changes. The rapid morphogenetic response of protonemal tissue to cytokinin (see Fig. 4) therefore appears to be an ideal starting point for such a study in *P. patens*.

We have isolated mRNA from tissue that had been grown for 9 days, following inoculation with a protonemal homogenate (Knight *et al.* 1988) and then treated with the cytokinin, benzylaminopurine. At the time of treatment, such tissue is composed almost entirely of protonemata, but will respond to cytokinin by producing very large numbers of buds within 36 h. To seek to identify early changes in gene expression, mRNAs were pooled from tissue that had been treated with cytokinin for 1, 2, 4 and 6 h. No morphological response is visible six hours following treatment with cytokinin. A control sample of mRNA was made from tissue that was treated in parallel with water instead of benzylaminopurine. A cDNA library was constructed in lambda gt11 and a differential screen was carried out using a probe enriched for cytokinin-induced sequences and a probe from uninduced tissue. Enrichment of the probe for induced sequences was achieved by three rounds of subtractive hybridisation by cDNAs derived from uninduced tissue. The cDNA from uninduced tissue was amplified by the polymerase chain reaction (PCR) and was photobiotinylated, before being hybridised to cDNA from induced tissue. A 10-fold excess of cDNA from uninduced tissue was used at each round of subtraction. The hybridisation mixture was then separated using the method of Sive and St. John (1988) and biotinylated hybrids were discarded. Following three rounds of hybridisation, the cDNA sample enriched for cytokinin-induced sequences was amplified by PCR.

In a screen of 8×10^4 plaques, 500 were picked as possible candidates for containing cytokinin-regulated sequences. Upon rescreening 200 of these, it was confirmed that 10 clones hybridised differentially to the two probes but all showed some signal with the uninduced probe, indicating quantitative rather than qualitative changes in gene expression. The 10 clones are now being analysed further using northern hybridisation to confirm that they do indeed represent cytokinin-induced messages.

The identification of developmentally-relevant genes using heterologous probes

The probing of flowering plant genomes with mammalian *myb* oncogenes, which code for proteins that function as transcriptional activators, has revealed that they contain homologues (Paz-Ares *et al.* 1987; Marocco *et al.* 1989). The function of these homologues is unknown and in some cases the only similarity between them and *myb* genes resides in the highly conserved DNA binding domain. We are investigating whether similar genes can be identified in *P. patens* and if so whether they show developmental regulation.

We have isolated mRNA samples from protonemal tissue, from similar tissue but treated with cytokinin, and from tissue from a mature *P. patens* culture that contained abundant gametophores. First strand cDNA was made from each mRNA sample, using a modified oligo dT primer. This contained a sequence that would anneal with a primer for use in PCR. The cDNA was amplified using PCR and, as the second primer, a 42-mer oligonucleotide probe coding for the highly conserved DNA binding domain from maize *C1* gene, a *myb* homologue (Jackson *et al.* 1991), was used. Gel analysis of each of the three reactions showed in each, a single 1.3 kb band. Southern hybridisation showed that this fraction hybridised to the original *C1* oligonucleotide probe. The intensity of the bands from the three samples is not equal. Although no difference could be observed between the protonemal samples with and without cytokinin, neither contained as much material as the sample of tissue from a mature culture and so it is possible that there is a *myb* homologue in *P. patens* that is developmentally regulated.

We are also attempting to identify homologues of genes known to be involved in the regulation of the cell cycle in fungi. The developmental control of the cell cycle is a crucial part of gametophyte development and if we are to understand, for example, why caulonemal apical cells divide at three times the rate of chloronemal cells, it is essential that we understand the control points in the cell cycle. We have preliminary evidence (Atkinson and Cove, unpublished data) from Southern hybridisation analysis that there is a homologue of the *Aspergillus nidulans nim* gene in *P. patens*.

References

ASHTON, N. W., BOYD, P. J., COVE, D. J. AND KNIGHT, C. D. (1988). Genetic analysis in

Physcomitrella patens. In *Methods in Bryology*. Proceedings of the Bryological Methods Workshop, Mainz (ed. J. M. Glime), pp. 59–72. Nichinan: Hattori Botanical Laboratory.

ASHTON, N. W., GRIMSLEY, N. H. AND COVE, D. J. (1979). Analysis of gametophyte development in the moss, *Physcomitrella patens*, using auxin and cytokinin resistant mutants. *Planta* **144**, 427–435.

BAKER, B., COUPLAND, G., FEDEROFF, N., STARLINGER, P. AND SCHELL, J. (1987). Phenotypic assay for excision of the maize controlling element Ac in tobacco. *EMBO J.* **6**, 1547–1554.

BOYD, P. J., GRIMSLEY, N. H. AND COVE, D. J. (1988). Somatic mutagenesis of the moss, *Physcomitrella patens*. *Mol. Gen. Genet.* **211**, 545–546.

COUPLAND, G., BAKER, B., SCHELL, J. AND STARLINGER, P. (1988). Characterization of the maize transposable element Ac by internal deletions. *EMBO J.* **7**, 3653–3659.

COVE, D. J. (1984). The role of cytokinin and auxin in protonemal development in *Physcomitrella patens* and *Physcomitrium sphaericum*. *J. Hattori Bot. Lab.* **55**, 79–86.

COVE, D. J. AND ASHTON, N. W. (1984). The Hormonal regulation of gametophyte development in bryophytes. In *The Experimental Biology of Bryophytes* (ed. A. F. Dyer and J. G. Duckett), pp. 177–201. London: Academic Press.

COVE, D. J. AND ASHTON, N. W. (1988). Growth regulation and development in Physcomitrella patens: an insight into growth regulation and development of bryophytes. *Bot. J. Linn. Soc.* **98**, 247–252.

COVE, D. J. AND KNIGHT, C. D. (1987). Gravitropism and phototropism in the moss, *Physcomitrella patens*. In *Developmental Mutants of Higher Plants* (ed. H. Thomas and D. Grierson), pp. 181–196. London: Cambridge University Press.

COVE, D. J., SCHILD, A., ASHTON, N. W. AND HARTMANN, E. (1978). Genetic and physiological studies of the effect of light on the development of the moss, *Physcomitrella patens*. *Photochem. photobiol.* **27**, 249–254.

GRIMSLEY, N. H. AND WITHERS, L. A. (1983). Cryopreservation of cultures of the moss *Physcomitrella patens*. *Cryo-Letters* **4**, 251–258.

HAGEN, G. (1988). The control of gene expression by auxin. In *Plant hormones and their role in plant growth and development* (ed. P. J. Davies), pp. 149–163. Dordrecht: Kluwer Academic Press.

JACKSON, D., CULIANEZ-MACIA, F., PRESCOTT, A., ROBERTS, K. AND MARTIN, C. (1991). Expression patterns of *myb*-genes from *Antirrhinum* flowers. *Plant cell* (in press).

JENKINS, G. I., COURTICE, G. R. M. AND COVE, D. J. (1986). Gravitropic responses of wild-type and mutant strains of the moss, *Physcomitrella patens*. *Plant Cell Env.* **9**, 637–644.

JENKINS, G. I. AND COVE, D. J. (1983*a*). Phototropism and polarotropism of primary chloronemata of the moss, *Physcomitrella patens*: responses of the wild-type. *Planta* **158**, 357–364.

JENKINS, G. I. AND COVE, D. J. (1983*b*). Phototropism and polarotropism of primary chloronemata of the moss, *Physcomitrella patens*: responses of mutant strains. *Planta* **158**, 432–438.

KNIGHT, C. D., COVE, D. J., BOYD, P. J. AND ASHTON, N. W. (1988). The isolation of biochemical and developmental mutants in *Physcomitrella patens*. In *Methods in Bryology*. Proceedings of the Bryological Methods Workshop, Mainz (ed. J. M. Glime), pp. 47–58. Nichinan: Hattori Botanical Laboratory.

MAROCCO, A., WISSENBACH, M., BECKER, J., PAZ-AREZ, J., SAEDLER, H., SALAMINI, F. AND ROHDE, W. (1989). Multiple genes are transcribed in *Hordeum vulgare* and *Zea mays* that carry the DNA-binding domain of the *myb* oncogenes. *Molec. gen. Genet.* **216**, 183–187.

MITRA, G. C. AND ALLSOPP, A. (1959*a*). Effects of kinetin, gibberellic acid and certain auxins on the development of shoot buds on the protonema of *Pohlia nutans*. *Nature* **183**, 974–975.

MITRA, G. C. AND ALLSOPP, A. (1959*b*). Effects of various physiologically active substances on the development of the protonema and bud formation in *Pohlia nutans* (Hedw.) Lindb. *Phytomorphology* **9**, 64–71.

MITRA, G. C., ALLSOPP, A. AND WAREING, P. F. (1959). Effect of light of various qualities on the development of the protonema and bud formation in *Pohlia nutans* (Hedw.) Lindb. *Phytomorphology* **9**, 47–55.

PAZ-AREZ, J., GHOSAL, D., WIENLAND, U., PETERSON, P. A. AND SAEDLER, H. (1987). The regulatory *C1* locus of *Zea mays* encodes a protein with homology to *myb* proto-oncogene products and with structural similarities to transcriptional activators. *EMBO J.* **6**, 3553–3558.

SAUL, M. W., SHILLITO, R. D. AND NEGRUTIU, I. (1988). Direct DNA transfer to protoplasts with and without electroporation. In *The Plant Molecular Biology Manual* (eds. S. B. Gelvin and R. A. Schilperoort), pp. 1–16. Netherlands: Kluwer Acad. Publ.

SCHAEFER, D., ZRYD, J.-P., KNIGHT, C. D. AND COVE, D. J. (1991). Stable transformation of the moss, *Physcomitrella patens. Molec. gen. Genet.* (in press).

SIVE, H. L. AND ST JOHN, T. (1988). A simple subtractive hybridization technique employing photoactivatable biotin and phenol extraction. *Nucl. Acids Res.* **16**, 10937.

WETTSTEIN, F. V. (1924). Morphologie und Physiologie des Formwechsels der Moose auf genetischer Grundlage. *Zeit. Indukt. Abst. Vereb.* **33**, 1–236.

Printed in Great Britain © Society for Experimental Biology 1991

MAPPING THE *ARABIDOPSIS* GENOME

BRIAN M. HAUGE, SUSAN HANLEY, JEROME GIRAUDAT* and HOWARD M. GOODMAN

Department of Genetics, Harvard Medical School and Department of Molecular Biology, Massachusetts General Hospital, Boston, MA 02114, USA

Summary

We are engaged in a project to assemble a complete physical map of the *Arabidopsis thaliana* (L.) Heynh. genome. The first stage of this project involved the analysis of approximately 20 000 random cosmid clones representing an 8- to 10-fold sampling redundancy. Using computer matching programs, these clones have been assembled into some 750 contigs, encompassing 90–95 % of the *Arabidopsis* genome. We are currently attempting to bridge the gaps by selecting the missing clones by hybridization. As a complement to this project we have constructed an RFLP map which currently contains 175 markers. The RFLP map provides contact points between the physical map and classical genetic map. Our main objective for undertaking this project is to simplify the cloning of genes where only the locus and not the product of the gene is known. In other words, the combined RFLP/physical map serves as a general cloning tool by simplifying the movement from the genetic locus to the cloned gene.

Introduction

One of the major challenges of molecular biology is the isolation of genes where the biochemical function of the gene product is unknown. In a number of plant species, genes controlling a wide range of fundamental developmental and metabolic processes have been identified by mutational analysis and placed on classical genetic linkage maps. Although the mutant phenotypes and genetic map locations are known, virtually nothing is known about the gene product. The cloning of these loci, therefore, rely solely on their mutant phenotype and genetic map position. The first step toward cloning the locus is to identify DNA probes residing within one to several cM of the gene of interest. Typically this is achieved by analyzing the meiotic segregation of hundreds or even thousands of restriction fragment length polymorphisms (RFLPs). Once a linked RFLP has been identified, cloning the gene necessitates bridging the intervening gap often spanning distances of millions of base pairs that typically separate genetically

*Present address: Institut des Sciences Vegetales, C.N.R.S., 91198 Gif sur Yvette, Cedex, France.

Key words: abscisic acid, *Arabidopsis*, fingerprinting, gibberellic acid, physical mapping, RFLPs.

linked markers in both mammals and higher plants. While mapping such large
regions is theoretically possible with the traditional technique of 'chromosome
walking' using overlapping cosmid or λ clones, in practice the procedure is
extremely labour intensive and illsuited for large projects where more than a few
steps are required.

Recently, interest has focused on strategies for constructing physical maps of
entire genomes. By definition, a physical map consists of a linearly ordered set of
DNA fragments encompassing the genome or region of interest. Physical maps are
of two types, macro-restriction maps and ordered clone maps. The former consists
of an ordered set of large DNA fragments generated by using restriction enzymes
whose recognition sequences are infrequently represented in the genome (Smith *et
al.* 1987). The macro-restriction map provides information about the organization
of DNA fragments at the level of the intact chromosome, thereby providing long
range continuity. As the name implies, an ordered clone map consists of an
overlapping collection of cloned DNA fragments. The DNA may be cloned into
any one of the available vector systems: yeast artificial chromosomes (YACs),
cosmids, phage or even plasmids. The major advantages of ordered clone maps is
that they are of high resolution and they directly provide the clones for further
study.

The benefits of having a physical map are two-fold. First, the map provides
access to any region of the genome that can be genetically identified. In other
words, the physical map serves as a cloning tool by facilitating the movement from
the genetic locus to the cloned gene. Given a mutation of known genetic map
location, the physical map can be used to easily isolate a collection of overlapping
clones encompassing the locus of interest. By eliminating the need for labour
intensive steps such as chromosome walking, researchers are free to focus their
efforts on the isolation and characterization of the gene of interest. Conversely,
the physical map allows for the movement from the cloned gene to the genetic
locus. Cloned genes can be easily integrated into the physical map and
consequently aligned with the genetic linkage map. In some cases it may be
possible to assign the clone to a defined genetic locus, thereby establishing its
biological function. Clearly, in most cases assignment of clones to known loci will
not be possible. Nonetheless, as an increasing number of genes are cloned and
molecular biological information is accumulated, the map can be used to
investigate the physical linkage of cloned genes, study the organization and
distribution of repetitive elements and address questions such as how physical
distance and genetic distance are correlated. In this context, the map provides the
framework for cataloging and integrating molecular biological information,
thereby serving as a starting point to investigate large scale genomic organization.
Ultimately, genome organization will be investigated at the nucleotide level.
Clearly, physical maps are the logical substrates for genome sequencing projects.

We are engaged in a project to construct a complete physical map of the
Arabidopsis thaliana genome that will ultimately consist of a fully overlapping
collection of cloned DNAs encompassing the five *Arabidopsis* linkage groups.

There are several reasons for using *Arabidopsis* as a model system for the study of plant biology. Its short life cycle, small size and large seed output make it well suited for classical genetic analysis (reviewed by Meyerowitz, 1987). Mutations have been described affecting a wide range of fundamental developmental and metabolic processes (reviewed by Estelle and Somerville, 1986). A genetic linkage map consisting of some 100 loci has been assembled (Koornneef, 1987) and an increasing number of cloned genes and RFLPs are available for the alignment of the genetic map with the physical map (Chang *et al.* 1988; Nam *et al.* 1989). For molecular biological studies, *Arabidopsis* offers the additional advantages of having a very small genome (70000 kb) and a remarkably low content of interspersed repetitive DNA (Leutwiler *et al.* 1984; Pruitt and Meyerowitz, 1986). The small, relatively simple genome greatly simplifies the cloning of interesting loci that have been identified by mutational analysis and makes the *Arabidopsis* genome ideally suited for a physical mapping project.

The first stage of the mapping project involves the characterization of random clones using the 'fingerprinting' strategy of Coulson, Sulston and co-workers (1986). At present more than 20000 random cosmid clones have been 'fingerprinted'. Our working data set consists of 17000 clones representing an 8- to 10-fold sampling redundancy. Using computer matching programs, these clones have been aligned into overlapping groups, referred to as contigs. The 17000 clones fall into some 750 contigs representing approximately 90–95 % of the *Arabidopsis* genome (Hauge *et al.* 1990). Ideally, as the random clone analysis proceeds and the sample size grows, the contigs will be expanded and ultimately merge to produce a map of the five *Arabidopsis* linkage groups. However, having fingerprinted 8–10 genomic equivalents we have reached the practical limit of the procedure. Success in completing the map now depends on the ability to close the gaps. Several approaches are being employed; multi-enzyme fingerprinting of clones residing at the end of the contigs and the unattached clones to enhance the statistical detection of overlaps; selecting linking clones by hybridization using end-probes from clones residing at the ends of contigs and using YACs as probes to bridge the gaps.

For the physical map to be of any utility it is necessary to align the physical map with the genetic map. Toward this goal, we have assembled a restriction fragment length polymorphism (RFLP) map of the *Arabidopsis* genome (Nam *et al.* 1989). The RFLP map, which currently consists of 175 markers, provides the contact points for alignment of the genetic and physical maps.

One of our major objectives for engaging in the mapping project is to simplify the cloning of genes when only the locus and not the product of the gene is known. In particular, we are using the maps to clone genes involved in plant hormone metabolism and their corresponding signal transduction pathways. We are interested in two of the hormones, abscisic acid (ABA) and the gibberellins (GA). Both hormones have long been implicated as important regulators affecting cell growth and development, seed germination, stem elongation and flower development; however, their precise role in regulating these processes is presently unclear. In all likelihood, the magnitude of pleiotropic effects reflect the complex interplay between the various plant hormones.

Consequently, it is difficult to establish the primary mode of action. As an approach to investigate the mechanism of hormone action, we are using the maps to clone genes involved in ABA and GA biosynthesis and their signal transduction pathways. When the molecular probes have been isolated, the pattern of expression of these genes will be examined during plant development, and in response to alterations in hormone levels achieved either by exogenous application or by the use of mutants. These experiments will be a starting point to investigate the role of hormones in specific developmental processes.

Physical mapping of the *Arabidopsis* genome

Several efforts have been initiated to assemble physical maps based on the fingerprinting of random clones (Coulson *et al.* 1986; Olson *et al.* 1986; Kohara *et al.* 1987). The fingerprints are generated by digesting randomly selected clones from primary libraries with one to several restriction enzymes. Following size fractionation by gel electrophoresis (either agarose or polyacrylamide), the lengths of the fragments are determined. The number and the size of the fragments constitute a unique signature or fingerprint of the cloned insert. For fingerprinting, it is unnecessary to generate a restriction map of the clone. The bands must however, be descriptive of the insert and the informational content of the fingerprint must be sufficient to make a reliable assignment of overlapping regions. Clones are said to be overlapping when the fingerprints of two clones are sufficiently similar.

The fingerprinting strategy that we are employing to construct the *Arabidopsis* map is essentially as described by Coulson, Sulston and co-workers (1986; Fig. 1). Briefly, DNA is isolated from randomly selected clones. We are working with random shear cosmid libraries having a mean insert size of 40 kb. The DNAs are digested with *Hin*dIII, which contains an average of 15 recognition sites per cosmid clone and the ends are simultaneously labeled with [^{32}P]ATP. Following termination of the reactions, the samples are subject to a second round of cleavage with an enzyme having a 4 bp specificity (*Sau*3A). Subsequent to size fractionation by electrophoresis on a 4 % denaturing polyacrylamide gel, the labeled bands are visualized by autoradiography. Fig. 2 shows an example of a typical autoradiogram containing the reaction products from 48 clones. The band that is common to all of the clones corresponds to the vector band, the intensity of which reflects the viability of the clone. It is readily apparent that individual clones are propagated nonuniformly. It is likely that the nonuniform growth leads to a certain level of sampling bias that has the potential to create gaps in the map. The main point that is illustrated in Fig. 2 is that individual clones give a distinct banding pattern or 'fingerprint' which serves as a unique signature for a given clone.

For subsequent analysis, the banding patterns are put into the computer using a scanning densitometer and an image-processing package (Sulston *et al.* 1989).

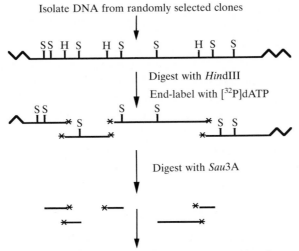

Fig. 1. Fingerprinting strategy (Coulson *et al.* 1986). The cloned insert is depicted by the solid lines and the vector sequences are represented by the zig-zag line. H, *Hind*III recognition site: S, *Sau*3A recognition sequence. The end-labeled digestion products are marked *.

Once the banding patterns have been entered into the computer, they are compared in pairs against the entire data set. Using clone matching programs the regions of probable overlap are determined and the clones are assembled into contigs (Coulson *et al.* 1986; Sulston *et al.* 1988). Fig. 3 shows an example of a contig generated by fingerprint analysis. It should be pointed out that the computer does not actually assemble the contigs. Assembly is performed interactively with a computer program (Coulson *et al.* 1986). Before the clones are joined, the reliability of the match is assessed by visually aligning the films and the overlap must be logically consistent.

The limitations of the fingerprinting strategy are two fold. First, the combination of enzymes used only permits the detection of overlaps in the range of 35–50 %. So in general, smaller overlaps will go undetected. Second, clones containing few or no *Hind*III sites will be unattached since a minimal number of bands is required for the statistical detection of overlapping regions. To circumvent these problems, the clones residing at the end of contigs and the unattached clones can be fingerprinted using several different combinations of enzymes. In doing so, we can establish joins that were previously undetected due to the statistical limitations since the match probability is now the product of the individual probabilities. In addition, it is unlikely that a given clone will have a nonrandom distribution of restriction sites for all of the enzymes employed. Therefore, by using several combinations of enzymes we can establish joins with the clones that were unattached as a result of having a minimal number of *Hind*III sites.

Random fingerprinting procedures are not expected to produce complete

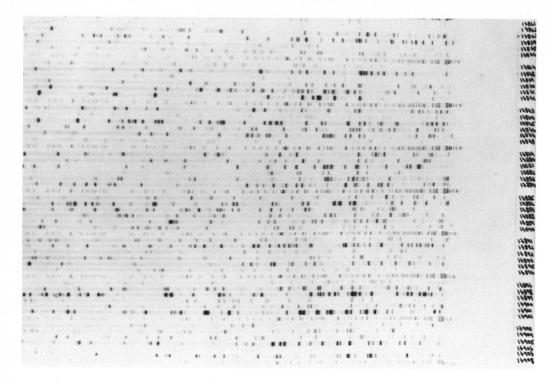

Fig. 2. Autoradiogram of a fingerprint gel. The autoradiogram shows the fingerprints of 48 randomly selected clones from a primary cosmid library. The common band originates from the vector. The end-labeled λ-Sau3A markers are repeated every 7 lanes.

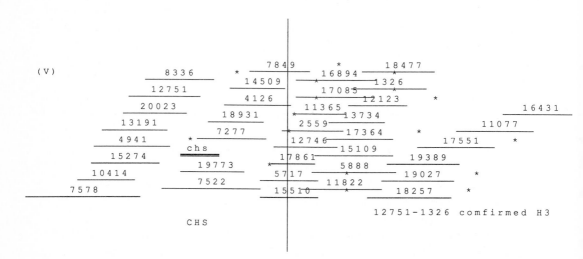

Fig. 3. Computer image of a contig. The solid lines depict individual clones, the length of which is proportional to the number of bands in the fingerprint. The asterisk denotes redundant clones that are not displayed. Assembly of the contigs has previously been described (Coulson et al. 1986; Sulston et al. 1988).

physical maps. Instead, the map will consist of many contigs composed of two or more overlapping clones. As the project progresses, the number of contigs decrease as the gaps are closed. After this point, one reaches the stage in the project where the rate of finding new joins becomes unacceptably low due to the scarcity of the remaining clones. Completion of the map now depends largely on the ability to bridge the remaining gaps. In an effort to close the gaps, we are simply selecting the missing clones by hybridization. The strategy is to label the clones residing at the ends of the contigs and the unattached clones and use the probes to select joining clones from an *Arabidopsis* yeast artificial chromosome (YAC) library. The use of YAC libraries is an important component of this analysis for two reasons. First, the large size of YAC clones means that fewer clones must be examined. Second, the yeast cloning system offers the potential to give a random or at least different representation of clones than is obtained using bacterial host/vector systems. This is important, since it is likely that many of the gaps in the contig map can be attributed to cloning bias, due either to the inability to clone certain sequences in cosmids or nonuniform growth of individual clones leading to sampling bias.

Although closing the gaps by hybridization is very straightforward, it is also labour intensive. We have 750 contigs so we need a minimum of 1500 hybridization probes. In an effort to minimize the labour, we are utilizing a multiplex hybridization strategy. Briefly, the end-clones and unattached clones are plated in a standard 96-well microtiter dish configuration. Probes are then made from pools of clones corresponding to rows and columns of the microtiter dish (Fig. 4; Evans and Lewis, 1989). The mixed-probes are then hybridized to ordered grids containing 2956 random YAC clones representing about 5 genomic equivalents. The hybridizations are repeated with mixed probes derived from each of the rows and columns. To determine which of the cosmids hybridizes to a given YAC, the common signals are identified. The hybridizing clone is identified by the intersection of a row and a column. For example, in Fig. 4, the cosmid clone at position 1A hybridizes to the YAC indicated by the arrows in panels A and B. By using the multiplex strategy outlined in Fig. 4, the number of hybridizations is reduced by 4.8-fold. Clearly, by using larger pools of probes the number of hybridizations can be further reduced. We are currently establishing the conditions for hybridization with mixed-probes and the feasibility of using this approach is being assessed by using characterized contigs.

As a complement to the strategy outlined above, we are performing the converse experiment, which is to use YACs as hybridization probes (Coulson *et al.* 1988). The strategy is to prepare an ordered grid of cosmids, which is representative of our contigs and the unattached clones. YACs are then isolated and used as hybridization probes. The YACs are selected on the basis of their hybridization to cosmids as described above or alternatively selected at random. The hybridization pattern of the cosmid grid is then used to establish linkage, as well as its position with respect to the cosmids (Coulson *et al.* 1988). This approach has been successfully used to reduce the number of contigs for the *C. elegans* map

Ordered arrays of
end-clones

Make mixed
probes from rows
and columns

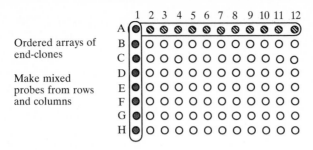

Probe ordered YAC libraries

A. Column 1   B. Row A

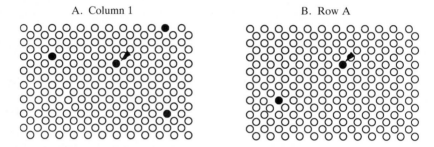

Fig. 4. Strategy for linking contigs by multiplex hybridization. Cosmid clones residing at the ends of contigs are archived in microtiter dishes and held as glycerol stocks at −80 °C. DNA is prepared from pools of 8 and 12 clones, corresponding to columns and rows respectively. Probes are prepared from the pools and hybridized to ordered arrays of random YAC clones. The hybridizations are repeated using pools from each of the rows and columns. The clone at the intersection of row A and column 1 hybridizes to the YAC indicated by the arrows in panels A and B.

from 700 to 170 (J. Sulston and A. Coulson, personal communication). It is significant that Coulson and co-workers (1988) find that sequences which are poorly represented in the cosmid libraries are represented in YAC libraries.

It is important to keep in mind that hybridization is not a sufficient criterion for declaring a join. One clear advantage of the YAC linking approach is that a given YAC, or overlapping set of YACs, is expected to hybridize to several cosmids in a contig. Therefore, the ordered cosmid map dictates a certain logic in the hybridization pattern that should minimize false joins resulting from hybridization to interspersed repetitive DNA.

RFLP mapping

An important component of the mapping project is the alignment of the physical map with the established genetic map. We have previously published an RFLP map consisting of 94 randomly distributed molecular markers (Nam *et al.* 1989). In addition, we have incorporated 17 markers from an independently constructed *Arabidopsis* RFLP map (Chang *et al.* 1988). The combination of the two maps provides good coverage of the genome; however, several gaps still exist. In an effort to close the gaps, we are adding additional markers to the map. Currently, our RFLP map contains 175 markers, several of the new markers map to previously unmarked regions of the genome. This analysis will be continued until coverage of the genome is complete. RFLPs are being identified using random clones as well as cloned genes.

Cloning of hormone biosynthetic and response loci

One of the main goals of this project is to utilize the maps to clone genes involved in the synthesis and response to plant hormones. We are focusing our efforts on two of the hormones, ABA and GA. Numerous mutants that are defective in the synthesis of GA have been characterized in *Arabidopsis* and genetically mapped to five loci (*ga-1*, *ga-2*, *ga-3*, *ga-4*, *ga-5*) (Koornneef and van der Veen, 1980). These mutants are all phenotypic dwarfs with reduced apical dominance and for many alleles GA is required for germination. Normal growth of these mutants is restored by exogenous application of GA. In addition, a sixth locus (*Gai*) which is GA-insensitive has been identified (Koornneef *et al.* 1985). By genetic criteria, the phenotypes are consistent with the possibility that Gai encodes a GA receptor or a second messenger.

Several 'ABA-insensitive' mutants (loci *abi-1*, *abi-2*, *abi-3*) have been identified which produce normal levels of endogenous ABA but display reduced sensitivity to ABA (Koornneef *et al.* 1984). Mutations at all three loci display reduced dormancy and ABA-inhibition of germination; however, with respect to the other responses, the effects of *abi-1* and *abi-2* are confined to vegetative growth, while *abi-3* primarily affects seed development (Koornneef *et al.* 1989). In contrast to ABA-deficient mutations (*aba* locus), the wild-type phenotype of the *abi* mutants cannot be restored by addition of exogenous ABA. Like the GA-insensitive mutant (*Gai*), the decreased sensitivity to ABA probably results from an alteration in the number or properties of the receptor(s) or other elements of the ABA signal transduction pathway.

Using the combined RFLP and physical maps, experiments are in progress to clone several of the loci described above. Clearly, the RFLP map provides the starting points for cloning. However, it is difficult to assign with confidence a precise position on the genetic map to an RFLP probe since only a few morphological markers per chromosome were used to align the genetic map with the RFLP map. In addition, due to the necessarily limited number of F_2 progeny used for the segregation analysis, we estimate that the resolution of the RFLP map is of the order of 2cM (Nam *et al.* 1989). So for each gene that we are interested in cloning, it is first necessary to map the locus with respect to the RFLP map. This is achieved using a 'fine structure mapping' approach (Fig. 5; Giraudat *et al.* 1989; Hauge *et al.* 1989). To enhance the resolution, we have constructed lines containing the mutation of interest flanked by linked markers. For the RFLP mapping, we examine the segregation of markers only among the F_2 progeny that have undergone a recombination event within a small genetically defined interval

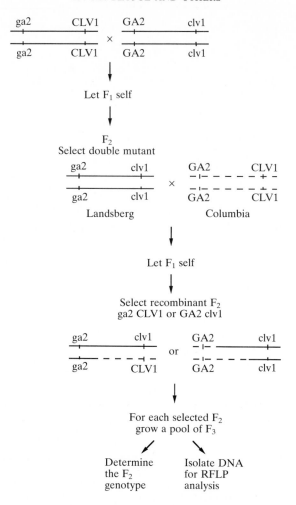

Fig. 5. Fine structure mapping strategy. Double mutants are constructed containing the mutation of interest flanked by a linked marker. In the example given, a *ga-2, clv-1* double mutant is constructed in a Landsberg *erecta* background. The double mutant is crossed to a Columbia line and the F_1 progeny are allowed to self. In the F_2 generation recombinants between the two markers are selected. The F_2s are selfed and the resultant F_3s are scored to uncover the F_2 heterozygotes and pooled to provide DNA for RFLP analysis.

(i.e. recombinant between the flanking markers and the gene of interest). Using this approach, we enhance the resolution of the RFLP map in the vicinity of the locus and minimize the number of progeny to be analyzed. Once a linked RFLP has been identified, the physical map can be used to localize the gene to one or to several clones. The gene can then be identified by complementation or alternatively, physical means.

The fine structure mapping is essentially complete for two loci, *abi-3* (Giraudat *et al.* 1989) and *ga-2* (Hauge *et al.* 1989). In both cases, using the combined RFLP/physical map we were able to localize the genes to a single cosmid.

Experiments to further localize the respective genes within the cosmids are in progress. We are also attempting to clone several of the other *ga* loci and the *Gai* locus. These experiments are in various stages of completion. The cloning of the *aba*, *abi-1* and *abi-2* loci is being done in collaboration with J. Giraudat (CNRS, Gif sur Yvette, France). Once the genes have been isolated they will be characterized using the various genetic and molecular biological tools that are available. Of particular interest, is the fact that the genes provide the starting point to elucidate the molecular mechanisms of plant hormone action.

DNA sequence analysis of the cosmid clones harboring *abi-3* and *ga-2* is near completion. In addition to the obvious goal, which is to determine the nucleotide sequence of the respective genes, we are using the clones to investigate strategies for large scale sequencing projects. We are pursuing several approaches; direct sequencing of cosmids in solution, sequencing of cosmids that have been immobilized onto solid supports, shotgun sequencing and multiplex sequencing.

References

CHANG, C., BOWMAN, J. L., DEJOHN, A. W., LANDER, E. S. AND MEYEROWITZ, E. M. (1988). Restriction fragment length polymorphism linkage map for *Arabidopsis thaliana*. *Proc. natn. Acad. Sci. U.S.A.* **85**, 6856–6860.

COULSON, A., SULSTON, J., BRENNER, S. AND KARN, J. (1986). Toward a physical map of the nematode *Caenorhabditis elegans*. *Proc. natn. Acad. Sci. U.S.A.* **83**, 7821–7825.

COULSON, A., WATERSTON, R., KIFF, J., SULSTON, J. AND KOHARA, Y. (1988). Genome linking with yeast artificial chromosomes. *Nature* **335**, 184–186.

ESTELLE, M. A. AND SOMERVILLE, C. R. (1986). The mutants of *Arabidopsis*. *Trends Genet.* **2**, 89–93.

EVANS, G. A. AND LEWIS, K. A. (1989). Physical mapping of complex genomes by multiplex analysis. *Proc. natn. Acad. Sci. U.S.A.* **86**, 5030–5034.

GIRAUDAT, J., HAUGE, B. M. AND GOODMAN, H. M. (1989). Progress toward the cloning of the *abi3* locus of *Arabidopsis*. Abstract; *The Genetics and Molecular Biology of* Arabidopsis. Bloomington, Indiana.

HAUGE, B. M., GALLANT, P., MALOLEPSZY, J. AND GOODMAN, H. M. (1989). Toward the cloning of the *ga-2* locus from *Arabidopsis*. Abstract; *The Genetics and Molecular Biology of* Arabidopsis. Bloomington, Indiana.

HAUGE, B. M., GIRAUDAT, J., HANLEY, S., HWANG, I., KOHCHI, T. AND GOODMAN, H. M. (1990). Physical mapping of the *Arabidopsis* genome and its applications. *J. Cell. Biol.* **14E**, 259.

KOHARA, Y., AKIYAMA, K. AND ISONO, K. (1987). The physical map of the whole *E. coli* chromosome: Application of a new strategy for rapid analysis and sorting of a large genomic library. *Cell* **50**, 495–508.

KOORNNEEF, M. (1987). Linkage map of *Arabidopsis thaliana* (2n=10). In O'Brien, S. J (ed.), *Genetic Maps*. Cold Spring Harbor Laboratory Press, Cold Spring Harbor, NY, pp. 742–745.

KOORNNEEF, M., ELGERSMA, A., HANHART, E. P., VAN LOENEN-MARTINET, C., VAN RIJN, L. AND ZEEVAART, J. A. D. (1985). A gibberellin insensitive mutant of *Arabidopsis thaliana*. *Physiologia Pl.* **65**, 33–39.

KOORNNEEF, M., HANHARD, C. J., HILHORST, H. W. M. AND KARSSEN, C. M. (1989). *In vivo* inhibition of seed development and reserve protein accumulation in recombinants of abscisic acid biosynthesis and responsiveness mutants in *Arabidopsis thaliana*. *Pl. Physiol.* **90**, 463–469.

KOORNNEEF, M., REULING, G. AND KARSSEN, C. M. (1984). The isolation and characterization of abscisic acid-insensitive mutants of *Arabidopsis thaliana*. *Physiologia Pl.* **61**, 377–383.

KOORNNEEF, M. AND VAN DER VEEN, J. H. (1980). Induction and analysis of gibberellin sensitive mutants in *Arabidopsis thaliana* (L.) Heynh. *Theor. appl. Genet.* **58**, 257–263.

LEUTWILER, L. S., HOUGH-EVANS, B. R. AND MEYEROWITZ, E. M. (1984). The DNA of *Arabidopsis thaliana. Molec. gen. Genet.* **194**, 15–23.

MEYEROWITZ, E. M. (1987). *Arabidopsis thaliana. A. Rev. Genet.* **21**, 93–111.

NAM, H-G., GIRAUDAT, J., DEN BOER, B., MOONAN, F., LOOS, W. D. B., HAUGE, B. M. AND GOODMAN, H. M. (1989). Restriction fragment length polymorphism linkage map of *Arabidopsis thaliana. Pl. Cell* **1**, 699–705.

OLSON, M. V., DUTCHIK, J. E., GRAHAM, M. Y., BRODEUR, G. M., HELMS, C., FRANK, M., MACCOLLIN, M., SCHEINMAN, R. AND FRANK, T. (1986). Random-clone strategy for genomic restriction mapping in yeast. *Proc. natn. Acad. Sci. U.S.A.* **83**, 7826–7830.

PRUITT, R. E. AND MEYEROWITZ, E. M. (1986). Characterization of the genome of *Arabidopsis thaliana. J. molec. Biol.* **187**, 169–183.

SMITH, C. L., ECONOME, J. G., SCHUTT, A., KLCO, S. AND CANTOR, C. R. (1987). A physical map of the *Escherichia coli* K12 genome. *Science* **236**, 1448–1453.

SULSTON, J., MALLETT, F., DURBIN, R. AND HORSNELL, T. (1989). Image analysis of restriction enzyme fingerprint autoradiograms. *Cabios* **5**, 101–106.

SULSTON, J., MALLETT, F., STADEN, R., DURBIN, R., HORSNELL, T. AND COULSON, A. (1988). Software for genome mapping by fingerprinting techniques. *Comput. applic. Biosci.* **4**, 125–132.

Printed in Great Britain © *Society for Experimental Biology 1991*   57

DEVELOPMENT OF A SYSTEM FOR EFFICIENT CHROMOSOME WALKING IN *ARABIDOPSIS*

ERWIN GRILL* and CHRIS SOMERVILLE

DOE Plant Research Laboratory, Michigan State Universtiy, East Lansing, MI 48824, USA

Summary

The small genome size of *Arabidopsis* and the low level of repetitive DNA sequences make this crucifer an attractive system for chromosome walking to isolate genes. Mapping of a mutant locus relative to restriction fragment length polymorphism (RFLP) markers provides the first step towards isolating the corresponding gene. The RFLP marker closest to the target gene serves as a starting point. The distance between gene and marker is generally in the range of 50–200 kb. In order to facilitate chromosome walking of this magnitude, we constructed a yeast artificial chromosome (YAC) library of *Arabidopsis*. Large fragments of *Arabidopsis* DNA were cloned into a YAC vector and transformed into yeast. The library contains more than 10 equivalents of the *Arabidopsis* genome. YACs containing sequences of RFLP markers of *Arabidopsis* revealed an average insert size of 150 kb. Thus, 1–3 (contiguous) YACs should be sufficient to clone genes from *Arabidopsis* by chromosome walking. In order to use the system to isolate genes involved in the signal transduction of abscisic acid, we fine-mapped the mutant loci *abi-1* and *abi-2*, which confer abscisic acid insensitivity, relative to RFLPs and isolated the corresponding YACs.

Introduction

Little progress has been made towards elucidating the molecular mechanisms of hormone action in plants. This situation reflects the difficulties involved in identifying and purifying components of the signal transduction pathway of plant hormones. As an alternative approach, 'reversed genetics' could be used to identify and characterize the single steps involved in signal transduction. Genes have been isolated from yeast, *Drosophila* and *Caenorhabditis* by virtue of a mutant phenotype and the gene products have been studied (e.g. Dietzel and Kurjan, 1987; Whiteway *et al.* 1989). The higher plant *Arabidopsis thaliana* (L.) Heynhold offers properties that make it well suited for genetic and molecular biological studies. The self-fertile crucifer is small and has a short generation time of several weeks (Redei, 1975). In addition, the *Arabidopsis* genome is the smallest so far known for any higher plant and appears to contain little repetitive

---

*Present address: Institut für Pflanzenwissenschaften, ETH-Zentrum, Sonneggstr. 5, 8092 Zurich, Switzerland.

Key words: chromosome walking, *Arabidopsis*, RFLP markers, YAC library.

DNA (Pruitt and Meyerowitz, 1986). Thus, *Arabidopsis* is a suitable organism for the isolation of genes by chromosome walking (Meyerowitz, 1987). RFLP markers serve as the starting points. Currently, there are 2 RFLP linkage maps available with a total of about 200 RFLP markers (Chang *et al.* 1988; Nam *et al.* 1989). Based on a genome size of 70 000 kb, the average distance between 2 adjacent RFLP markers is approximately 350 kb. Thus, the closest RFLP marker is expected to be within 50–250 kb of a target gene for chromosome walking.

For chromosome walking of this magnitude the YAC system (Burke *et al.* 1987) offers advantages. In YACs heterologous DNA fragments up to 800 kb in size were stably maintained in yeast (Burke *et al.* 1987; Coulson *et al.* 1988; Brownstein *et al.* 1989; Silverman *et al.* 1989). YACs have been successfully used to link up cosmids of the *Caenorhabditis* genome (Coulson *et al.* 1988) and have been used for chromosome walking (Silverman *et al.* 1989; Garza *et al.* 1989).

We and others (Guzman and Ecker, 1988; Ward and Jen, 1990), have constructed a YAC library of *Arabidopsis*. Our goal is to isolate genes involved in the signal transduction of abscisic acid (ABA). *Arabidopsis* plants insensitive to ABA have been isolated and characterized by Koornneef *et al.* (1984), and the mutant loci were designated *abi-1*, *abi-2*, and *abi-3*. The ABA insensitive phenotypes appear to reflect a defective transfer of the hormone signal.

Results and Discussion

For construction of the YAC library of *Arabidopsis* we modified the YAC vector pYAC4 (Burke *et al.* 1987) by introducing a cloning site (*Bam*HI) which is flanked by two T3 phage promoters in pYAC 41 or T3/T7 phage promoters in pYAC 45 (Grill and Somerville, 1990). The cloning site is located in the intron of the *sup 4* gene and allows testing for insertional inactivation by a color assay in suitable yeast strains. The cloning procedure is presented schematically in Fig. 1. High relative molecular mass DNA of *Arabidopsis*, ecotypes Columbia and Landsberg *erecta*, was isolated from leaf protoplasts with a yield of 10 μg DNA gram^{-1}. Pulsed-field electrophoresis demonstrated that the relative molecular mass of the isolated DNA was $\geqslant 1500 \times 10^6 M_r$. The DNA was partially digested with restriction enzymes (*Bam*HI or *Sau*3A) to fragments of about 100–400 kb and ligated into the cloning site of phosphatase-treated YAC vector. After size selection by pulse-field electrophoresis to remove excess vector DNA and small ligation products (<120 kb), the DNA was transformed into yeast. In a typical experiment we obtained 2500 transformed yeast colonies with 5 μg size-selected DNA. More than 99 % of the transformants appear to contain an insert, based on the insertional inactivation of the *sup 4* gene. In several experiments we have isolated more than 10 000 independent yeast clones which are individually maintained in microtiter plates.

A subset of the library was probed with RFLP markers of *Arabidopsis* by colony hybridization. Several YAC clones were identified with a frequency of 1 out of 800 clones per marker. The chromosomes from these identified clones were isolated

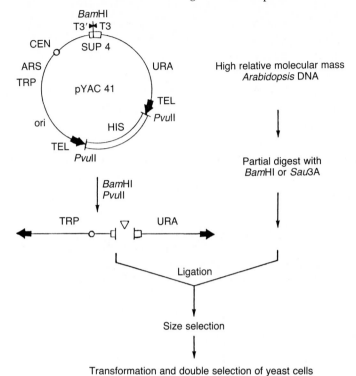

Fig. 1. Scheme of the cloning of high relative molecular mass DNA from *Arabidopsis* in yeast artificial chromosomes (YACs). For details see text.

and analyzed by pulse-field separation. The average insert size of the YACs was 150 kb, ranging from 70 kb to 220 kb (Grill and Somerville, 1991). The YACs seem to be stably propagated, since we could not observe heterogeneously sized YAC molecules from individual clones, and the RFLP sequences of *Arabidopsis* appeared to be faithfully maintained in the yeast. These results established that the library will be useful for chromosome walking.

As mentioned above, in most cases it will be necessary to walk distances of 50–250 kb on the chromosome to the target gene. Given the size of cloned *Arabidopsis* DNA fragments it is conceivable that simply identifying large YACs (150–220 kb) containing the RFLP marker proximal to the target gene will frequently be sufficient to clone the gene itself. This situation is particularly likely if the locus of interest maps close [less than 1 cM; 1 cM corresponds to approximately 140 kb (Meyerowitz, 1987)] to the adjacent RFLP marker. In this case, several YACs with the RFLP marker should be identified and examined for the presence of the target gene. This is accomplished by mapping end-probes of the YAC inserts relative to the target gene (see below). In some instances, the cloned fragments will not be large enough to contain both the RFLP marker and the target gene. Additional YACs can be identified that overlap with the distal sequences of the previous YACs after generating end-specific probes. We have been able to produce end-specific riboprobes of YAC inserts from total yeast

DNA by a combination of inverse polymerase chain reaction (PCR) (Ochman *et al.* 1988) and subsequent generation of labelled transcripts using the phage promoter located at the insert/vector junction (Grill and Somerville, 1991). Inverse PCR has also been used to generate end-probes from YACs not containing phage promoters (Silverman *et al.* 1989; Garza *et al.* 1989), but purified YAC DNA seems to be required to suppress unspecific amplification. Several YAC clones of the *Arabidopsis* library were identified after colony hybridization with the labelled riboprobes.

In order to persue our goal of isolating genes responsible for ABA signal

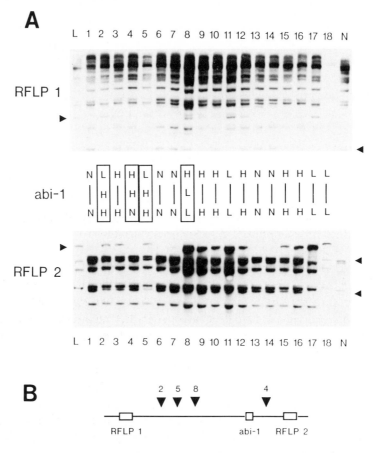

Fig. 2. Mapping of the *abi-1* locus relative to RFLP markers. (A) DNA of individual F_2 plants (1–18) from a cross between the ecotypes Landsberg-*erecta* (L), homozygous for *abi-1*, and Niederzenz (N) was isolated and scored for the inheritance of the linked RFLP markers 1 and 2 in a Southern blot. The arrowheads indicate the RFLP between the two ecotypes. In addition, the inheritance of the *abi-1* marker was determined in the progeny of these F_2 plants. The genetic background of these 3 markers (RFLP 1: upper row; *abi-1*: center row; RFLP 2 lower row) is indicated by L, N and H (for heterozygous). Recombination events are marked by boxes (F_2 plants 2, 4, 5 and 8). See also text. (B) Schematic drawing of the arrangement of the *abi-1* locus relative to the RFLP markers 1 and 2. The chromosome break points of the recombinants in orientation to the markers are indicated by the numbered arrowheads.

transduction, we fine-mapped the *abi-1* and *abi-2* loci relative to RFLPs. The homozygous mutant (*abi-1* or *abi-2*) in the genetic background of Landsberg-*erecta* (L) (Koornneef *et al.* 1984) was crossed with the ecotype Niederzenz (N). Plants of these crosses were allowed to self-pollinate and DNA from the subsequent F_2 plants was individually prepared from leaf material. Seeds from the same F_2 plants were scored for ABA insensitivity. In Fig. 2A, a low resolution mapping of the *abi-1* locus is shown. The DNA of the parental ecotypes (L, N) and of individual F_2 plants (lane 1–18) was cut with restriction enzymes and the Southern blot was probed with two linked RFLP markers 1 and 2. The arrowheads indicate the DNA fragment polymorphism between L and N. The F_2 DNA was scored for homozygous (L or N) or heterozygous inheritance (H) of the RFLP. The scoring is shown below and above the corresponding blot. Between them, the scoring of the *abi-1* phenotype of the F_3 progeny is indicated (L for homozygous *abi-1*, N for homozygous wild-type, and H for heterozygous). A vertical line indicates that the genetic background of the *ABI* locus was identical to the one of RFLP 1 and 2. In four of these F_2 plants recombinants were observed (the scoring of RFLP1 differs from that of RFLP 2), and in these cases the *abi-1* marker has either the genetic background of RFLP 1 or 2. Thus, the *ABI* gene appears to be located on the chromosome between the 2 RFLP markers, as outlined schematically in Fig. 2B. We mapped *abi-1* and *abi-2* with a resolution of 0.3 cM by analyzing 150 F_2 plants in each case and determined the proximal RFLP markers. YACs of the chromosomal regions were isolated and characterized, and are currently subcloned into binary vectors in order to identify the target genes after *Agrobacterium*-mediated transfer into *Arabidopsis*.

We thank E. Meyerowitz and H. Goodman and collaborators for providing RFLP probes, L. Danhof for excellent technical assistance, and M. Kertesz for reading the manuscript. This work was supported by the National Science Foundation (DMB 8351595) and by a fellowship to E.G. from the Deutsche Forschungsgemeinschaft.

References

BROWNSTEIN, B. H., SILVERMAN, G. A., LITTLE, R. D., BURKE, D. T., KORSMEYER, S. J., SCHLESSINGER, D. AND OLSON, M. V. (1989). Isolation of single-copy human genes from a library of yeast artificial chromosome clones. *Science* **244**, 1348–1351.

BURKE, D. T., CARLE, G. F. AND OLSON, M. V. (1987). Cloning of large segments of exogenous DNA into yeast by means of artificial chromosome vectors. *Science* **236**, 806–812.

CHANG, C., BOWMAN, J. L., DEJOHN, A. W., LANDER, E. S. AND MEYEROWITZ, E. M. (1988). Restriction fragment length polymorphism linkage map for *Arabidopsis thaliana*. *Proc. natn. Acad. Sci. U.S.A.* **85**, 6856–6860.

COULSON, A., WATERSTON, R., KIFF, J., SULSTON, J. AND KOHARA, Y. (1988). Genomic linking with yeast artificial chromosomes. *Nature* **335**, 184–186.

DIETZEL, C. AND KURJAN, J. (1987). The yeast SCG1 gene: a Gα-like protein implicated in the a- and α-factor response pathway. *Cell* **50**, 1001–1010.

GARZA, D., AJIOKA, J. W., BURKE, D. T. AND HARTL, D. L. (1989). Mapping the *Drosophila* genome with yeast artificial chromosomes. *Science* **246**, 641–646.

GRILL, E. AND SOMERVILLE, C. (1991). Chromosome walking in *Arabidopsis* using yeast artificial chromosomes. *Molec. gen. Genet.* in press

GUZMAN, P. AND ECKER, J. R. (1988). Development of large DNA methods for plants: molecular cloning of large segments of *Arabidopsis* and carrot DNA into yeast. *Nucl. Acids Res.* **16**, 11 091–11 105.

KOORNNEEF, M., REULING, G. AND KARSSEN, C. M. (1984). The isolation and characterization of abscisic acid-insensitive mutants of *Arabidopsis thaliana*. *Physiology Pl.* **61**, 377–383.

MEYEROWITZ, E. M. (1987). *Arabidopsis thaliana*. *A. Rev. Genet.* **21**, 93–111.

NAM, H. G., GIRAUDAT, J., DEN BOER, B., MOONAN, F., LOOS, W. B., HAUGE, B. AND GOODMAN, H. (1989). Restriction fragment length polymorphism linkage map of *Arabidopsis thaliana*. *Pl. Cell* **1**, 699–705.

OCHMAN, H., GERBER, A. S. AND HARTL, D. L. (1988). Genetic applications of an inverse polymerase chain reaction. *Genetics* **120**, 621–623.

PRUITT, R. E. AND MEYEROWITZ, E. M. (1986). Characterization of the genome of *Arabidopsis thaliana*. *J. molec. Biol.* **187**, 169–183.

REDEI, G. P. (1975). *Arabidopsis* as a genetic tool. *A. Rev. Genet.* **9**, 111–127.

SILVERMAN, G. A., YE, R. D., POLLOCK, K. M., SADLER, J. E. AND KORSMEYER, S. J. (1989). Use of yeast artificial chromosome clones for mapping and walking within human chromosome segment 18q21.3. *Proc. natn. Acad. Sci. U.S.A.* **86**, 7485–7489.

WARD, E. R. AND JEN, G. C. (1990). Isolation of single-copy-sequence clones from a yeast artificial chromosome library of randomly-sheared *Arabidopsis thaliana* DNA. *Pl. molec. Biol.* **14**, 561–568.

WHITEWAY, M., HOUGHAN, L., DIGNARD, D., THOMAS, D. Y., BELL, L., SAARI, G. C., GRANT, F. J., O'HARA, P. AND MACKAY, V. L. (1989). The *STE4* and *STE18* genes of yeast encode potential α and β subunits of the mating factor receptor-coupled G protein. *Cell* **56**, 467–477.

Printed in Great Britain © *Society for Experimental Biology 1991*

DEVELOPMENT OF AN EFFICIENT TRANSPOSON TAGGING SYSTEM IN *ARABIDOPSIS THALIANA*

CAROLINE DEAN[1], *CHRISTINA SJODIN*[1,*], *IAN BANCROFT*[1], *EMILY LAWSON*[1], *CLARE LISTER*[1], *STEVE SCOFIELD*[2] *and JONATHAN JONES*[2]

[1] IPSR Cambridge Laboratory, JI Centre, Colney Lane, Norwich, UK
[2] Sainsbury Laboratory, JI Centre, Colney Lane, Norwich, UK

Summary

We are currently developing a transposon tagging system in *Arabidopsis thaliana* using the maize transposable elements *Ac* and *Ds*. In order to make the system as efficient as possible, four different antibiotic resistance markers have been tested for their usefulness in monitoring excision and reinsertion of transposons in the *Arabidopsis* genome. Owing to the low transposition frequency of wild-type *Ac* in *Arabidopsis* we have also tested a number of modifications to the *Ac* element. Deletion of the CpG-rich region in the transposase 5′ untranslated leader was found to significantly increase the activity of *Ac* in *Arabidopsis*. In our first non-targeted tagging experiment, 200 individuals with an inherited transposed *Ac* element have been collected and their progeny are being screened in families of twelve for segregation of mutant phenotypes. A two-element system, using *Ac* and *Ds*, is also being developed and a way of stabilising the *Ds*-induced mutations by counter selecting plants carrying the transposase source is being investigated. The *Agrobacterium* T-DNA insertions carrying the different transposons are currently being mapped onto the *Arabidopsis* RFLP map. Inverse polymerase chain reaction (IPCR) is being used to generate flanking DNA probes, that will be used on the RFLP blots to map the T-DNA relative to the other known markers. Once the first few have been mapped the first targeted tagging experiment will be initiated.

Introduction

Transposon tagging was first used in eukaryotes to isolate the *Drosophila* white gene (Bingham *et al.* 1981). The attraction of using transposons in gene isolation is that the function and expression of the gene product need not be known, allowing genes identified only by their mutant phenotype, to be cloned. In plant systems, endogenous transposons have been well characterised from *Zea mays* (maize), e.g. *Ac, Ds, Spm/En, Mu* (McClintock, 1951; Peterson, 1953, Robertson, 1978;

*Present address: Department of Molecular Genetics, Uppsala Genetic Centre, Swedish University of Agricultural Sciences, Sweden.

Key words: transposon, *Arabidopsis*, cloning.

Federoff *et al.* 1983; Federoff, 1983; Doring and Starlinger, 1986) and *Antirrhinum majus* (snapdragon), e.g. *Tam3* (Sommer *et al.* 1985; Coen and Carpenter, 1986) and the list of genes being cloned using these transposons in these species is growing steadily (Federoff *et al.* 1984; Martin *et al.* 1985; O'Reilly *et al.* 1985; Paz-Ares *et al.* 1986; Schmidt *et al.* 1987; Theres *et al.* 1987). Many plant species have, as yet, no identified endogenous transposon system. In order to use transposons to clone genes from these species, heterologous transposons have been introduced into them using *Agrobacterium tumefaciens* transformation (Chilton *et al.* 1977). There are now many reports of the activity of the maize transposable element *Ac* in heterologous systems (Baker *et al.* 1986; van Sluys *et al.* 1987; Yoder *et al.* 1988; Knapp *et al.* 1988; Finnegan *et al.* 1988; Jones *et al.* 1989; Schmidt and Willmitzer, 1989). The maize element *Spm/En* and the snapdragon element *Tam3* have also been shown to be active in tobacco (Masson and Federoff, 1989; Martin *et al.* 1989).

We and others (Schmidt and Willmitzer, 1989; Masterson *et al.* 1989) are developing a transposon tagging system in *Arabidopsis thaliana*. The usefulness of an insertional mutagenesis system in *Arabidopsis* has already been demonstrated using *Agrobacterium* T-DNA and the seed transformation procedure (Feldman and Marks, 1987; Feldman *et al.* 1989). The seed transformation procedure has the advantage that it does not have a tissue-culture stage, thereby avoiding the problems of background mutations caused by somaclonal variation. The transformation frequency with the seed transformation procedure, however, is exceedingly low. Use of transposons instead of T-DNA as the insertional mutagen would increase the ease and efficiency of generating new mutants in *Arabidopsis*. The transposons can be engineered to give an optimal rate of transposition and the transformants can be screened for tissue-culture-induced mutations before the screen for transposon-induced mutations takes place.

Van Sluys *et al.* (1987) and Schmidt and Willmitzer (1989) have demonstrated the activity of the maize autonomous element *Ac* in *Arabidopsis thaliana*. The activity of the element in *Arabidopsis* was significantly lower than had been observed in tobacco and tomato (Baker *et al.* 1986; Yoder *et al.* 1988; Jones *et al.* 1989). In order to increase the efficiency of a transposon tagging system in *Arabidopsis* using the *Ac* element or a two-element *Ac/Ds* system we have investigated a number of parameters that are described below. These include testing different antibiotic resistance markers, to monitor excision and reinsertion of the transposon and testing modifications of the *Ac* and *Ds* elements that may increase the transposition frequency. We are currently using these introduced transposons in both targeted and non-targeted tagging experiments.

Monitoring *Ac* activity in *Arabidopsis*

Resistance to the antibiotic kanamycin was first used to select for excision of *Ac* from the tobacco genome by Baker *et al.* (1987). The *Ac* element was cloned into the 5′ untranslated leader region of an NPTII fusion, rendering the fusion inactive until the element had excised. Jones *et al.* (1989) have also monitored the activity of *Ac* in tobacco using a similar approach but have replaced the kanamycin resistance fusion (NPTII) with a streptomycin resistance marker, SPT. Use of a streptomycin selection to monitor transposon activity has the advantage over most other antibiotic resistance markers in that sensitive cells bleach but do not die on streptomycin-containing media, provided they are supplied with a carbon source (Maliga *et al.* 1988) and resistance is cell-autonomous. This allows both somatic and germinal activity of the elements to be monitored. Transgenic tobacco plants carrying an SPT::*AC* fusion were generated and seeds from the transgenic plants were plated on streptomycin-containing media. Somatic excision of the transposon was visualised as green sectors on bleached cotyledons (shown schematically in Fig. 1).

Fully streptomycin-resistant seedlings were also present in the progeny of the primary tobacco transformants. These represented individuals in which *Ac* had

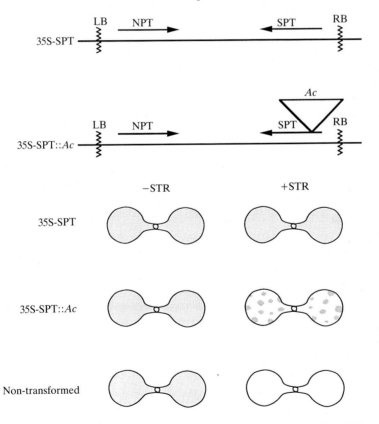

Fig. 1. Streptomycin resistance assay to monitor excision of *Ac*. The 35S-SPT and the 35S-SPT::*Ac* constructs are shown at the top of the figure. The NPT fusion is used in the initial selection for transformants. The 35S-SPT or 35S-SPT::*Ac* fusions are assayed by plating progeny from the transformants onto streptomycin-containing media. The expected phenotypes of the seedlings on streptomycin-free and streptomycin-containing media are shown below. Stippled regions indicate green areas and white regions indicate bleached areas of the cotyledons.

excised in cells forming the germ line prior to fertilisation. This resulted in an intact copy of the SPT fusion and a transposed *Ac* being transmitted through to the progeny. This is referred to as germinal excision of the transposon and it is in progeny from these individuals in which recessive mutations, due to the insertion of *Ac*, should appear.

Use of the 2′ SPT fusion in *Arabidopsis*

The streptomycin fusion used by Jones *et al.* (1989) was composed of a 2′T-DNA promoter (Velten *et al.* 1984), the streptomycin phosphotransferase coding region from Tn5 (Mazodier *et al.* 1985) and a 3′ flanking region from another T-DNA gene, octopine synthase (DeGreve *et al.* 1983). This was carried on an *Agrobacterium* binary vector that also conferred resistance to kanamycin and was referred to as pJJ2668 (Jones *et al.* 1988). Here it will be referred to as the 2′ SPT construct. The same plasmid, but also carrying the maize *Ac* element in the 5′ untranslated region of the SPT fusion, was termed pJJ2853, referred to here as the 2′ SPT::*Ac*(Cla) construct.

These plasmids were introduced into *Arabidopsis thaliana* using the root transformation procedure described by Valvekens *et al.* (1988). To establish the conditions for the streptomycin assay with *Arabidopsis* seedlings, non-transformed seed of the ecotype Landsberg *erecta* were plated on a variety of media with a range of streptomycin concentrations. These media were GM (germination media, Valvekens *et al.* 1988) with either 1% sucrose, 1% glucose, 0.1% glucose or no glucose, with 0, 25, 50, 100, 200 mg l^{-1} streptomycin phosphate. On GM 1% glucose, a streptomycin concentration of 200 mg l^{-1} was required to achieve complete bleaching of the sensitive seedlings. However, even at this concentration the sensitive seedlings could be easily rescued by subsequent transfer to streptomycin-free media. On GM media containing 0.1% glucose or no glucose a streptomycin concentration of 25 mg l^{-1} effectively bleached non-transformed seedlings. The bleached seedlings, however, did not fully expand on the GM with reduced sugar and did not recover so effectively when subsequently rescued onto streptomycin-free media.

Seed from ten independent Landsberg *erecta* transformants carrying the 2′-SPT construct were plated on either GM 1% glucose, streptomycin 200 mg l^{-1} or GM 0.1% glucose, streptomycin 25 mg l^{-1}. These transformants had either one or two T-DNA insertions carrying both the nos–NPTII and the 2′-SPT markers as judged by the segregation data after plating the seeds on kanamycin-containing media. PCR analysis was used to confirm the integrity of the T-DNA. The streptomycin resistance phenotype of the progeny from the different transformants plated on the different media was, however, very poor. The cotyledons from the seedlings containing the 2′SPT fusion were pale green and the first leaves had a variegated phenotype due to incomplete detoxification of the streptomycin. We believe this poor resistance phenotype was due to the low level of expression of the 2′SPT fusion in *Arabidopsis* cotyledons. This phenotype was not improved by the inclusion of 10 μM NAA in the media, previously shown to stimulate expression of the 2' promoter in transformed plants (Langridge *et al.* 1989).

The 2'-SPT::Ac(Cla) construct used by Jones *et al.* (1989) contained a mutation (GGTTCC to AGATCGATCT conferring a Cla1 site) approximately 80 bp downstream of the transposase polyadenylation site (Jones *et al.* 1990*b*). Twenty transformants carrying this construct were generated but since the resistance phenotype from the 2'-SPT construct was so poor the streptomycin selection could not be used to monitor the activity of the introduced *Ac*(Cla) element. Instead, DNA was isolated from pooled progeny of the transformants carrying the 2'-SPT::*Ac* (Cla) construct and amplified using oligonucleotides homologous to the 2' promoter and the streptomycin phosphotransferase coding region in PCR experiments. If the *Ac*(Cla) element had excised, a PCR product of 663 bp would have been expected. This is 99 bp larger than the corresponding PCR fragment from the 2'-SPT construct due to the *Ac* flanking sequences, cloned in with the element, remaining after excision. Despite all the control fragments appearing as expected, a 663 bp fragment, indicating excision of *Ac*, was never observed. Jones *et al.* (1990*a*) have recently reported that the Cla1 mutation reduces the level of *Ac* excision at least 10-fold in tobacco. Since the level of excision of the wild-type *Ac* element is already very low in *Arabidopsis* (Schmidt and Willmitzer, 1989) a further 10-fold drop in activity would inevitably lead to difficulties in detecting excision.

Other excision markers tested in *Arabidopsis*

Since the 2'-SPT construct gave a poor level of resistance to streptomycin several other different antibiotic resistance markers were tested in *Arabidopsis* for use as markers to monitor transposon activity. The kanamycin resistance marker used in previous studies (Baker *et al.* 1987; Schmidt and Willmitzer, 1983) was not used as this was the marker (1'NPT11) being used to select for the introduction of constructs into *Arabidopsis* using *Agrobacterium tumefaciens* transformation. This marker was found to be considerably more efficient than a 1' hygromycin phosphotransferase (HPT) marker in root transformation experiments using the *Arabidopsis thaliana* ecotype Landsberg *erecta*.

The other fusions tested were: a 35S (CaMV promoter)–streptomycin phosphotransferase fusion (35S–SPT), a nos–spectinomycin adenyltransferase fusion (nos–SPEC) conferring resistance on spectinomycin (Hollingshead and Vapnek, 1985; Svab *et al.* 1990; Jones, J., Scofield, S., Bishop, G. and Harrison, K., unpublished) and a 35S–phosphinothricine acetyltransferase fusion (35S–BAR) conferring resistance on bialaphos and Basta (Thompson *et al.* 1987), De Block *et al.* 1987; Jones, J., Scofield, S., Bishop, G. and Harrison, K., unpublished). Unlike the 2'-SPT fusion the 35S–SPT, nos–SPEC and 35S–BAR fusions conferred clear resistant phenotypes on the progeny of *Arabidopsis*. However, the streptomycin resistance assay was the only one that enabled the somatic excision of *Ac* to be monitored. Seedlings with a range of variegated

phenotypes were observed in progeny of transformants carrying the 35S–SPT::*Ac*
fusion. Green sectors on the seedlings on streptomycin-containing media arising
from excision of *Ac* from the 35S–SPT fusion had very sharp boundaries and the
cell-autonomy of the resistance phenotype was confirmed by microscopical
examination of variegated seedlings. Cells containing fully developed green
chloroplasts could be seen adjacent to cells in which the chloroplasts had remained
white and undifferentiated (K. Pyke, personal communication). On the other
hand, there were very few discrete sectors arising in seedlings containing
nos–SPEC:*Ac* plated on spectinomycin, larger sectors with diffuse edges being
more commonly observed. On bialaphos, seedlings containing a 35S–BAR::fu-
sion showed three phenotypes; full resistance, varying degrees of intermediate
resistance or sensitivity (Sjodin, C., Jones, J. and Dean, C., unpublished). These
and other data in tobacco (Jones, J., Scofield, S., Bishop, G. and Harrison, K.,
unpublished) indicate that the spectinomycin and bialaphos resistance phenotypes
are not fully cell-autonomous. With resistance markers whose phenotype is not
fully cell-autonomous it is difficult to distinguish between individuals showing
germinal excision events (with a corresponding transposed element) and those
showing a high level of somatic activity. Further analysis of *Ac* and *Dc* activity in
Arabidopsis was therefore carried out exclusively with the streptomycin resistance
assay.

Ac activity in *Arabidopsis*

The level of somatic and germinal excision of a wild-type *Ac* element cloned in
both orientations with respect to the 35S–SPT fusion has now been followed in 50
independent *Arabidopsis* transformants through two generations. In the progeny
of the primary transformant (the T_2 generation) the germinal excision frequency
of the wild-type *Ac* element was between 0.07 and 0.58 %, which agrees with
frequencies reported by Schmidt and Willmitzer (1989). The frequency of somatic
and germinal excision increased in homozygous as compared to heterozygous lines
showing that, as in tobacco, there is a positive dosage effect of *Ac* (Jones *et al.*
1989). We have also observed a significant increase in the frequency of somatic and
germinal excision of an *Ac* element following deletion of 537 bp (brought about by
digestion of the *Ac* element with *Nae*I and subsequent religation) from the CpG-
rich 5′ untranslated leader region (Lawson, E., Scofield, S., Sjodin, C., Jones, J.
and Dean, C., unpublished). In progeny of the transformants carrying this deleted
Ac the germinal excision frequency increased to 5 %. The basis of this increase is
currently being investigated.

We have collected 200 individuals picked as 'full greens', i.e. fully streptomycin
resistant which have inherited an excision event and are currently screening their
progeny for the segregation of recessive mutations caused by insertion of the *Ac*
element. We do not yet know the genomic location of the T-DNA inserts carrying
the *Ac* elements thus this screen is termed non-targeted transposon tagging.

Development of a two-element transposon system using *Ac* and *Ds*

One of the major disadvantages of using an autonomous element in tagging experiments is the somatic reversion of the mutation. If the transposon is lying in a gene whose gene product is non-cell autonomous then somatic excision of the transposon from that gene might yield enough gene product to mask the mutant phenotype. Since our main interest is in the isolation of genes involved in floral induction, many of which may confer non-cell autonomous phenotypes, we have also, along with many others (Lassner *et al.* 1989; Hehl and Baker, 1989; Masterson *et al.* 1989), been establishing a two-element transposon system. In this system the two components are:

(1) a non-autonomous *Ds* element; (2) a transposase source that provides transposase *in trans* to activate the *Ds* but cannot itself move.

The two-element system would also have a means by which the transposase source can be efficiently counter selected after transactivation of the *Ds* element so as to prevent any somatic reversion of the *Ds* element. Thus any mutations caused by the insertion of *Ds* would be stable.

The components of the two-element system we have developed are summarised in Fig. 2. Two different *Ds* elements have been cloned into the 5′ untranslated leader of the 355S–SPT fusion. One of the *Ds* elements was constructed by causing a frameshift in the transposase open reading frame (by digestion, Klenow treatment with dNTPs and religation at the *Eco*RI site in the *Ac* element). The other *Ds* element was constructed by replacing the *Xho*I–*Hin*dIII internal fragment of *Ac* with a *Bgl*II–*Hin*dIII fragment carrying a 35S–HPT fusion. This enables selection for re-insertion of the element into the genome following excision.

The transposase sources were constructed by removing the 3′ 200 bp from a wild-type *Ac* element and from an *Ac* element carrying deletion of the CpG-rich region (ΔNaeI) at the beginning of the transposase transcript. These can provide the transposase protein but do not contain the *in cis* sequences required for transposition. Transposase fusions to heterologous promoters are also being tested as efficient transposase sources by our colleague Dr George Coupland and his group. Linked to the transposase source are either a 2′-*iaaH* or a 2′-GUS fusion (Jefferson, 1989). We have tested these fusions as ways to efficiently counterselect or screen seedlings for loss of the transposase source after *Ds* transactivation. The *Agrobacterium* T-DNA gene, *iaaH* encodes the last gene product required for the T-DNA induced auxin biosynthetic pathway that converts indole acetamide to indole acetic acid. Plants expressing this gene product die when they are plated on indole acetamide or its analogue naphthalene acetamide (Nam *et al.* 1989; Klee *et al.* 1987). We have incorporated this selection into our tagging strategy, which is outlined in Fig. 3. Obtaining a good seed yield after backcrossing is difficult in *Arabidopsis*. In order to make the system as efficient as possible, it is important to minimize the number of steps where crossing is required. Plants containing a *Ds* element in a 35S–SPT fusion are crossed with plants containing an unlinked transposase source, which is itself tightly linked to the *iaaH* fusion or the GUS

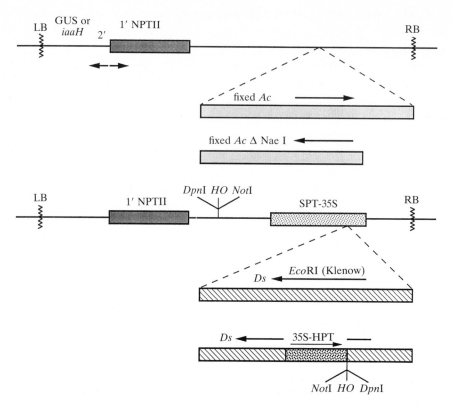

Fig. 2. Components of the two element transposon system. The different constructs introduced into *Arabidopsis* to act as transposase sources and as *Ds* elements are shown. LB and RB are the left and right borders of the introduced T-DNAs. 1′NPTII is the fusion conferring kanamycin resistance. 35S–SPT confers streptomycin resistance and 35S–HPT confers hygromycin resistance. The term 'fixed *Ac*' refers to an *Ac* element in which the 3′ 200 bp have been deleted.

fusion. F_1 plants are allowed to self and F_2 seed collected. These are then plated on media containing streptomycin, hygromycin and either naphthalene acetamide (NAM) or X-glucuronide. Streptomycin, hygromycin and NAM resistant seedlings are then rescued from the plates and allowed to self. Selection of fully streptomycin resistant progeny ensures that the *Ds* has excised and selection of hygromycin resistant progeny ensures that the element has re-inserted into the genome (at least for those progeny heterozygous for *Ds*). Selection of NAM-resistant or GUS minus plants means that only plants where the transposase source has been lost due to segregation will be carried through the selection. Thus any mutation caused by insertion of the *Ds* element will be stable and can be screened for in the progeny of the rescued plants.

Targeted *versus* non-targeted tagging experiments

Several reports have shown that *Ac* and *Ds* transpose preferentially to closely linked sites (Greenblatt, 1984; Jones *et al.* 1990*b*). We are currently investigating whether *Ac* transposes to closely linked sites in *Arabidopsis*. If this is the case, then in order to make the system as efficient as possible the strategy for tagging a specific gene, whose map position is known, needs to include starting from a linked *Ac* or *Ds* element. This strategy has been termed targeted transposon tagging. A desirable resource for targeted transposon tagging would be series of lines containing mapped *Ds* elements, ideally one approximately every 5–10 cM

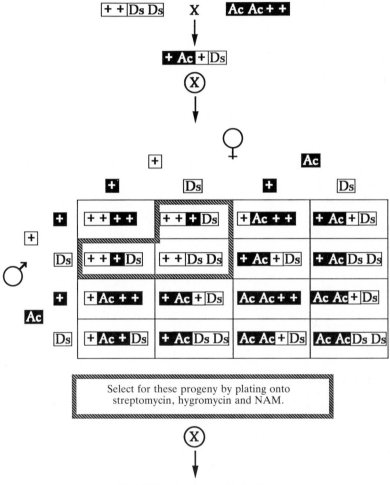

Fig. 3. Selection of individuals containing a transposed *Ds* and lacking a transposase source. Punnett square to show the different F$_2$ progeny obtained from a cross of a line homozygous for a *Ds* insertion with a line homozygous for an unlinked fixed *Ac*. The progeny highlighted by the hatched box would be insensitive to NAM media because they would not contain the fixed *Ac* tightly linked to the 2′-iaaH fusion. Selection of fully streptomycin-resistant and hygromycin-resistant individuals within that group of seedlings would select those individuals containing a transposed *Ds*. Selfed progeny of these plants would then be screened for segregating mutant phenotypes.

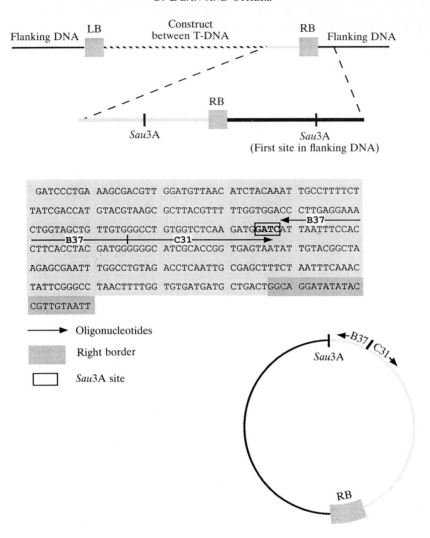

Fig. 4. Inverse polymerase chain reaction to generate DNA probes flanking the right border of the *Agrobacterium* T-DNA. Genomic DNA from transformed plants is digested with *Sau*3A and then circularised by self-ligation. Inverse PCR reactions (Ochman *et al.* 1988) are then carried out using oligonucleotides B37 and C31, amplifying the sequences flanking the T-DNA insertion.

covering the entire genome. An international collaboration is currently underway to generate this resource. The T-DNA insertions carrying the introduced transposons are being mapped onto the *Arabidopsis* RFLP map (Chang *et al.* 1988; Nam *et al.* 1989). In order to achieve this we are generating DNA probes flanking the T-DNA using the inverse polymerase chain reaction (IPCR) technique. This is summarised in Fig. 4.

If the IPCR fragments show restriction fragment length polymorphisms (RFLP) then they can be used to map the position of the T-DNA in relation to the other RFLP markers.

Conclusions

Use of the streptomycin resistance marker to monitor both somatic and germinal activity of the transposons has greatly enhanced our understanding of the behaviour of the transposable elements *Ac* and *Ds* in *Arabidopsis*. *Arabidopsis* lines containing modifications of the elements that significantly increase the transposition frequency have now been made. Once some of the T-DNA lines carrying the transposons have been mapped onto the RFLP map targeted tagging experiments will be initiated. Non-targeted tagging experiments are currently underway and so it should not be long before the first *Arabidopsis* gene is cloned using transposon tagging.

References

BAKER, B., COUPLAND, G., FEDEROFF, N., STARLINGER, P. AND SCHELL, J. (1987). Phenotypic assay for excision of the maize controlling element *Ac* in tobacco. *EMBO J.* **6**, 1547–1554.

BAKER, B., SCHELL, J., LORZ, H. AND FEDEROFF, N. (1986). Transposition of the maize controlling element 'Activator' in tobacco. *Proc. natn. Acad. Sci. U.S.A.* **83**, 4844–4848.

BINGHAM, P. M., LEVIS, R. AND RUBIN, G. M. (1981). Cloning of DNA sequences from the *white* locus of *D. melanogaster* by a novel and general method. *Cell* **25**, 693–704.

CHANG, C., BOWMAN, J. L., DEJOHN, A. W., LANDER, E., MEYEROWITZ, E. M. (1988) Restriction fragment length polymorphism linkage map for *Arabidopsis thaliana. Proc. natn. Acad. Sci. U.S.A.* **85**, 6856–6860.

CHILTON, M.-D., DRUMMOND, M. H., MERLO, D. J., SCIAKY, D., MONTOYA, A., GORDON, M. P. AND NESTER, E. W. (1977). Stable incorporation of plasmid DNA into higher plant cells: the molecular basis of crown gall tumorigenesis. *Cell* **11**, 263–274.

COEN, E. S. AND CARPENTER, R. (1986). Transposable elements in *Antirrhinum,* generators of genetic diversity. *Trends Genet.* **2**, 292–296.

DE BLOCK, M., BOTTERMAN, J., VANDEWIELE, M., DOCKX, J., THOEN, C., GROSSELE, V., MOVVA, N. R., THOMPSON, C., VAN MONTAGU, M. AND LEEMANS, J. (1987). Engineering herbicide resistance in plants by expression of a detoxifying enzyme. *EMBO J.* **6**, 2513–2518.

DEGREVE, H., DHAESE, P., SEURINCK, J., LEMMERS, S., VAN MONTAGU, M. AND SCHELL, J. (1983). Nucleotide sequence and transcript map of the *Agrobacterium tumefaciens* Ti plasmid encoded octopine synthase gene. *J. molec. appl. Genet.* **1**, 499–501.

DORING, H. P. AND STARLINGER, P. (1986). Molecular genetics of transposable elements in plants. *A. Rev. Genet.* **20**, 175–200.

FEDEROFF, N. (1983). Controlling elements in maize. In *Mobile Genetic Elements* (ed. J. Shapiro), pp. 1–63, Academic Press NY.

FEDEROFF, N., WESSLER, S. AND SHURE, M. (1983). Isolation of the transposable maize controlling elements *Ac* and *Ds. Cell* **35**, 235–242.

FEDEROFF, N. V., FURTEK, D. B. AND NELSON, O. E. (1984). Cloning of the bronze locus in maize by a simple and generalizable procedure using the transposable controlling element *Activator (Ac). Proc. natn. Acad. Sci. U.S.A.* **81**, 3825–3829.

FELDMAN, K. A. AND MARKS, D. (1987). *Agrobacterium* mediated transformation of germinating seeds of *Arabidopsis thaliana*: a non-tissue culture approach. *Molec. gen. Genet.* **208**, 1–9.

FELDMAN, K. A., MARKS, M. D., CHRISTIANSON, M. L. AND QUATRANO, R. S. (1989). A dwarf mutant of *Arabidopsis* generated by T-DNA insertional mutagenesis. *Science* **243**, 1351–1354.

FINNEGAN, E. J., TAYLOR, B. H., DENNIS, E. S. AND PEACOCK, W. J. (1988). Transposition of the maize transposable element *Ac* in maize seedlings and in transgenic tobacco. *Molec. gen. Genet.* **212**, 505–509.

GREENBLATT, I. (1984). A chromosomal replication pattern deduced from pericarp phenotypes resulting from movements of the transposable element, Modulator, in maize. *Genetics* **108**, 471–485.

HEHL, R. AND BAKER, B. (1989). Induced transposition of *Ds* by a stable *Ac* in crosses of transgenic tobacco plants. *Molec. gen. Genet.* **217**, 53–59.

HOLLINGSHEAD, S. AND VAPNEK, D. (1985). Nucleotide sequence analysis of a gene encoding a streptomycin/spectinomycin adenyltransferase. *Plasmid* **13**, 17–30.

JONES, J. D. G., CARLAND, F. M., MALIGA, P. AND DOONER, H. (1989). Visual detection of transposition of the maize element Activator (*Ac*) in tobacco seedlings. *Science* **224**, 204–207.

JONES, J. D. G., CARLAND, F., HARPER, L., LIM, E. AND DOONER, H. (1990*a*). Genetic properties of the maize transposon *Activator (Ac)* in tobacco. In *Plant Gene Transfer,* UCLA Symp. Molec. cell. Biol., New Series, vol. 129 (ed. Lamb C. J. and Beachy R. N.), pp. 59–64, Alan R. Liss, NY.

JONES, J., CARLAND, F., LIM, E., RALSTON, E. AND DOONER, H. (1990*b*) Preferential transposition of the maize transposon *Activator (Ac)* to linked chromosomal locations in tobacco. *The Plant Cell* **2**, 709–721.

JEFFERSON, R. A. (1989). The GUS reporter gene system. *Nature* **343**, 837–838.

KLEE, H. J., HORSCH, R. B., HINCHEE, M. A., HEIN, M. B. AND HOFFMAN, N. L. (1987). The effects of overproduction of two *Agrobacterium tumefaciens* T-DNA auxin biosynthetic gene products in transgenic petunia plants. *Genes & Dev.* **1**, 86–96.

KNAPP, S., COUPLAND, G. M., UHRIG, H., STARLINGER, P. AND SALAMINI, F. (1988). Transposition of the maize transposable element *Ac* in Solanum tuberosum. *Molec. gen. Genet.* **213**, 285–290.

LANGRIDGE, W. H. R., FITZGERALD, K. J., KONCZ, C., SCHELL, J. AND SZALAY, A. A. (1989). Dual promoter of *Agrobacterium tumefaciens* mannopine synthase genes is regulated by plant growth hormones. *Proc. natn. Acad. Sci. U.S.A.* **86**, 3219–3223.

LASSNER, M. W., PALYS, J. M. AND YODER, J. I. (1989). Genetic transactivation of *Dissociation* elements in transgenic tomato plants. *Molec. gen. Genet.* **218**, 25–32.

MALIGA, P., SVAB, Z., HARPER, E. AND JONES, J. (1988). Improved expression of streptomycin resistance in plants due to a deletion in the streptomycin phosphotransferase coding sequence. *Molec. gen. Genet.* **214**, 456–459.

MARTIN, C., CARPENTER, R., SOMMER, H., SAEDLER, H. AND COEN, E. S. (1985). Molecular analysis of instability in flower pigmentation of *Anthirrhinum majus* following isolation of the pallida locus by transposon tagging. *EMBO, J.* **4**, 1625–1630.

MARTIN, C., PRESCOTT, A., LISTER, C. AND MACKAY, S. (1989). Activity of the transposon *Tam 3* in *Anthirrhinum* and tobacco: possible role of DNA methylation. *EMBO J.* **8**, 997–1004.

McCLINTOCK, B. (1951). Chromosome organisation and genic expression. *Cold Spring Harb. Symp. quant. Biol.* **16**, 13–47.

MASSON, P. AND FEDEROFF, N. (1989). Mobility of the maize Supressor-mutator element in transgenic tobacco cells. *Proc. natn. Acad. Sci. U.S.A.* **86**, 2219–2223.

MASTERSON, R. V., FURTEK, D. B., GREVELDING, C. AND SCHELL, J. (1989). A maize *Ds* transposable element containing a dihydrofolate reductase gene transposes in *Nicotiana tabacum* and *Arabidopsis thaliana. Molec. gen. Genet.* **219**, 461–466.

MAZODIER, P., GOSSART, P., GIRAUD, E. AND GASSER, F. (1988). Completion of the nucleotide sequence of the central region of Tn5 confirms the presence of three resistance genes. *Nucl. Acids Res.* **13**, 195–203.

NAM, H. G., GIRAUDAT, J., DEN, BOER, B., MOONAN, F., LOOS, W, D. B., HAUGE, B. AND GOODMAN, H. (1989). Restriction fragment length polymorphism map of *Arabidopsis thaliana. The Plant Cell* **1**, 699–705.

OCHMAN, H., GERBER, A. S. AND HARTL, D. L. (1988). Genetic application of an inverse polymerase chain reaction. *Genetics* **120**, 621–623.

O'REILLY, C., SHEPHERD, N. S., PEREIRA, A., SCHWARZ-SOMMER, Z., BERTRAM, I., ROBERTSON, D. S., PETERSON, P. A. AND SAEDLER, H. (1985). Molecular cloning of the *al* locus of *Zea mays* using the transposable elements *En* and *Mul. EMBO J.* **4**, 877–882.

PAZ-ARES, X., WIENAND, U., PETERSON, P. A. AND SAEDLER, H. (1986). Molecular cloning of the *C* locus of *Zea mays*: a locus regulating the anthocyanin pathway. *EMBO J.* **5**, 829–833.

PETERSON, P. (1953) A mutable pale green locus in maize. *Genetics* **38**, 682–683.

ROBERTSON, D. (1978) Characterisation of a mutator system in maize. *Mutation Res.* **51**, 21–28.

SCHMIDT, R. J., BURR, F. A. AND BURR, B. (1987). Transposon tagging and molecular analysis of the maize regulatory locus *opaque-2. Science* **238**, 960–963.

SCHMIDT, R. AND WILLMITZER, L. (1989). The maize autonomous element *Activator (Ac)* shows a minimal germinal excision frequency of 0.2 %–0.5 % in transgenic *Arabidopsis thaliana* plants. *Molec. gen. Genet.* **220**, 17–24.

SOMMER, H., CARPENTER, R., HARRISON, B. J. AND SAEDLER, H. (1985). The transposable element *Tam3* of *Antirrhinum majus* generates a novel type of sequence alteration upon excision. *Molec. gen. Genet.* **199**, 225–231.

SVAB, Z., HARPER, E. C., JONES, J. D. G. AND MALIGA, P. (1990). Aminoglycoside-3'-adenyltransferase confers resistance to spectinomycin and streptomycin in *Nicotiana tabacum. Pl. Molec. Biol.* **14**, 197–205.

THERES, N., SCHEELE, T. AND STARLINGER, P. (1987). Cloning of the Bz-2 locus of *Zea mays* using the transposable element *Ds* as a gene tag. *Molec. gen. Genet.* **209**, 193–197.

THOMPSON, C. J., MOVVA, N. R., TIZARD, R., CRAMERI, R., DAVIES, J., LAUWEREYS, M. A. AND BOTTERMAN, J. (1987). Characterisation of the herbicide resistance gene *bar* from *Streptomyces hyroscopicus. EMBO J.* **6**, 2519–2523.

VALVEKENS, D., VAN MONTAGU, M. AND VAN LIJSEBETTENS, M. (1988). *Agrobacterium tumefaciens* mediated transformation of *Arabidopsis thaliana* root explants by using kanamycin selection. *Proc. natn. Acad. Sci. U.S.A.* **85**, 5536–5540.

VAN SLUYS, M. A., TEMPE, J. AND FEDEROFF, N. (1987). Studies on the introduction and mobility of the maize Activator element in *Arabidopsis thaliana* and *Daucus carota. EMBO J.* **6**, 3881–3889.

VELTEN, J., VELTEN, R., HEIN, R. AND SCHELL, J. (1984). Isolation of a dual plant promoter from the Ti plasmid of *Agrobacterium tumefaciens. EMBO J.* **3**, 2723–2730.

YODER, J. I., PALYS, J., ALPERT, K. AND LASSNER, M. (1988). *Ac* transposition in transgenic tomato plants. *Molec. gen. Genet* **213**, 291–296.

Printed in Great Britain © Society for Experimental Biology 1991

A MODEL FOR CELL-TYPE DETERMINATION AND DIFFERENTIATION IN PLANTS

M. DAVID MARKS, JEFF ESCH, PAT HERMAN,
SHAN SIVAKUMARAN and DAVID OPPENHEIMER

School of Biological Sciences, University of Nebraska, Lincoln, NE 68588-0118, USA

Summary

We are using trichome formation on the plant *Arabidopsis thaliana* as a model for the study of plant cell determination and differentiation. Several of the genes that are required for trichome formation are defined by mutations. Two mutations, *ttg* and *gl1*, prevent the initiation of trichome differentiation. Thus, these mutations define products that are involved in the signalling of trichome determination. Other mutations, *gl2*, *gl3*, *dis1*, and *dis2*, define genes that are involved in trichome maturation.

Our immediate goal has been to isolate the genes defined by these mutations and determine the role that they play in trichome formation. Our general goals are (1) to identify counterparts to these genes that are involved in other cell type determination and differentiation processes; (2) to manipulate cell development by altering the normal expression of these genes; and (3) to determine if this information can be used to improve crop plants.

Presently, most of our progress has centered on the *GL1* gene, which has been isolated and characterized. We have found that *GL1* is a *myb*-related gene that is uniquely required for trichome initiation. As in other plants, *Arabidopsis* has a family of *myb*-related genes. We are currently investigating the possibility that some of these other *myb*-related genes are also uniquely required for other types of cell determination events.

Introduction

Plant morphogenesis is largely a process of cell determination and differentiation. Understanding how morphogenesis is controlled at the molecular level will not only satisfy scientific curiosity, but may also generate new ideas for improving crop plants. Although it has long been known that certain concentrations of plant hormones can induce certain tissues to differentiate, the actual underlying mechanisms that mediate the process are largely unknown. Our goal has been to identify genes that mediate specific cell determination and differentiation events. To this end we have chosen as a model system, the development of plant hairs (called trichomes) on *Arabidopsis thaliana* (L.) Heynh. The use of *Arabidopsis* as a model system has been well documented (Somerville *et al.* 1985; Meyerowitz and Pruitt, 1985). It offers the ability to isolate genes by

Key words: trichome formation, *Arabidopsis*, *myb*-related genes, cell differentiation.

chromosome walking and by gene tagging (Guzman and Ecker, 1988; Feldmann *et al.* 1989). Trichome formation provides an excellent model for plant cell determination and differentiation because, (1) the development of trichomes occurs on the surface of the plant where it can be easily observed, (2) the process is relatively simple compared to other plant cell differentiation events, and (3) it is possible to isolate mutants altered in trichome differentiation because trichomes are not required for plant growth and development.

The availability of trichome mutants allows many of the genes uniquely required for trichome development to be identified. The only apparent effect of most of these mutants is altered trichomes. Furthermore, the mutations result in several different phenotypes suggesting that the affected genes are responsible for different aspects of trichome formation. Our short-term goal is to identify the normal function of these genes. The next step will be to determine if there are counterparts to these genes that carry similar functions during the development of other cell types. It is probable that many embryo lethal mutations are in genes required for the determination and differentiation of essential cell-types. The isolation of such genes from crop plants may allow direct manipulation of embryo development and of other developmental processes resulting in crop improvement.

Development of trichomes on normal and mutant plants
Normal trichomes

The trichomes of *Arabidopsis* are composed of single cells that develop on the protodermal tissue of young leaves and stems. The mature leaf trichomes have two to four branches, whereas stem trichomes are unbranched (Fig. 1). Individual cells destined to become trichomes distinguish themselves by rapidly enlarging outward (Fig. 2). The trichomes of the stem elongate away from the surface as a single spike. As the trichome matures, the cell wall thickens and acquires a papillate

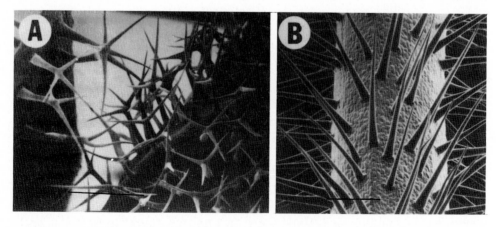

Fig. 1. Normal leaf and stem trichomes. (A) mature leaf trichomes. (B) mature stem trichomes. Bar=200 μm.

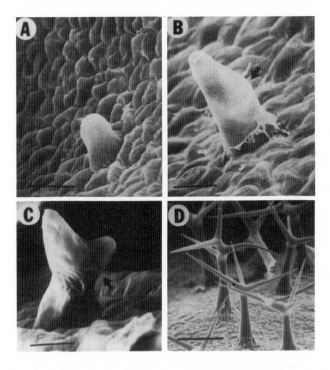

Fig. 2. Development of trichome on the leaf. (A) initiation of a trichome on immature leaf protodermal tissue. (B) elongating trichome showing the initiation of a branch (arrow). (C) elongating trichome with growing branch (arrow). (D) field of mature trichomes. (A–C) bar=10 μm; (D) bar=100 μm.

cuticle. Formation of leaf trichomes is similar except that secondary weakenings occur in the cell wall of the developing trichome. These areas serve as regions of secondary growth which mature into branches. The nuclei of the trichomes enlarge and accumulate up to ten times the amount of DNA found in the guard cells (Jerry Melagarno, Rhode Island College, RI, USA, personal communication). The nature of this amplification process is unknown.

We are currently working with six trichome mutants whose phenotypes fall into three groups. A characterization of these mutants by scanning electron microscopy (SEM) has led to some insights on the possible role that the wild-type alleles of the mutated genes may play during trichome development.

dis1 *and* dis2

The mutants *dis1* and *dis2* have distorted trichomes (Freenstra, 1978); the mature trichomes have an over-inflated appearance and are curved instead of erect. SEM analysis of trichomes in various stages of formation indicate that the defect is apparent at the earliest stage of development (Fig. 3). When the trichomes first initiate, they overexpand. The tendency to overexpand persists throughout trichome development as seen during the formation of branches. The trichome stalk often overexpands at the apparent branch initiation sites instead of

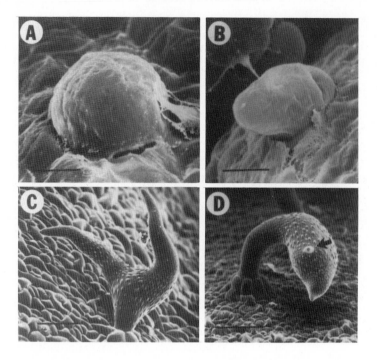

Fig. 3. Development of *dis* trichomes on the leaf. (A) initiation of a *dis* trichome. Bar=5 μm. (B) growing *dis* trichome showing typical swelling. Bar=10 μm. (C) mature *dis* trichome with two branches. Bar=35 μm. (D) an unbranched *dis* trichome with one aborted branch (arrow). Bar=40 μm.

forming a branch, and when branches do form they also overexpand. Several possible defects could lead to the *dis* trichome phenotype. Some component(s) required for strengthening the cell wall may not be synthesized, or the enzymes required for softening the cell wall may be overactive. Alternately, there could be a defect in the regulation of the osmotic pressure during cell expansion. The *DIS1* and *DIS2* gene products may participate directly or have a regulatory role during trichome development. We are using a chromosome walking strategy to isolate *DIS1*. The characterization of these genes and their products should lead to a better understanding of the control of cell expansion during differentiation.

gl2 *and* gl3

The trichomes of the mutants gl2 and gl3 (Koornneef *et al.* 1982), in contrast to the *dis* phenotype, have an under-inflated appearance. The mature trichomes of these plants are erect but have a narrower shaft and fewer branches than normal. In addition, the number of trichomes is reduced. This is especially true of the first pair of leaves, which are often almost completely glabrous. The SEM analysis of trichomes in different stages of development indicate that the defect is evident at the earliest visual stage of initiation (Fig. 4). The developing trichomes appear to initiate properly but fail to fully expand. It is unknown whether the reduced

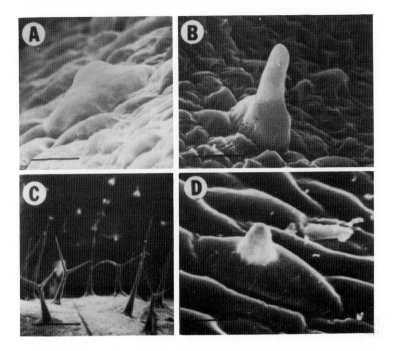

Fig. 4. Development of *gl3* trichomes. (A) initiation of a *gl3* leaf trichome. Bar=10 μm. (B) elongating *gl3* trichome. Bar=10 μm. (C) mature field of *gl3* trichomes on the leaf. Bar=180 μm. (D) aborted trichome on mature stem tissue. Bar=10 μm.

number of branches is due to a defect in the process leading to secondary weakenings in the cell wall or a failure of branches to form at the initiation sites. The former may be true because aborted branches have not been observed. The reduced number of trichomes on these mutants appears, at least in part, to result from a failure of some trichomes to expand once they have been initiated. This phenomenon is especially prevalent on the stems, which often have a greatly reduced number of trichomes. As with the *dis* mutants, the defect could be structural, enzymatic, or osmotic in nature.

We are actively pursuing the isolation of *gl2*. We have obtained two cosmid clones from Howard Goodman (Harvard Medical School, Boston MA, USA, cosmids sent courtesy of Brian Hauge) that should contain the wild-type gene. Overlapping DNA fragments from these cosmids are being cloned into the pBIN19 *Agrobacterium* Ti plasmid binary vector (Bevan, 1984). These constructions are being used to transform *gl2* explants (Valvekens *et al.* 1988). Fragments that restore normal trichome development will be further characterized to localize the gene.

ttg *and* gl1

Most alleles of the *ttg* and *gl1* loci result in almost completely glabrous plants

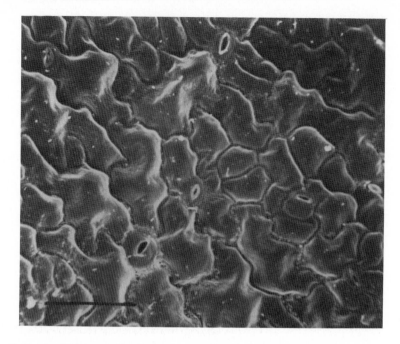

Fig. 5. Lack of trichome development on *gl1* leaf surface. Bar=35 μm.

(Koornneef, 1981; Koornneef *et al.* 1982). Aside from a few trichomes on the leaf margin and petiole, these mutants are devoid of trichomes. The *ttg* mutant also lacks seed coat pigmentation, seed mucilage, and appears to have roots that grow abnormally slowly (Koornneef, 1981; Esch and Marks unpublished data). This multitude of functions for the *TTG* gene product indicates that it may be an important regulatory gene (Koornneef, 1981). SEM analysis of *ttg* and *gl1-1* reveals that these mutations result in a block in the initiation of trichome formation (Fig. 5). Both of these loci most likely encode regulatory proteins that are required for trichome cell-type determination.

Isolation of *GL1*

The actual starting point for this work was mutant *gl1-43* (Marks and Feldmann, 1989). This mutant was isolated from the first population of plants transformed by the *Agrobacterium*-mediated seed transformation procedure (Feldmann and Marks, 1987). Genetic complementation tests with the other trichome mutants revealed that the mutant was an allele of the *gl1* locus. Linkage analysis indicated that a T-DNA insert was closely linked to the mutation. We therefore screened a *gl1-43* genomic library with a T-DNA probe. Southern hybridization and restriction enzyme mapping of the selected clones revealed that the insertion site contained multiple copies of the T-DNA that were tandemly linked in both direct and inverted orientations. One genomic clone was found to contain one of the junctions between the T-DNA and plant DNA. A 3 kb DNA fragment that flanked the T-DNA insert was used as a probe to screen a normal genomic library in the hope of isolating the uninterrupted *GL1* gene.

Two overlapping clones were isolated from the wild-type library. Southern hybridization analysis indicated that the T-DNA insertion site was in the middle of clone NT2. To localize the *GL1* gene, DNA fragments from NT2 were subcloned into the pBIN19 T-DNA vector. These constructions were used to transform *gl1* root explants. Positive molecular complementation was defined as the ability of a fragment to restore normal trichomes to the shoots of regenerating transformed *gl1* explants (Herman and Marks, 1989). This analysis allowed the *GL1* gene to be localized to a 4.5 kb DNA fragment (Marks *et al.* unpublished). Analysis of the DNA sequence of this fragment revealed the presence of 4 significant open reading frames (ORFs) in the same orientation. A computer search has shown that 3 of these ORFs encode amino acid sequences that show homology to the *myb* class of DNA binding proteins (Marks *et al.* unpublished).

GL1 as a *myb*-related gene

myb genes were first found associated with avian proto-oncoviruses that cause myeloblastosis (Moscovici, 1975; Klempnauer *et al.* 1982). *myb* genes of vertebrates are involved in the control of blood cell differentiation (Lipsick and Baluda, 1986). The Myb protein can be divided into two domains, the amino-terminal domain, which contains the DNA binding region and the carboxy-terminal domain, which participates in protein–protein interactions (Sakura *et al.* 1989). The existence of plant genes containing the Myb DNA binding domain has previously been reported (Paz-Ares *et al.* 1987; Marocco *et al.* 1989). However, the DNA binding domain of the animal Myb proteins contains three imperfect repeats of approximately 60 amino acid residues (Klempnauer and Sippel, 1987) whereas, Myb-related proteins from plants contain only the second and third repeat.

The structure of the *GL1* gene is similar to that of *C1* (*Zea mays* regulatory gene involved in controlling seed pigmentation;Paz-Ares *et al.* 1987). It also only contains two repeats toward the amino terminus and the predicted amino terminus of *GL1* is two residues longer than that of *C1*. The predicted exon/intron junctions between the first and second exon and the second and third exon are the same as those in *C1* (unpublished data). The putative carboxy-terminal domain lacks significant homology to *C1*. The lack of homology in this region is also found between *C1* and several other maize and barley *myb* homologues that have been characterized.

To date we have been unable to identify a mRNA encoded by the *GL1* gene. This is not unexpected since only a few cells develop into trichomes and it is known that many other mRNAs that encode regulatory DNA binding proteins in other organisms are unstable. Furthermore, toward the putative 3' end of the gene are two AUUUA sequences that function in other organisms to destabilize the mRNA (Malter, 1989). We have not yet tried polymerase chain reaction (PCR) amplification of cDNA (Vrieling *et al.* 1988) which may allow us to identify the mRNA.

Ongoing experiments with *Gl1*

Even without a mRNA for the gene, we are in a position to ask some interesting questions about the function of *GL1*. We have placed the putative *GL1* promoter in front of the *GUS* (β-glucuronidase) reporter gene (Jefferson *et al.* 1987). If the level of *GUS* expression is sufficient, we may be able to detect cells before they begin the differentiation process. We will also use this construct to transform the *gl1-1* mutant, which lacks the *GL1* gene (see below). It will be interesting to see if we will be able to detect GUS-positive cells in these plants. We may be able to test for the presence of negative feedback by comparing the levels of GUS activity in the transformed normal and mutant plants.

Is *GL1* activity, in itself, sufficient for trichome initiation? To address this question we have placed the CaMV 35S promoter (Odell *et al.* 1985) in front of the putative start codon. This construct will be electroporated into protoplasts derived from the epidermis. In some systems it has been found that expression of a single gene can convert one cell type into another. For example, transformation of the *myoD* gene into fibroblast cells induces them to differentiate into myoblasts (Davis *et al.* 1987). In other systems two or more genes act as positive regulators. For example, anthocyanin production in seedlings requires *R1* gene activity and either *Pl* gene activity or light (Ludwig *et al.* 1990). If *GL1* by itself does not induce cell differentiation then it is possible that the *TTG* product may also be required.

gl1 mutants

The insertional mutant *gl1-43* has a novel trichome phenotype. This mutant lacks trichomes on the stem, but has normal leaf trichomes. The position of the T-DNA insert is located toward the 3′ end of the gene. The insert interrupts a short open reading frame of 62 amino acids. It is possible that this region encodes a protein domain that is required for trichome initiation in the stem, or that the *GL1* transcript undergoes alternate splicing and the insert disrupts *GL1* transcript processing in the stem. Finally, it is possible that the region contains an enhancer-like sequence that is required for *GL1* expression in stem tissue. To determine which of these possibilities is correct we have placed the putative promoter region of the *GL1* gene in front of the *GUS* reporter gene. If we detect *GUS* expression in stem trichomes then it is likely that the 3′ region does not contain an enhancer needed for stem expression. Instead, it would be likely that the insert has interrupted a functional reading frame or disrupted RNA processing.

The *gl1-1* mutant has allowed an important biological question to be addressed. Is the *GL1* gene required for other functions besides trichome formation? It is possible that the *gl1-43* mutation is only inactivating a region of the gene required for trichome formation. Our analysis of the *gl1-1* allele has revealed that this mutation is due to a deletion. This deletion extends from approximately 900 bp upstream of the putative start codon to over 1000 bp beyond the 5′ end of the clone that complemented the mutation. Thus, the deletion of the entire *GL1* region strongly suggests that the only function of the *GL1* gene is in trichome development.

myb-related genes in *Arabidopsis*

Do genes required for trichome formation have counterparts responsible for other types of cell determination events? To test this hypothesis for *GL1* we are attempting to identify other *myb*-related genes that are involved in controlling the initiation of cell differentiation events. Previous work by others has shown that maize and barley both contain *myb*-related gene families (Marocco *et al.* 1989). Our preliminary data from genomic Southern analysis suggest that *Arabidopsis* also contains an extensive *myb*-related gene family. Under permissive

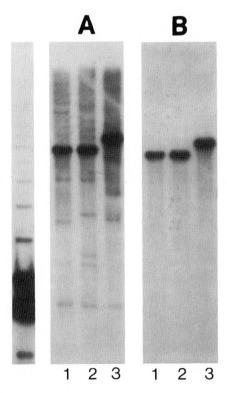

Fig. 6. Southern blot of *Arabidopsis* DNA probed with *myb* probe. DNA isolated from three ecotypes of *Arabidopsis* (lane 1=Columbia, lane 2=Landsberg; lane 3=Wassilewskija), was cleaved with *Sst*I and resolved on a 0.7% agarose gel. The gel was blotted onto a nylon filter and hybridized in 50% formamide and 5×SSC (1×=0.15 M sodium chloride and 0.015 M sodium citrate) at 30°C to a probe containing the first and second putative exons of the *GL1* gene. The filters were washed in 2×SSC at (A) 45°C and (B) 65°C. The far left lane contains the 1 Kb ladder (BRL, Gaithersbug, MD) that was resolved on the same gel.

hybridization conditions numerous DNA fragments hybridize to a *GL1* probe containing the conserved *myb*-related DNA binding domain (Fig. 6A). Under more stringent conditions only the *GL1* sequence hybridized (Fig. 6B). To confirm these results we have isolated 50 genomic clones that hybridized under permissive conditions to a probe containing the first and second exons of the *GL1* gene. DNA sequence analysis of two clones confirmed that they encode the *myb* DNA binding domain. DNA sequence analysis is continuing on the other clones. We will attempt to determine if other *myb*-related genes from *Arabidopsis* are involved in cell-type determination.

Conclusions

We have found that the *GL1* gene, which is required for trichome determination, is a *myb*-related gene. Many questions still remain about the function of this gene. For example: does *GL1* activate a developmental gene cascade?, and how is *GL1* activated? It is known that plants contain extensive *myb*-related gene families and it is highly likely that members of the families are involved in other cell differentiation events. What we learn about *GL1* will aid in our efforts to understand the action of the other *myb*-related genes. Furthermore, our studies on the other trichome genes should also provide important information about the molecular mechanisms that underlie cell-type determination and differentiation.

D.G.O. is supported by the University of Nebraska Centre for Biotechnology postdoctoral fellowship programme. Major funding for this work was provided by an Individual Investigator Award from the McKnight Foundation and by grant DCB-8801949 from the National Science Foundation.

References

Bevan, M. W. (1984). Binary *Agrobacterium* vectors for plant transformation. *Nucl. Acids Res.* 12, 8711–8721.

Davis, R. L., Weintraub, H. and Lassar, A. B. (1987). Expression of a single transfected cDNA converts fibroblasts to myoblasts. *Cell* 51, 987–1000.

Feldmann, K. A. and Marks, M. D. (1987). *Agrobacterium*-mediated transformation of germinating seeds of *Arabidopsis thaliana*: A non-tissue culture approach. *Molec. gen. Genet.* 208, 1–9.

Feldmann, K. A., Marks, M. D., Christianson, M. L. and Quantrano, R. S. (1989). A dwarf mutant of *Arabidopsis thaliana* generated by T-DNA insertional mutagenesis. *Science* 243, 1351–1354.

Freenstra, W. J. (1978). Contiguity of linkage groups I and IV as revealed by linkage relationships of two newly isolated markers *dis1* and *dis2*. *Arabidopsis Information Service* 15, 35–38.

Guzman, P. and Ecker, J. R. (1988). Development of large DNA methods for plants: Molecular cloning of large segments of *Arabidopsis* and carrot DNA into yeast. *Nucl. Acids Res.* 16, 11 091–11 105.

Herman, P. L. and Marks, M. D. (1989). Trichome development in *Arabidopsis thaliana*. II. Isolation and complementation of the *GLABROUS1* gene. *Plant Cell* 1, 1051–1055.

Jefferson, R. A., Kanvanagh, T. A. and Bevan, M. W. (1987). GUS fusion: B-glucuronidase as a sensitive and versatile gene fusion marker in higher plants. *EMBO J.* **6**, 3901–3907.

Klempnauer, K.-H., Gonda, T. J. and Bishop, J. M. (1982). Nucleotide sequence of the retroviral leukemia gene v-*myb* and its cellular progenitor c-*myb*: the architecture of a transduced oncogene. *Cell* **31**, 453–463.

Klempnauer, K.-H. and Sippel, A. E. (1987). The highly conserved amino-terminal region of the protein encoded by the v-*myb* oncogene functions as a DNA-binding domain. *EMBO J.* **6**, 2719–2725.

Koornneef, M. (1981). The complex syndrome of *ttg* mutants. *Arabidopsis Information Service* **18**, 45–51.

Koornneef, M., Dellaert, L. W. M. and van der Veen, J. H. (1982). EMS and radiation-induced mutation frequencies at individual loci in *Arabidopsis thaliana*. *J. Hered.* **74**, 265–272.

Lipsick, J. S. and Baluda, M. A. (1986). The *myb* oncogene. In *Gene amplification and analysis*, vol. 4., Oncogenes (ed. Papas, T. S. and Vande Woude, G. F.), pp. 73–98. Elsevier Science Publishing Co., New York.

Ludwig, S. E., Bowen, B., Beach, L. and Wessler, S. R. (1990). A Regulatory gene as a novel visible marker for maize transformation. *Science* **247**, 449–450.

Malter, J. S. (1989). Identification of an AUUUA-specific messenger RNA binding protein. *Science* **246**, 664–666.

Marks, M. D. and Feldmann, K. A. (1989). Trichome development in *Arabidopsis thaliana*. I. T-DNA tagging of the *GLABROUS1* gene. *Plant Cell* **1**, 1043–1050.

Marocco, A., Wissenbach, M., Becker, D., Paz-Ares, J., Saedler, H., Salaminin, F. and Rohde, W. (1989). Multiple genes are transcribed in *Hordeum vulgare* and *Zea mays* that carry the DNA binding domain of the *myb* oncoproteins. *Molec. gen. Genet.* **216**, 183–187.

Meyerowitz, E. M. and Pruitt, R. E. (1985). *Arabidopsis thaliana* and plant molecular genetics. *Science* **229**, 1214–1218.

Moscovici, C. (1975). Leukemic transformation with avian myeloblastosis virus: present status. *Curr. Topics Microbiol. Immunol.* **71**, 79–101.

Odell, J. T., Nagy, F. and Chua, N-H. (1985). Identification of DNA sequences required for activity of the cauliflower mosaic virus 35S promoter. *Nature* **313**, 810–812.

Paz-Ares, J., Ghosal, D., Wienand, U., Peterson, P. A. and Saedler, H. (1987). The regulatory c1 locus of *Zea mays* encodes a protein with homology to *myb* proto-oncogene products and with structural similarities to transcriptional activators. *EMBO J.* **6**, 3553–3558.

Sakura, H., Kanei-Ishii, C., Nagase, T., Nakagoshi, H., Gonda, T. and Ishii, S. (1989). Delineation of three functional domains of the transcriptional activator encoded by the c-*myb* proto-oncogene. *Proc. natn. Acad. Sci. U.S.A.* **86**, 5758–5762.

Somerville, C. R., McCourt, P., Caspar, T., Estelle, M. and Keith, K. (1985). *Arabidopsis thaliana* as a model system for plant genetics and molecular biology. In *Plant Genetics* (ed. M. Freeling), pp. 651–660. Alan R. Liss, New York.

Valvekens, D., Van Montagu, M. and Van Lifsebettens, M. (1988). *Agrobacterium*-mediated transformation of *Arabidopsis thaliana* root explants by using kanamycin selection. *Proc. natn. Acad. Sci. U.S.A.* **85**, 5536–5540.

Vrieling, H., Niericker, M. J., Simons, J. W. and van Zeeland, A. A. (1988). Nucleotide sequence determination of point mutations at the mouse HPRT locus using *in vitro* amplification of HPRT mRNA sequences. *Mutation Res.* **198**, 99–106.

Printed in Great Britain © Society for Experimental Biology 1991

GENETIC CONTROL OF PATTERN FORMATION DURING FLOWER DEVELOPMENT IN *ARABIDOPSIS*

JOHN L. BOWMAN and ELLIOT M. MEYEROWITZ

Division of Biology 156-29, California Institute of Technology, Pasadena, California 91125, USA

Summary

Arabidopsis flowers develop from groups of undifferentiated cells on the flank of an inflorescence meristem. The cells in these flower primordia must somehow assess their position within the primordium and differentiate accordingly to produce a flower with a precisely defined pattern of organ types and positions. The molecular mechanisms by which this is accomplished are largely unknown. We are studying a set of genes whose mutations give homeotic phenotypes in *Arabidopsis* flowers. A genetic model to explain the specification of organ identity by combinatorial action of the products of these homeotic genes is presented, along with several aspects that are not readily addressed by the model. The recent cloning of one of the *Arabidopsis* homeotic genes, and an additional homeotic gene from *Antirrhinum*, has provided an opportunity for molecular tests of our genetic model. So far, the molecular data are in accord with the genetic model.

Introduction

Flowers begin their development as groups of undifferentiated cells, which comprise the floral meristem. The cells in these flower primordia then divide and differentiate to produce a flower with a precisely defined pattern of organ positions and types. During this process, cells within the primordium must somehow 'assess' their relative position and differentiate accordingly. The mechanisms by which these cells sense and determine their position are largely unknown. It appears that some processes utilized in animals are not involved. There is no cell migration in higher plants, and maternal deposition of positional information is not required, since most plant cells are totipotent. As a method of understanding the molecular mechanisms involved in pattern formation in the flower, our laboratory is studying genes whose wild-type products are required for proper pattern formation in *Arabidopsis thaliana* (L.) Heynh. flowers (Pruitt *et al.* 1987; Bowman *et al.* 1988; Bowman *et al.* 1989; Yanofsky *et al.* 1990; Bowman *et al.* 1991).

Flower formation in *Arabidopsis* can be considered as a series of developmental steps. The first of these, termed floral induction, is the switch from vegetative growth to reproductive growth. This involves a reorganization of the apical meristem in response to both environmental and internal signals such as day length, temperature, and age (Vaughan, 1955; Miksche and Brown, 1965; Drews

Key words: flower development, *Arabidopsis*, homeotic genes.

and Goldberg, 1989). Following floral induction, the reproductive (or inflorescence) meristem produces one to three cauline (stem) leaves, followed by an
indeterminate number of individual flowers in a phyllotactic spiral (Smyth *et al.*
1990). Rapidly enlarging individual flower primordia (floral meristems), which
exhibit determinate growth, are produced on the flank of the apical meristem.
These cells divide to produce four whorls of organ primordia in a sequential
manner, and in a well-defined whorled pattern (Smyth *et al.* 1990). Each of the
organ primordia then differentiates into one of the four types of floral organs,
depending upon its position within the flower. These events do not occur as
discrete steps, but rather as a continuous process, and it is possible that individual
genes are involved in more than one of the steps. Mutations disrupting various
stages in these processes have been isolated in *Arabidopsis*, providing genetic tools
for elucidating the mechanisms of flower development (Koornneef *et al.* 1983;
Pruitt *et al.* 1987; Bowman *et al.* 1988; Komaki *et al.* 1988; Haughn and
Sommerville, 1988; Bowman *et al.* 1989; Meyerowitz *et al.* 1989; Hill and Lord,
1989; Okada *et al.* 1989; Kunst *et al.* 1989; Bowman *et al.* 1991). Our laboratory is
primarily concerned with processes after floral induction, therefore those
mutations affecting events following inflorescence meristem formation will be the
focus of this review. Since apparently homologous mutations exist in *Antirrhinum
majus* (Stubbe, 1966; Sommer *et al.* 1990; Carpenter *et al.* 1990; Schwarz-Sommer
et al. 1990), similarities and differences in flower development between the two
systems will be noted throughout.

Flower structure

The radially symmetric *Arabidopsis* flower, which is typical of the Brassicaceae,
is composed of four concentric whorls of organs (Fig. 1D): the first or outer whorl
is occupied by four green sepals; four white petals are found in the second whorl
positions, which are alternate and interior to the first whorl positions; four long
medial (with respect to the inflorescence meristem) and two short lateral stamens,
each composed of a filament capped with a pollen bearing anther, occupy the third
whorl; and a gynoecium composed of a two-chambered ovary topped with a short
style and capped with stigmatic papillae occupies the fourth whorl (Smyth *et al.*
1990). Each gynoecium contains approximately 30–50 ovules born in rows along
the margins of fusion of the carpels. The flowers develop in a raceme so that a
single plant has a series of flowers in different stages of development, the older
flowers and mature fruits toward the base, and the younger flowers at the apex.
The vegetative growth of *Arabidopsis* is characterized by the production of several
rosette leaves in a phyllotactic spiral, each with a secondary meristem in their axil
(Smyth *et al.* 1990). After floral induction, before the individual flowers are
formed in a phyllotactic spiral, one to three (average=2.05) cauline leaves are
produced, each also with a secondary meristem in its axil (Smyth *et al.* 1990). Each
of the secondary meristems is capable of producing its own cauline leaves and
flowers.

The development of individual flowers, which is similar to that of *Cheiranthus cheiri* (Payer, 1857; Sattler, 1973) and *Brassica napus* (Polowick and Sawhney, 1986), has been described in detail and stages defined by Smyth *et al.* (1990) (Fig. 1A–D). Briefly, flower development commences when a group of cells emerges to form a buttress on the flank of the inflorescence meristem (stage 1; Fig. 1A). As these cells divide, an indentation separates the flower primordium from the inflorescence meristem (stage 2; Fig. 1A). Sepal buttresses arise (stage 3; Fig. 1A) on the flanks of the flower primordium. The abaxial sepal primordium arises first, followed by the adaxial and then the lateral sepal primordia. Shortly after these primordia grow to overlay the remaining flower meristem (stage 4), the second (petal) and third (stamen) whorl primordia become visible (stage 5). The sepals' continued growth causes them to enclose the developing bud entirely (stage 6; Fig. 1B). The rapidly growing stamen primordia (in the third whorl positions) develop distinct filament and anther regions (stage 7; Fig. 1C) and locules appear soon after in the anthers of the long (medial) stamens (stage 8). Petal elongation marks the beginning of stage 9 and they reach the height of the short stamens by stage 10. During these stages the gynoecium develops from those cells interior to the stamens. Development of the gynoecium begins as the remaining floral meristem, which is dome shaped at this time, grows into a cylinder due to cell division at the periphery of the dome. Stigmatic papillae appear on the rim of the cylinder at stage 11 (Fig. 1D) about the same time that nectaries, which began their development during stage 9, mature at the base of the lateral stamens. All floral organs continue to grow, with the petals reaching the level of the long stamens by stage 12. Development through stage 12 takes approximately 13 days under growth conditions of 25°C in 24 h illumination. After stage 12, the bud opens and the petals and stamens continue to elongate until anthesis occurs. Later

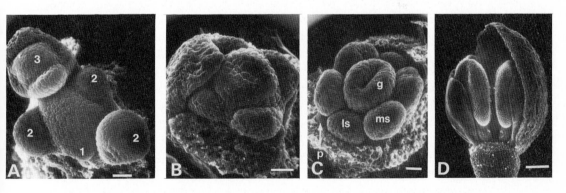

Fig. 1. SEM micrographs depicting the early development of wild-type *Arabidopsis* flowers grown at 25°C. (A) Apical meristem and flower primordia development through stage 3. Stages of the buds are indicated. Bar=10 μm. (B) Stage 6 bud. Three sepals, two medial and one lateral have been removed. Bar=10 μm. (C) Stage 7 bud. All four sepals have been removed to expose the petal (p), medial stamen (ms), and lateral stamen (ls) primordia and the gynoecial cylinder (g). Bar=10 μm. (D) Stage 11 bud. Three sepals have been removed revealing the inner three whorls of organs. Bar=100 μm.

stages of development, through seed maturation, have been described by Müller (1961).

Antirrhinum majus flowers also consist of four concentric whorls of organs and are produced in a raceme of spiral phyllotaxis, but are bilaterally symmetric (Stubbe, 1966). They are much larger than *Arabidopsis* flowers, and each is produced in the axil of a bract, an organ type not present in *Arabidopsis*. Those organs closest to the bract are referred to as lower while those on the opposite side are referred to as upper. The first whorl consists of five free green sepals, and the second whorl of five petals, which are fused for part of their length forming a tube with five lobes. The lobes of the upper two petals are distinct in shape from the lower three. Five stamens comprise the third whorl, but in wild-type flowers the uppermost is aborted in early develoment. The fourth whorl is occupied by an ovary composed of two carpels. The prominent zygomorphy of *Antirrhinum* flowers allows a class of mutations not identified in *Arabidopsis*, those in which the symmetry of the flower is no longer bilateral, but rather radial. The production and development of floral organ primordia is sequential; the outer whorl primordia form first and the central last, as is observed in *Arabidopsis* (Awasthi *et al*. 1984). Vegetative growth of *Antirrhinum* is characterized by pairs of leaves in decussate phyllotaxis. Following the transition from vegetative to inflorescence meristem, bracts are produced in a phyllotactic spiral with a flower in the axil of each.

Classes of mutations

One approach to elucidating the mechanisms by which pattern formation occurs is to identify mutations that specifically disrupt developmental processes, and to study the wild-type products of genes identified by such mutations. Mutations have been isolated that disrupt the major events of flower development after floral induction, and these can be divided into three classes: (1) mutations that disrupt the formation of individual flower primordia, (2) mutations that affect the generation of organ primordia within the flower and (3) mutations that alter the fate of organ primordia. Many of the mutations affecting the formation of floral organ primordia also alter their fate. These two processes are concurrent and will therefore be discussed together.

Mutations in genes involved in the production of flower primordia and the transition from the inflorescence meristem to the flower meristem

The production of individual flower primordia and the transition from inflorescence to floral meristem are defined by three types of mutations, many of which are not well-characterized. Thus, only brief descriptions are presented. *Arabidopsis* plants homozygous for the *pin-formed* mutation (Goto *et al*. 1987) fail to produce individual flower primordia even though the transition from vegetative to inflorescence meristem appears to have occurred. An inflorescence stem is produced but no flowers are formed, resulting in a pointed and flowerless stem emerging from the rosette leaves. The *sterilis* mutation of *Antirrhinum* is similar in phenotype, with an inflorescence stem producing floral bracts with no flowers in their axils (Stubbe, 1966).

Arabidopsis plants homozygous for a strong *leafy* mutation fail to make the transition from inflorescence meristem to floral meristem (Haughn and Sommerville, 1988; Detlef Weigel, David Smyth, and Elliot Meyerowitz, unpublished). Flower primordia appear to form properly in a phyllotactic spiral, but rather than differentiating into a flower, the primordium behaves as if it is an inflorescence meristem, remaining indeterminate and producing its own cauline leaves, and with each of these leaves, an associated secondary meristem. This process can be repeated in each of the secondary meristems resulting in an often highly branched structure consisting mostly of cauline leaves. Late in development, carpelloid organs are produced towards the terminal portion of the branched structure. Mutations similar in phenotype in *Antirrhinum* include *floricaula*, *squamata*, and *squamosa* (Stubbe, 1966; Carpenter *et al.* 1990; Schwarz-Sommer *et al.* 1990).

The final type of mutation identified in this class is one in which the inflorescence meristem displays determinate growth and produces a terminal flower instead of continuing to produce the normal phyllotactic spiral of flowers. The *centroradialis* mutation in *Antirrhinum* displays this phenotype (Stubbe, 1966). Candidates for genes with similar functions in *Arabidopsis* include *triple flower* (Alvarez *et al.* 1990) and *premature termination of inflorescence-1* (Shannon *et al.* 1990). Homozygous plants of either of these *Arabidopsis* mutants produce a terminal structure composed of organs of more than one flower after a limited number of normal flowers are produced.

Mutations in genes involved in the generation of organ primordia and their subsequent specification

Three types of genes belong to this category: those that alter organ primordium formation, those that alter the specification of organ primordia (homeotic genes), and those that disrupt both processes. Mutations at two loci, *clavata1* and *clavata2* (*clv1* and *clv2*), appear to fall in the first category (Koornneef *et al.* 1983; Bowman *et al.* 1988). Homozygous, recessive mutations at either of the *CLAVATA* loci, result in the production of extra floral organs, although the organ identity in the individual whorls is not altered. The extra organs may occur in any whorl, but a gynoecium of four carpels is the most frequent floral phenotype. The number of extra organs in *clv1-1* flowers increases when the plants are grown at $16°C$ (Bowman *et al.* 1988). The phenotypic effects of these mutations are not restricted to the flower, however, since in addition to the floral organ phenotype, the spiral phyllotaxis of the rosette leaves and flower primordia is disrupted. Thus, these loci appear to be involved in a fundamental process that controls the number and position of organ primordia throughout the plant.

The homeotic mutations that alter floral organ indentity have been characterized in greater detail than the mutants in the other classes. Each of the

Table 1. *Summary of phenotypes of* Arabidopsis *flowers*

	Whorl			
	first (medial/lateral)	second	third	fourth
wild type	Sepals	Petals	Stamens	Carpels
apetala2-1	Leaves	Staminoid petals	Stamens	Carpels
apetala2-2	Carpels/absent	Absent	Absent/ stamens	Carpels
apetala2-1/2	Carpels/leaves	Absent	Stamens	Carpels
apetala3-1	Sepals	Sepals	Carpels	Carpels
pistillata-1	Sepals	Sepals	Absent	Extra carpels
pistillata-3	Sepals	Sepals	Carpels	Carpels
agamous-1,2,3	Sepals	Petals	Petals	Another flower
superman-1	Sepals	Petals	Stamens (stamens)	Reduced carpels

floral homeotic mutations described in *Arabidopsis* alters the identity of two adjacent whorls of organs, and thus fall into three categories: (1) those that affect whorls one and two, (2) those that affect whorls two and three, and (3) those that affect whorls three and four. There is more than one locus that alters the fate of the second and third whorls. These types of mutants have also been identified in *Antirrhinum* (Stubbe, 1966; Carpenter *et al.* in press; Sommer *et al.* 1990; Schwarz-Sommer *et al.* submitted) as well as in a number of other species (Masters, 1869; Penzig, 1890–4; Meyer, 1966; Meyerowitz *et al.* 1989). Mutations in which the identity of only a single whorl of organs is altered have not been documented in either *Arabidopsis* or *Antirrhinum*. Many of the homeotic loci of *Arabidopsis* seem to be involved in the production of organ primordia as well as specifying organ fate. The phenotypes of the homeotic mutants of *Arabidopsis* described below are for plants grown at 25 °C in 24 h illumination, unless otherwise noted, and are summarized in Table 1.

Mutations at the *APETALA2* (*AP2*) locus alter the identity of the outer two whorls of organs. A wide allelic series of recessive *ap2* mutations have been isolated in *Arabidopsis* (Koornneef *et al.* 1983; Komaki *et al.* 1988; Bowman *et al.* 1989; Meyerowitz *et al.* 1989; Kunst *et al.* 1989; Bowman *et al.* 1991), three examples of which are shown in Fig. 2A–C. Each appears to be a loss of function mutation because they are all recessive, and trans-heterozygotes between alleles have a phenotype intermediate between the homozygous phenotypes.

Plants homozygous for the mildest allele, *ap2-1* (Koornneef *et al.* 1983; Bowman *et al.* 1989), produce flowers in which the first whorl organs are leaves rather than sepals (Fig. 2A). These leaf-like organs have stellate trichomes on their adaxial surfaces and have stipules at their bases, both characteristics of leaf development. In addition, these organs senesce on the time course of leaves, rather than sepals. Occasionally, a secondary flower develops in the axil of one of these outer whorl leaves. This can be interpreted as another leaf-like characteristic, since each leaf

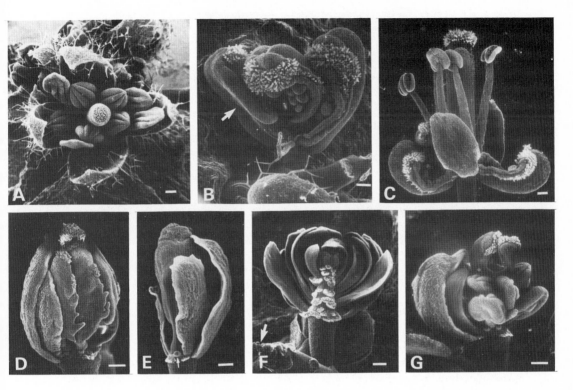

Fig. 2. SEM micrographs of *Arabidopsis* homeotic flower mutants. All plants were grown at 25°C. (A) *apetala2-1*. Cauline leaf-like organs are present in the outer whorl while staminoid petals are visible in the second whorl. (B) *apetala2-2*. A stamen-carpel mosaic organ is indicated (arrow) in a medial first whorl position and the central gynoecium is not fused properly. (C) *apetala2-1/2*. Note the difference in phenotype between the medial (carpel-like) and lateral (leaf-like) first whorl organs. (D) *apetala3-1*. Note the second whorl sepal and the carpeloid third whorl organs. Some first and second whorl organs have been removed. (E) *pistillata-1*. A second whorl sepal, which is smaller than the first whorl sepal, and the abnormal gynoecium are visible. The small dome of cells at the base of the flower (arrow) is a nectary. Several first and second whorl organs have been removed. (F) *agamous-1*. Cross section through a flower showing the indeterminate growth. The apical meristem is visible in the lower left (arrow). (G) *superman-1*. Some first whorl sepals and second whorl petals have been dissected off to reveal what appears to be two whorls of stamens and some carpeloid tissue interior to the stamens. Bar=100 μm.

on a wild-type plant is associated with a secondary meristem in its axil. In *ap2-1* flowers, the second whorl organs are staminoid petals (Fig. 2A) or morphologically normal stamens, which dehisce and sometimes have nectaries at their base (Fig. 3). The third and fourth whorls of *ap2-1* flowers are normal.

ap2-2 homozygotes (Meyerowitz *et al.* 1989; Bowman *et al.* 1991) produce flowers in which the medial first whorl organs are solitary carpels and the lateral first whorl organs, when present (about 50 % of the time), are leaf-like (Fig. 2B). Both the lateral and medial first whorl organs may have sectors of stamen tissue at their outer margins. In *ap2-2* flowers, second whorl organs are completely missing,

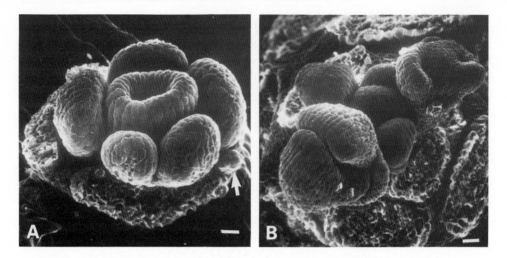

Fig. 3. (A) An *ap2-1* flower at an early stage of development (stage 7). When grown at 29°C, *ap2-1* flowers often fail to form second whorl organs (Bowman *et al.* 1989); this is due to a failure to form second whorl primordia. The remaining third whorl organs are also in ectopic positions (compare with Fig. 1C). The small group of cells indicated (arrow) arises from the base of an outer whorl organ which would develop into a cauline leaf-like organ and therefore is a stipule. Bar=10 μm. (B) Apex of an *ap2-8* plant. Flowers of *ap2-8* homozygotes are similar in phenotype to those of *ap2-2* homozygotes. The medial first whorl primordia are enlarged and the lateral first whorl primordia arise lower on the flower primordium compared to those of wild-type (Fig. 1A). Bar=10 μm.

as are most of the third whorl organs (90% missing). When the third whorl organs are present, they develop into morphologically normal stamens. The central gynoecium often fails to fuse properly, and when it does fuse, its orientation is altered by 90° with respect to that seen in wild-type plants. The transheterozygote *ap2-1/ap2-2* displays a phenotype intermediate between these two extremes (Fig. 2C; Bowman *et al.* 1991). The medial first whorl organs are phylloid carpels, while the lateral first whorl organs are leaf-like. The second whorl organs are always absent. The third whorl organs, which are normal stamens, are present in near-normal numbers. The central gynoecium is also phenotypically normal.

Two notable features of the *ap2* phenotype are (1) that only the outer two whorls of organs have their identity altered (the third and fourth whorl organs, when present, are phenotypically normal with respect to cell type) and (2) that the formation of the pattern of organ primordia is altered. When organs are absent, it is due to the failure to initiate an organ primordium rather than to the aborted development of an organ primordium (Fig. 4A; Bowman *et al.* 1989). When several organ primordia are not formed, the positions of the remaining primordia are abnormal. For example, in *ap2-2* flowers the medial first whorl primordia are enlarged relative to wild-type flowers, while the lateral first whorl primordia develop in a position lower on the flower meristem than they do in the wild-type

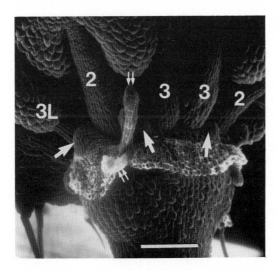

Fig. 4. Medial view of the base of an *ap2-1* flower grown at 29 °C; the first whorl organs have been removed. Second whorl organs when present (they are missing about 3/4 of the time, see Fig. 3A) often develop into stamens. All second and third whorl positions visible are occupied by stamens and are indicated as: second (2), medial third (3), and lateral third (3L) whorls stamens. The domes and other outgrowths of cells therefore do not represent aborted organ primordia. The domes at the base of the stamens (large arrows), both second and third whorl, are nectaries and have stomata on their apices. The other filamentous outgrowths (small double arrows) arise from the base of the first whorl organs that were cauline leaf-like and therefore these outgrowths are stipules. Bar = 100 μm.

(Fig. 4B; Bowman *et al.* 1991). The positions of third whorl organ primordia may also be abnormal if fewer than six are formed (Fig. 4A; Bowman *et al.* 1991).

All of the *ap2* alleles that we have tested (4 alleles) are temperature sensitive, suggesting that either the *ap2* gene product is inherently temperature sensitive or that the developmental process in which it is involved is temperature sensitive (Bowman *et al.* 1989; Bowman *et al.* 1991). The second whorl organs of *ap2-1* flowers are nearly phenotypically normal petals at 16 °C while they are either stamens or fail to form at 29 °C. To determine the temperature-sensitive developmental stage for the formation and specification of the second whorl organs, we performed temperature shift experiments (Bowman *et al.* 1989). The temperature sensitive period (tsp) for each of the processes was found to begin in stage two, when the flower primordium becomes distinct from the inflorescence meristem, but prior to the formation of any organ primordia, and continues into stage four, by which time the sepal primordia have formed, but no other primordia are yet visible (see Fig. 1A). This suggests that the processes of organ primordium formation and organ specification are concurrent, and is consistent with AP2 having a role early in the formation and differentiation of the outer two whorls of organs. A generalized *ap2* phenotype of carpels, stamens, stamens, carpels may be derived from the allelic series.

A similar mutation, *ovulata*, has been described in *Antirrhinum* in which the identity of the outer two whorls has been altered, the outer whorl organs from sepals to carpels and the second whorl organs from petals to stamens (Carpenter *et al*. 1990). In contrast to the *ap2* alleles described, the *ovulata* mutation is dominant with the heterozygote having carpelloid outer whorl organs and staminoid petals in the second whorl. When homozygous, the outer whorl organs are very carpel-like, while the second whorl organs are morphologically normal stamens. In contrast to *ap2* mutations, organs are rarely missing in *ovulata*, even in homozygotes. The isolation of other alleles of *OVULATA* is necessary to determine if this is an allele- or locus-specific difference, since the nine well-described *ap2* alleles in *Arabidopsis* vary in this respect. A semi-dominant mutation at the *AP2* locus in *Arabidopsis* has also been isolated; when heterozygous it displays a phenotype less severe than *ap2-1*, and when homozygous, flowers consist of little more than a two carpelled gynoecium (Uli Mayer, Gerd Jürgens, and Detlef Weigel, unpublished).

Two *Arabidopsis* mutations, *pistillata* (*pi*) and *apetala3* (*ap3*), cause the second whorl organ primordia to develop into sepals rather than petals, and the third whorl organ primordia to develop into carpels instead of the wild-type stamens (Koornneef *et al*. 1983; Bowman *et al*. 1989; Hill and Lord, 1989). In *ap3* flowers the organ primordia are all formed properly, but their subsequent differentiation is altered (Fig. 2D). The *ap3-1* allele is temperature sensitive; the third whorl organs are phenotypically normal stamens at 16°C, and are solitary carpels at 29°C (Bowman *et al*. 1989). The second whorl organs are sepals at 29°C, but are slightly sepaloid petals at 16°C. The temperature sensitive period (tsp) of *ap3* for the specification of the third whorl organs begins in stage 5, at which time all organ primordia have formed, but have not yet begun to visibly differentiate, and ends early in stage 7, when the third whorl organ primordia begin to visibly differentiate into stamens in wild-type flowers (Bowman *et al*. 1989). The tsp of *ap3* for the second whorl organs also begins in stage 5, but extends into stage 7, and possibly later (Bowman *et al*. 1989). *pi* alleles range in phenotype from ones indistinguishable from *ap3-1* (*pi-3*; Bowman *et al*. 1991), to ones in which third whorl organs fail to form (*pi-1*; Bowman *et al*. 1989; Hill and Lord, 1989). The development of *pi-1* flowers deviates from that of wild-type at the time during which the third whorl organ primordia are formed (stage 5; Bowman *et al*. 1989). In *pi-1* flowers the cells that would normally give rise to the third whorl organ primordia appear to be incorporated into the gynoecial cylinder causing the cylinder to be abnormal in size and shape, and thus producing a flower in which no third whorl organs are present, and in which the fourth whorl gynoecium is abnormal (Fig. 2E; Bowman *et al*. 1989; Hill and Lord, 1989). The *pi/ap3* phenotype can thus be summarized as: sepal, sepal, carpel, carpel.

Mutations in at least three loci (*deficiens*, *globosa*, and *sepaloidea*) in *Antirrhinum* give a phenotype similar to the *ap3* and *pi* mutations of *Arabidopsis* (Stubbe, 1966; Sommer *et al*. 1990; Carpenter *et al*. 1990). Some *deficiens-621* (*def-621*) flowers are similar in phenotype to *ap3* flowers but some display a more extreme phenotype, in which the third whorl develops into a ring of fused carpels, whereas the fourth whorl fails to develop (Carpenter *et al*. 1990). This extreme phenotype resembles *pi-1* flowers in the respect that only three whorls of organs develop. Since the *def-621* mutation is caused by the insertion of an active transposon, somatic reversions are often observed in flowers. Somatic reversion events in second whorl organs can produce small sectors of petal tissue in the sepals occupying second whorl positions suggesting that the fate of those cells is not irreversibly determined until late in development (Sommer *et al.* 1990; Carpenter *et al.* in press). Small sectors from somatic excision events are not seen in third whorl organs. This is consistent with the temperature shift experiments in *Arabidopsis*, in which the tsp of *ap3* for the third whorl organs is a short well-defined developmental period, whereas the tsp of *ap3* for the second whorl organs extends until late in development (Bowman *et al.* 1989).

agamous-1 (as well as *ag-2* and *ag-3*) flowers of *Arabidopsis* consist of multiple whorls of sepals and petals, and display indeterminate growth (Fig. 2F; Koornneef *et al.* 1983; Bowman *et al.* 1989; Yanofsky *et al.* 1990; Bowman *et al.* 1991). Observations on developing flowers elucidates the developmental basis of the phenotype (Bowman *et al.* 1989). The first two whorls of organ primordia are produced correctly and differentiate properly, but the third whorl organ primordia, although they are produced in the correct positions, differentiate into another whorl of petals rather than the wild-type stamens. The cells which would normally give rise to the gynoecial cylinder behave instead as if they constituted another flower primordium; this type of developmental defect has been termed heterochronic. This internal flower primordium repeats the pattern and the cycle continues indeterminately, causing the phenotype commonly referred to as a 'double flower'. The organs of the internal flowers may be mosaics of petal and sepal tissue, with the petal tissue always at the margin, and the sepal tissue central. Thus the *ag* phenotype may be summarized as (sepal, petal, petal)$_n$ with respect to organ identity or $(1, 2, 3)_n$ in terms of organ position.

In *Antirrhinum*, the *pleniflora* mutation is similar in phenotype to *agamous*, except that the organs produced interior to the third whorl organs of *pleniflora* flowers may be carpelloid in addition to being sepalloid and petaloid (Carpenter *et al.* 1990).

Another mutation of *Arabidopsis* which has phenotypic effects in the third and fourth whorls, is *superman-1* (Fig. 2G; Uli Mayer, Gerd Jürgens, and Detlef Weigel, unpublished). In *superman-1* flowers, the first and second whorls develop normally into four sepals and four petals, respectively. Interior to these outer two whorls, 9–12 organ primordia are formed in a pattern that is not well-defined, either spatially or numerically (J. Bowman and Detlef Weigel, unpublished). Most of the primordia that form develop into stamens although the more peripheral may be petalloid and the more central may be carpelloid. The fate of the cells interior to the staminoid organs is variable. Nearly normal gynoecia, free carpelloid organs, thread-like filamentous structures, and no organ development at all have been observed. Further characterization of this allele as well as the generation of an allelic series for this locus should provide a clearer understanding of the role of *SUPERMAN* in organ primordium formation and specification of identity. A mutation with a similar phenotype has been described in an abstract by Schultz and Haughn (1990); allelism tests between this mutation and *superman-1* have yet to be performed.

Genetic interactions between the homeotic genes

Double and triple mutant combinations have been constructed between many of the homeotic mutations described for *Arabidopsis* (Bowman *et al.* 1989; Bowman *et al.* 1991), but, unfortunately, not for the similar *Antirrhinum* mutations. These genetic experiments have provided evidence for the interaction of the *AP2* and *AG* gene products in establishing organ identity and number as well as the *AP2* and *PI* gene products in the establishment of organ number.

ag-1 ap3-1 and *ag-1 pi-1* flowers consist entirely of sepals and display the indeterminate growth seen in *ag* flowers (Fig. 5A; Bowman *et al.* 1989). The pattern of organ primordia observed in *ap3-1 ag-1* flowers is indistinguishable from that of *ag-1* flowers: $(1, 2, 3)_n$ (Bowman *et al.* 1989). In contrast, the pattern of organ primordia formed in *ag-1 pi-1* flowers is $(1, 2)_n$; the third whorl organ primordia are absent just as they are in *pi-1* flowers (Bowman *et al.* 1989).

ap2-2 pi-1 flowers comprise a central gynoecium usually with four carpels and an occasional lateral first whorl leaf-like organ (Fig. 5B; Bowman *et al.* 1991). Observations on developing flowers provide insight into the basis of the phenotype. Lateral first whorl organ primordia are initiated about half of the time, as seen in *ap2-2* flowers, and develop into cauline leaf-like organs. Medial first whorl organ primordia are initiated but are congenitally fused to each other, and grow to form a cylinder into which all of the remaining flower meristem is incorporated. The four carpels in the mature flower could arise from a fusion of the two first whorl primordia, which differentiate into carpels, and the cells that would ordinarily give rise to the two carpels of the central gynoecium. No second or third whorl organs are present, but nectaries may arise between the lateral first whorl positions and the central gynoecium. *ap2-2 ap3-1* flowers (Bowman *et al.* 1991) are similar to *ap2-2 pi-1* flowers except that there are occasional third whorl organ primordia, which differentiate into carpels, and the medial first whorl carpels are often separate from the central carpels. Thus, the effects on the specification of organ identity of the two mutations in each of the four double mutants described are essentially additive in the double mutant flowers.

In contrast, non-additive effects in the specification of organ identity are seen in *ap2-2 ag-1* flowers (Fig. 5C; Bowman *et al.* 1991). The outer whorl of *ap2-2 ag-1* flowers is composed of carpelloid leaves in the medial positions while, about half of the time, cauline leaf-like organs develop in the lateral positions. The second and third whorl organs are petaloid stamens; hybrid organs with both an overall morphology as well as epidermal cellular morphology intermediate between that of wild-type petals and wild-type stamens. These organs have rudimentary locules containing pollen, but fail to dehisce. The number of second and third whorl organ primordia formed (7.6) is intermediate between that formed in wild-type flowers

(10) and *ap2-2* flowers (0.25). These organ primordia arise in a pattern resembling the wild-type, with some of the primordia occupying the same positions as organ primordia in wild-type flowers, but the remaining primordia are in abnormal positions. This pattern repeats indefinitely, producing a double flower with the following order of organs: (carpelloid leaves, petaloid stamens, petaloid stamens)$_n$ and $(1, 2^*, 3^*)_n$ in terms of organ primordia with the * denoting ectopic positions and incorrect numbers. Thus the *ag-1* mutation suppresses two aspects of the *ap2-2* phenotype: the carpelloidy of the medial first whorl organs and the failure of second and third whorl organ formation. That the *ag-1* mutation affects the first and second whorl in an *ap2-2* mutant background, but it has no such effects alone, strongly suggests that the *AG* and *AP2* gene products interact at some level.

Fig. 5. SEM micrographs of double and triple mutant flowers. (A) *ap3-1 ag-1*. All organs are sepals. (B) *ap2-2 pi-1*. Flowers consist of only a central gynoecium usually composed of four carpels. (C) *ap2-2 ag-1*. Medial first whorl organs are carpelloid leaves (c/l) with stigmatic tissue visible along the margin and the second and third whorl organs are petaloid stamens (p/s). (D) *ap2-2 pi-1 ag-1*. All organs are carpeloid leaves. Both trichomes and sigmatic tissue are evident. Bars=100 μm.

Flowers of plants mutant for all three types of homeotic genes, *ap2-2 pi-1 ag-1* homozygotes, consist of many whorls of carpelloid leaves (Fig. 5D; Bowman *et al.* 1991). Analysis of developing flowers shows that lateral first whorl organ primordia are formed about half of the time, and that they arise lower on the flower pedicel than the similar primordia in wild-type flowers. These develop into cauline leaf-like organs. The lower position and leaf-like differentiation of the lateral first whorl primordia are characteristic of all flowers with the *ap2-2* genotype (*ap2-2*, *ap2-2 pi-1*, *ap2-2 ap3-1*, *ap2-2 ag-1*, and *ap2-2 pi-1 ag-1*; Bowman *et al.* 1991). The medial first whorl primordia develop in the correct positions, although they are enlarged relative to the wild-type. Interior to the medial first whorl organ primordia, two more primordia are formed in lateral positions. This is followed by the production of two additional primordia in medial positions that are interior to all of the previously produced primordia. This process continues indefinitely, generating many whorls of alternating organs. Each of these primordia develops into a carpelloid leaf-like organ. These pairs of organs could be interpreted as a repeating pattern of (medial first whorl organs)$_n$ or (1)$_n$ if each successive internal flower is rotated 90° with respect to the next outer one. That the *pi-1* mutation prevents the production of second whorl organs in an *ap2-2 ag-1* background suggests that the products of these genes interact during the process of organ primordium formation; however, it is not possible to distinguish whether this interaction is between AP2 and PI or AG and PI or both.

ap2-1 ag-1 pi-1 and *ap2-1 ag-1 ap3-1* flowers consist entirely of cauline leaf-like organs with virtually no carpellody (Fig. 6; Bowman *et al.* 1991). The *ap2-1* allele is the mildest mutant allele isolated so far. Thus, an apparent low level of AP2 activity, in the absence of the other types of homeotic gene products, causes floral organs that would otherwise develop into carpelloid leaves, as in *ap2-2 ag-1 pi-1* flowers, to develop into cauline leaf-like organs, complete with stipules, stellate trichomes, and an occasional floral meristem in the axil. Therefore, it is possible that the ground state of organs produced by the floral meristem is carpelloid leaves

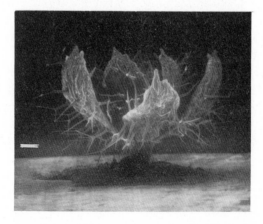

Fig. 6. *ap2-1 pi-1 ag-1* flower. All organs are cauline leaf-like with prominant stellate trichomes on their abaxial surfaces and stipules at their bases. Bar=100 μm.

or that there is another as yet unidentified gene involved in carpel formation that the *AP2* gene product represses.

Several of the genes involved in floral morphogenesis encode transcription factors

During floral morphogenesis, cells in the developing flower primordia must assess their position relative to nearby cells, or assess it globally, and differentiate accordingly. Two types of genes are likely to play a role in these processes: those that encode transcription factors and those that encode components of signal transduction pathways. Indeed, many genes identified as being involved in morphogenesis of other organisms such as *Drosophila melanogaster* and *Caenorhabditis elegans* encode these types of proteins. The recent cloning of several of these genes in *Arabidopsis* and *Antirrhinum* and their identification as putative transcription factors suggests that this is also the case for the genes controlling morphogenesis in plants.

The AGAMOUS *and* DEFICIENS *genes encode putative DNA-binding proteins*

The recent molecular cloning of the *AGAMOUS* gene (Yanofsky *et al.* 1990) from *Arabidopsis* and the *DEFICIENS* gene (Sommer *et al.* 1990) from *Antirrhinum* have been described. In each, the predicted protein sequence contains a region with remarkable similarity to the DNA-binding region of transcription factors from yeast and humans, MCM1 (Passmore *et al.* 1988) and SRF (Norman *et al.* 1988), respectively. The region of homology is limited to the DNA-binding and dimerization domains in SRF, and does not extend to the rest of the predicted protein sequence. This region of homology has been designated as the MADS-box, referring to four of the proteins known to contain it: MCM1, AGAMOUS, DEFICIENS, and SRF (Schwarz-Sommer *et al.* 1990). A family of genes homologous to the *AGAMOUS* gene has been isolated in *Arabidopsis*, some of which appear to be preferentially expressed in flowers (Yanofsky *et al.* 1990). Likewise, a family of genes homologous to the *DEFICIENS* gene has been identified in *Antirrhinum*, two of which correspond to the *GLOBOSA* and *SQUAMOSA* genes (Schwarz-Sommer *et al.* 1990). It will be of interest to see if there is a family of genes related to *DEFICIENS* in *Arabidopsis* and one related to *AGAMOUS* in *Antirrhinum*. The roles of each of the MADS-box genes during floral morphogenesis are very different: the *AGAMOUS* gene is involved in specifying the identity of the third and fourth whorl organs as well as the determinate growth of the flower, the *DEFICIENS* and *GLOBOSA* genes are required for the proper specification of organ indentity in the second and third whorls, and the *SQUAMOSA* gene is involved in the switch from vegetative meristem to inflorescence meristem. Thus, there appears to be a family of related genes that act to direct many steps of floral morphogenesis. A similar situation has been observed in animals where a single DNA-binding motif has been found to be used in different regulatory pathways.

The AGAMOUS and DEFICIENS genes are expressed in complex spatial and temporal patterns

The localization of *AG* mRNA in developing *Arabidopsis* flowers has been analyzed by tissue *in situ* hybridization experiments (Yanofsky *et al.* 1990; Gary Drews, J. Bowman and E. Meyerowitz, unpublished). *AG* mRNA is detectable as early as stage three, at which time the outer whorl organ primordia have formed, but no other organ primordia are visible (Gary Drews, J. Bowman, and E. Meyerowitz, in preparation). Expression is localized to those cells that will give rise to the third and fourth whorl organs, and continues until late in development, past stage 12, in both the third and fourth whorl organs (Yanofsky *et al.* 1990; Gary Drews, J.B., and E.M., in preparation). RNA blot analysis of dissected floral organs performed with the *DEFICIENS* gene in *Antirrhinum* shows that *DEFICIENS* expression is concentrated in those whorls that are affected in *deficiens* mutants (Schwarz-Sommer *et al.* 1990). A low level of expression was also detected in first and fourth whorl organs, where the *deficiens* mutation does not have an obvious phenotypic effect (Schwarz-Sommer *et al.* 1990). That the mRNA of each is preferentially expressed in those floral organs that are altered when the genes are mutated, implies that the genes are spatially as well as temporally regulated.

The processes of specification of organ identity and organ primordia formation are concurrent

Each of the homeotic flower genes of *Arabidopsis* appears to have a role early in the development of the flower, when the organ primordia are being generated. Evidence for this comes from phenotypic effects, temperature sensitive periods (tsps), and mRNA expression data. The tsp for the second whorl organs of *ap2-1* flowers begins in stage two (when no organ primordia are visible yet; Bowman *et al.* 1989), *AG* mRNA is detectable at stage three (when the first whorl organ primordia become visible; Gary Drews, J.B., and E.M., in preparation), the *pi-1* mutation has phenotypic effects at stage five (when the second and third whorl organ primordia are forming; Bowman *et al.* 1989), and the tsp for the third whorl organs of *ap3-1* flowers begins in stage five (Bowman *et al.* 1989). The fates of the cells that will give rise to the organs of the second whorl are at least in part determined by the *AP2*, *PI*, and *AP3* gene products. In *ap2-1* flowers the second whorl organs can be stamens, although in *ap3* and *pi* flowers the second whorl organs are sepals rather than the wild-type petals. In the temperature sensitive alleles, *ap2-1* and *ap3-1*, the fates of the cells that will give rise to the second whorl organs can be manipulated by altering the growth temperature. The tsp for the specification of identity of the second whorl organs in *ap2-1* flowers begins in stage two and extends into stage four, during which time the first whorl organ primordia are formed. In contrast, the tsp for the specification of identity of the second whorl organs in *ap3-1* flowers begins in stage five, at which time all the organ primordia have formed, and extends at least into stage seven, when the third whorl primordia have begun to differentiate, but the second whorl organ primordia have not yet begun their visible differentiation. This suggests that the specification of organ identity is initiated before the visible formation of the organ primordia and the process is not complete until after the formation of the organ primordia. Mutations at the *AP2* locus can disrupt both of these processes in the second whorl and the tsps for each process encompass approximately the same developmental time period (Bowman *et al.* 1989).

Although the identity of the second whorl primordia is not irreversibly determined at stage five, their possible fates may be restricted (Bowman *et al.* unpublished). For example, the cells that give rise to the second whorl organs may develop as either sepal/petal or stamen/carpel during stages 2–4, depending upon the action of the *AP2* gene product. The fate of those same cells to develop into petal or sepal is determined slightly later in development, depending upon the action of the *AP3* gene product. After stage four the option to differentiate into a stamen may no longer be open, but the cells are not irreversibly committed to differentiate into those of a petal.

That most of the homeotic flower mutations of *Arabidopsis* disrupt organ primordium formation, but most of those identified thus far in *Antirrhinum* do not, is the most obvious phenotypic difference between the two systems. The more severe phenotypes of *deficiens* alleles do resemble those of *pi-1* flowers, in that they consist of only three whorls of organs: sepals, sepals, carpels (Carpenter *et al.* 1990). The number of carpels composing the *pi-1* gynoecium is variable, ranging from 2 to 5 with an average of 2.7 (Bowman *et al.* 1989). The carpels of *deficiens* flowers are also variable in number (Sommer *et al.* 1990; Carpenter *et al.* 1990). Klemm (1927) described a *globifera* mutant (which is the same allele as *deficiens*) in which the gynoecium appeared to be a fusion of both the third and fourth whorls, and could be composed of up to seven carpels. Thus, the *deficiens* mutation in *Antirrhinum* is similar in phenotype to that of *pi-1* flowers in *Arabidopsis* in terms of organ loss as well as altered organ identity. The extra organs of *pleniflora* mutants may be carpelloid, and do not form an inner flower with the normal *Antirrhinum* floral structure (Carpenter *et al.* 1990). In comparison, the extra organs of *ag* flowers of *Arabidopsis*, are flowers (Bowman *et al.* 1989). Finally, leaf-like organs or loss of organ primordia, both prominent features of the *ap2* phenotype in *Arabidopsis*, have not been observed in *ovulata* flowers (Carpenter *et al.* 1990; Schwarz-Sommer *et al.* 1990), the *Antirrhinum* equivalent. Since a limited number of alleles of these loci have been analyzed, a wider allelic series of the *Antirrhinum* mutations might resolve these differences. Alternatively, there might be a fundamental difference between the two plant species in the mechanisms of floral morphogenesis.

Models of floral morphogenesis

Many early models of flower development rely on communication between developing organs in adjacent whorls to sequentially specify the identity of the organs in each whorl (Heslop-Harrison, 1963; McHughen, 1980; Green, 1988). This is clearly not the case in *Arabidopsis* since the proper specification of any whorl of organs is not dependent on the correct differentiation of either the next outer or the next inner whorl of organs (Bowman *et al.* 1989). For example, normal petals can form in *ap2-1* flowers in the second whorl when grown at 16 °C despite the fact that the outer whorl organs develop as leaves rather than the wild-type sepals. Conversely, wild-type petals form in the second whorl of *ag* flowers despite the third whorl organ primordia differentiating incorrectly into petals rather than stamens. Examples exist for each of the other floral organ types as well. It is clear that there are few constraints on the identity of the organs in any particular whorl. For example, all four floral organ types in addition to leaf-like organs can be observed in second and third whorl positions in various single, double, and triple mutant flowers of *Arabidopsis*. This also seems to be the case for *Antirrhinum* although the lack of any reported double and triple mutant phenotypes leaves open the possibility of some constraints in this species.

The major hormones known to act in plants do not seem to be involved in specifying organ identity in any simple way in *Arabidopsis*. Application of exogenous gibberellic acid, indole acetic acid, or kinetin has no effect on phenotype of wild-type or homeotic mutant flowers (Bowman *et al.* 1989; Bowman, unpublished). Additionally, mutants that fail to produce or respond properly to ethylene, auxins, gibberellins, or abscisic acid develop phenotypically normal flowers (Koornneef *et al.* 1985; Bleecker *et al.* 1988; King, 1988).

Models, in which the flower primordium is thought of as concentric fields of cells with each field being specified by a combination of gene products, have been proposed (Brieger, 1935; Holder, 1979; Bowman *et al.* 1989; Kunst *et al.* 1989; Carpenter *et al.* 1990; Schwarz-Sommer *et al.* 1990; Bowman *et al.* 1991). In this type of model, the flower primordium is divided into radially concentric fields, each of which comprises of two adjacent whorls. Our model (Bowman *et al.* 1991) for *Arabidopsis* has three fields (Fig. 7): field A comprises the first and second whorls and is identified by AP2; field B comprises whorls two and three, which is specified by AP3 and PI; field C is made of whorls three and four and is identified by AG. The combination of homeotic gene products present in each whorl unambiguously identifies each whorl, and the combination of gene products is responsible for specifying the fate of the cells within the whorl. It follows that wild-type whorl one (occupied by sepals in wild-type) is specified by AP2, whorl two (occupied by petals in wild-type) by the combination of AP2 and AP3/PI, whorl three (occupied by stamens in wild-type) by the combination of AP3/PI and AG, and whorl four (occupied by carpels in wild-type) by AG. For example, if AP2 is the only product present in a particular whorl, then the organ primordia of that whorl develop into sepals, and if AP3/PI is present in addition to AP2 then petals would develop, and so on. We also proposed that the *AG* and *AP2* gene products act antagonistically, based on the genetic data that the *ag-1* mutation has phenotypic effects in the first and second whorls in an *ap2-2* background, but not in the *ag* single mutant (Bowman *et al.* 1991). Thus, in an *ap2* background the

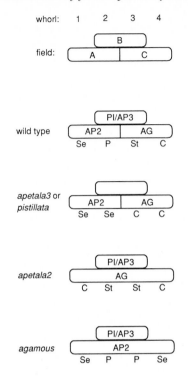

Fig. 7. Simple model depicting how the three classes of floral homeotic genes in *Arabidopsis* could specify the identity of each of the four whorls of floral organs. Details of the model are in the text. Se=sepals, P=petals, St=stamens, and C=carpels.

phenotypic effects of the wild-type *AG* gene product expand to encompass both fields A and C, all four whorls. And conversely, in an *ag* background the phenotypic effects of the wild-type *AP2* gene product expand into field C, encompassing all whorls.

If one or more of these gene products are missing, the distribution of the remaining gene products determines organ identity. For example, if the *AP3* and/or *PI* gene products are missing, leaving only the *AP2* and *AG* gene products in their wild-type domains, the outer two whorls develop into sepals, while the inner two develop into carpels. If both the *AG* and *AP3* gene products are missing, the phenotypic effects of the *AP2* wild-type product are observed in all four whorls. Since AP2 is the only functional homeotic gene product, a flower in which all organs are sepals develops. Similar arguments can be made for *ag*, *ap2*, *ag pi*, *ap2 ap3*, and *ap2 pi* flowers. In *ap2 ag* flowers both the *AP2* and *AG* gene products are not functional, leaving only the *AP3/PI* products in field B. Thus the distribution of gene products in each whorl is different from any distribution in wild-type flowers. The production of both petals and stamens is dependent on the *AP3/PI* gene products, the decision to produce a petal or a stamen, however, depends on the combinatorial action of *AP2* and *AG* gene products, respectively.

Those organs produced in field B (whorls two and three) of *ag ap2* flowers are neither wild-type petals nor wild-type stamens, but intermediate in phenotype between petals and stamens (Bowman *et al.* 1989; Bowman *et al.* 1991). In those whorls in which none of the three types of homeotic gene products are functional, whorls one and four of *ap2 ag* flowers as well as all positions in *ap2 ag pi* flowers, carpelloid leaves are produced (Bowman *et al.* 1989; Bowman *et al.* 1991). The models proposed on the basis of single mutant phenotypes in *Antirrhinum* (Carpenter *et al.* 1990; Schwarz-Sommer *et al.* 1990) are consistent with, and are subsets of, this model.

The model successfully predicts the basic phenotypes of the singly, doubly, and triply mutant flowers. There are, however, a few details in *Arabidopsis* that are not explained by this simplistic model (see Bowman *et al.* 1991, for a comprehensive discussion). For example, in *ag-1* homozygotes (in the double and triple mutants) both carpelloid and staminoid organs develop. If the *AG* gene product is responsible for directing the development of carpels, then the stigmatic tissue and the ovules observed in *ap2-2 ag-1* and *ap2-2 ag-1 pi-1* flowers pose a quandary. If there is residual *AG* activity in *ag-1* flowers, then some carpelloid tissue would be expected. Therefore, one possible solution would be that the *ag-1* allele is hypomorphic but not null. The molecular data do not support this hypothesis, however. *ag-1* and *ag-2* flowers are very similar in phenotype. The *ag-2* mutation is caused by a 35 kilobase insertion of T-DNA into an intron towards the 5' end of the gene while the *ag-1* mutation is caused by a single base change at a splice site acceptor site also towards the 5' end of the gene (Yanofsky *et al.* 1990). An alternative hypothesis is that the *AG* product merely sets up a prepattern on which other genes, whose expression may not be strictly dependent on *AG*, act to direct carpel differentiation in wild-type flowers. These other genes could be regulated by general factors, but in the absence of the *AG* and *AP2* gene products, could be expressed ectopically. The second and third whorl organs of *ap2-2 ag-1* flowers are actually much more stamen-like than petal-like, with locules full of pollen and a general morphology of wild-type stamens. It may thus be that the proper differentiation of stamens is also not directly dependent on the *AG* gene product. The slight petallody of the organs could be due to residual *AP2* activity, since the *ap2-2* mutation may not be null. Another complication of the *ag* phenotype is the occurrence of the internal flowers. One interpretation is that there are not any fourth whorl organs in *ag* flowers; those cells that would ordinarily develop into the gynoecium instead develop into another flower primordium. This primordium repeats the process producing organ primordia in the pattern seen in the first three whorls and another flower primordium internal to the three whorls of organ primordia. This continues indeterminately such that no fourth whorl primordia are ever produced. Therefore, the process of specification of organ identity starts over after the third whorl and the specification of the fourth whorl becomes irrelevent.

A second complication is the presence of mosaic organs. Stamen-carpel mosaics are observed in the first whorl of *ap-2* flowers (Bowman *et al.* 1991) and sepal-petal mosaics are seen in the internal flowers in *ag* (Bowman *et al.* 1989). The stamen-carpel mosaic organs consist of sectors of wild-type stamen and wild-type carpel tissue with the stamen tissue always at the margins of the organs and the carpel sectors occupying the central region. The outer organ primordia of *ap2-2* flowers encompass a larger proportion of the flower primordium than do the first whorl organ primordia in wild-type flowers. It might be the case that the enlarged primordia of *ap2-2* flowers contain cells from more than a single whorl or field. The shape of the primordium is such that the cells at the margins could come from cells that would ordinarily form second or third whorl organs while those in the central portion of the primordia arise from the region that usually gives rise to the outer whorl (Fig. 4B). This would explain the consistent spatial relationship of the stamen and carpel sectors. Since the processes of specification of organ identity and organ primordium formation are concurrent (see above) and the *AP2* gene product has a role in each of these, it is conceivable that the two processes might not be precisely coordinated in *ap2* mutants. A similar argument could be made for the mosaic organs in *ag* flowers. In summary, the model presented for the specification of organ identity based on simple combinatorial distributions of homeotic gene products, successfully predicts the specification of organ identity except in cases when organ number or position is drastically altered and does not address the issues of organ primordia formation.

How the products and phenotypic effects of the homeotic genes are localized to specific whorls is also unknown. It is true that the mRNA of both *AG* and *DEFICIENS* is preferentially present in those whorls that are altered in mutants, implying that transcriptional (or RNA stability) control is involved (Yanofsky *et al.* 1990; Schwarz-Sommer *et al.* 1990; Gary Drews, J.B., and E.M., in preparation). Additionally, the spatial pattern of the *AG* mRNA is altered in an *ap2* background (Gary Drews, J.B., and E.M., in preparation), as predicted by our model. This not only implies that there must be an initial prepattern, which regulates the expression of the homeotic genes, but also that their final pattern of expression depends on interactions between the homeotic gene products. Thus, the mechanism of establishment of initial homeotic gene expression is different from the maintenance of their expression. One can envision a variety of types of interactions between prepatterns and between the homeotic genes. For *AG*, one might imagine the initial expression pattern to be uniform, with subsequent modification by *AP2*. There is, however, no evidence of *AG* being expressed in all four whorls initially (Gary Drews, J.B., and E.M., in preparation) although the method of detection used, *in situ* hybridization to tissue sections, might be unable to detect very low levels of expression. Since the whorls of organ primordia are generated sequentially, the outer whorl first and the fourth whorl last, it is possible that this developmental feature is utilized in the pattern formation process. *AP2* expression probably precedes that of *AG*, since the tsp determined for the *ap2-1* allele begins in stage two (Bowman *et al.* 1989), whereas *AG* mRNA is not detectable until stage three (Gary Drews, J.B., and E.M., in preparation). The *AP2* gene product could become established in those cells destined to give rise to the outer whorls before *AG* expression commences. This might preclude the expression of *AG* in these whorls, since in an *ap2* mutant background, *AG* is expressed in the outer whorls. *AG* could then be expressed in those cells destined to become the third and fourth whorl primordia, consequently either reducing the expression of *AP2*, or suppressing the activity of the *AP2* gene product in some manner in the inner whorls. The spatial regulation of the *AP3* and *PI* gene products could also be dictated by temporal aspects in the first, second, and third whorls. The phenotypic effects of their wild-type products could be masked in the fourth whorl by an, as yet, unidentified gene. The phenotype of *superman-1* flowers (Uli Mayer, Gerd Jürgens, J.B., and Detlef Weigel, unpublished) suggests that the *SUPERMAN* gene product has a role in defining the patterns of activity of the other homeotic genes. One interpretation of *SUPERMAN* is that both the second and third whorl regions are expanded at the expense of the fourth whorl. A candidate for an even earlier regulator of the homeotic genes is the *LEAFY* gene (Haughn and Sommerville, 1988; David Smyth, Detlef Weigel, J.B., and E.M., unpublished). *leafy* mutants exhibit a phenotype similar to but not identical to that of *ap2-2 ag pi-1* triple mutants (see above; Bowman *et al.* 1991) suggesting that it might be involved in the initial appearance of the homeotic gene products. Other genes must be involved as well, since the phyllotaxy of the triple mutant flowers and *leafy* mutants differs. It should be noted that the mRNA and protein expression of the homeotic genes does not have to be an all-or-nothing phenomenon in those whorls in which they have phenotypic effects. That the *DEFICIENS* gene may be expressed in all four whorls, although at a low level in those whorls in which no phenotypic effects are observed, suggests that a pattern in which there are threshold levels of expression required to produce phenotypic effects is possible, as suggested by Schwarz-Sommer *et al.* (1990). A model similarly relying on threshold levels of gene activity to produce phenotypic effects has been proposed for specification of the *hairy achenes* character in *Microseris* inflorescences (Bachmann, 1983). Thus, a characteristic aspect of plant development, that of the sequential production of organs from a group of meristematic cells, could play a pivotal role in the generation of patterns in flower development.

Evolutionary considerations

In addition to the apparently homologous mutations in *Antirrhinum*, mutations described in hundreds of other species of plants (including *Matthiola*, *Cheiranthus*, *Petunia*, and *Primula*; see Masters, 1869; Penzig, 1890–4; Meyer, 1966; Meyerowitz *et al.* 1989) resemble the homeotic flower mutations of *Arabidopsis*. Thus, the mechanisms used in flower development of *Arabidopsis* and *Antirrhinum* are likely to be utilized in other flowering plants as well. Flowering plants appeared recently in the fossil record (120–140 million years ago; lower cretaceous) and rapidly became the dominant class of plants (80–90 million years ago; upper cretaceous). Due to the sparse nature of the early fossil record it is difficult to ascertain the relative order of appearance of the floral organ types, but it is apparent that their evolution was rapid since the four basic types were present even in early flowers (120–140 million years ago; Friis and Crepet, 1987). It has been suggested that the floral organs are modified leaves and that carpels evolved from sporophylls of primitive vascular plants (Stebbins, 1976). When all three types of flower homeotic genes are mutant in *Arabidopsis*, a flower in which all organs are leaves or carpelloid leaves, which could be interpreted as sporophylls, is produced. This suggests that the three classes of genes represented by the known *Arabidopsis* homeotic genes may be responsible for the majority of fundamental features associated with the flower.

The two cloned homeotic genes (*AGAMOUS* in *Arabidopsis* and *DEFICIENS* in *Antirrhinum*; Yanofsky *et al.* 1990; Sommer *et al.* 1990), both putative transcription factors, are involved in distinct developmental processes in pattern formation in the flower, and share homology in their DNA-binding motif. It will be of interest to see if these genes are related to other floral homeotic genes. If so, it is possible that much of the genetic machinery required for the development of flowers may have evolved from a small number of progenitor genes. Early duplication and divergence could produce a small number of gene families whose members direct the development of the basic floral structure.

Future prospects

To understand the mechanisms by which flower development occurs, it will be necessary to analyze each of these genes, and probably many others, at the molecular level. Both *Antirrhinum* and *Arabidopsis* are amenable to approaches to clone genes of which nothing more than their phenotype is known. Several endogenous transposons in *Antirrhinum* make gene tagging by transposon mutagenesis an attractive method in this system (Coen and Carpenter, 1986). The large floral organs also make feasible the construction of organ-specific cDNA libraries. Differential screening or subtractive hybridizations can be used to identify organ specific clones or enrich libraries for genes of interest (Sommer *et al.* 1990). In contrast, the genome of *Arabidopsis* contains no known active transposons. Insertional mutagenesis can, however, be performed using the T-DNA of *Agrobacterium* (Feldmann and Marks, 1987; Feldmann *et al.* 1989; Yanofsky *et al.* 1990; Feldmann and Meyerowitz, 1991) In addition, the introduction of heterologous transposons may allow gene tagging by transposon mutagenesis (Van Sluys *et al.* 1987; Masterson *et al.* 1989; Schmidt and Willmitzer, 1989). Although the genome of *Arabidopsis* has no known active transposons, others of its characteristics, such as its small size (70000–100000 kilobases; Leutwiler *et al.* 1984) and the near-absence of dispersed repetitive elements (Pruitt and Meyerowitz, 1986), can be exploited to clone genes by chromosome walking. RFLP maps with over 250 total markers have been constructed to provide starting points (Chang *et al.* 1988; Nam *et al.* 1989); walks of several hundred kilobases are already feasible if genomic libraries based on phage P1 or yeast artificial chromosome vectors are used. In addition, the possibility of physically mapping the entire *Arabidopsis* genome is being pursued. This would provide immediate access to any locus that can be mapped, consequently expanding the realm of plant molecular biology.

The small flowers of *Arabidopsis* preclude experiments relying on the isolation of large amounts of floral organ specific mRNA. Other attributes of *Arabidopsis*, however, such as its rapid generation time and the ability to be transformed by *Agrobacterium* (Lloyd *et al.* 1986; Masterson *et al.* 1989; Schmidt and Willmitzer, 1989), allow other types of analyses, such as the production of double and triple mutants, and the misexpression of cloned homeotic genes, which are at present impracticible in *Antirrhinum*. Thus, both *Arabidopsis* and *Antirrhinum* have distinct advantages that complement each other. Since the flowers of each system are similar in structure, it is likely that their general processes of flower development will also be similar. The advantages of each system can then be applied to the solution of a common problem: the mechanism of flower development.

We thank Uli Mayer and Gerd Jürgens (University of Munich, Munich, Germany) for the gift of *superman-1* and the semi-dominant *ap2* allele and David Smyth (Dept. of Genetics and Developmental Biology, Monash University, Melbourne, Australia) for the gift of several *leafy* alleles. We also thank members of the Meyerowitz laboratory (Laura Brockman, Gary Drews, Tom Jack, Leslie Sieburth, and Detlef Weigel) for discussions and critical review of the manuscript and Pat Coen of the Caltech Electron Microscope Facility for technical advice. The Meyerowitz laboratory's work on *Arabidopsis* flowers is supported by National Science Foundation grant DCB-8703439 and J.L.B. was supported by National Institutes of Health training grant 5T32-GM07616.

References

ALVAREZ, J., GULI, C. L. AND SMYTH, D. R. (1990). Mutations affecting inflorescence development in *Arabidopsis thaliana*. Abst. *Fourth International Conference on Arabidopsis Research*, Vienna, 101.

AWASTHI, D. K., KUMAR, V. AND MURTY, Y. S. (1984). Flower development in *Antirrhinum majus* (Scrophulariaceae) with a comment upon corolla tube formation. *Bot. Mag. Tokyo* **97**, 13–22.

BACHMANN, K. (1983). Evolutionary genetics and the genetic control of morphogenesis in flowering plants. *Evol. Biol.* **16**, 157–208.

BLEECKER, A. B., ESTELLE, M. A., SOMERVILLE, C. AND KENDE, H. (1988). A dominant mutation confers insensitivity to ethylene in *Arabidopsis thaliana*. *Science* **241**, 1086–1089.

BOWMAN, J. L., SMYTH, D. R. AND MEYEROWITZ, E. M. (1989). Genes directing flower development in *Arabidopsis*. *The Plant Cell.* **1**, 37–52.

BOWMAN, J. L., SMYTH, D. R. AND MEYEROWITZ, E. M. (1991). Genetic interactions among floral homeotic genes of *Arabidopsis*. *Development* **112**, 1–20.

BOWMAN, J. L., YANOFSKY, M. F. AND MEYEROWITZ, E. M. (1988). *Arabidopsis thaliana*: A review. *Oxford Surv. Pl. molec. cell. Biol.* **5**, 57–87.

BRIEGER, F. G. (1935). The developmental mechanics of normal and abnormal flowers in *Primula*. *Biol. J. Linn. Soc.* **147**, 126–130.

CARPENTER, R., LUO, D., DOYLE, S. AND COEN, E. S. (1990). Floral homeotic mutations produced by transposon mutagenesis in *Antirrhinum majus*. *Genes and Dev.* **4**, 1483–1493.

CHANG, C., BOWMAN, J. L., DEJOHN, A. W., LANDER, E. S. AND MEYEROWITZ, E. M. (1988). Restriction fragment length polymorphism linkage map for *Arabidopsis thaliana. Proc. natn. Acad. Sci. U.S.A.* **85**, 6856–6860.

COEN, E. S. AND CARPENTER, R. (1986). Transposable elements in *Antirrhinum majus*: Generators of genetic diversity. *Trends Genet.* **2**, 292–296.

DREWS, G. N. AND GOLDBERG, R. B. (1989). Genetic control of flower development. *Trends Genet.* **5**, 256–261.

FELDMANN, K. A. AND MARKS, M. D. (1987). *Agrobacterium* mediated transformation of germinating seeds of *Arabidopsis thaliana*: a non-tissue culture approach. *Molec. gen. Genet.* **208**, 1–9.

FELDMANN, K. A., MARKS, M. D., CHRISTIANSON, M. L. AND QUATRANO, R. S. (1989). A dwarf mutant of *Arabidopsis* generated by T-DNA insertion mutagenesis. *Science* **243**, 1351–1354.

FELDMANN, K. A. AND MEYEROWITZ, E. M. (1991). Tagging floral structure genes. In *Genetics and Breeding of Ornamental Species* (ed. J. N. M. Mol) Kluwer, Amsterdam, (in press).

FRIIS, E. M. AND CREPET, W. L. (1987). Time of appearance of floral features. In *The Origins of Angiosperms and their Biological Consequences* (ed. E. M. Friis, W. G. Chaloner, and P. R. Crane) pp. 145–179. Cambridge University Press, Cambridge.

GOTO, N., STARKE, M. AND KRANZ, A. R. (1987). Effect of gibberellins on the flower development of the *pin-formed* mutant of *Arabidopsis thaliana. Arabidopsis Information Service* **23**, 66–71.

GREEN, P. B. (1988). A theory for inflorescence development and flower formation based on morphological and biophysical analysis in *Echeveria. Planta* (Berl.) **175**, 153–169.

HAUGHN, G. W. AND SOMMERVILLE, C. R. (1988). Genetic control of morphogenesis in *Arabidopsis. Dev. Genet.* **9**, 73–89.

HESLOP-HARRISON, J. (1963). Sex expression in flowering plants. In *Meristems and Differentiation, 16th Brookhaven Symposium in Biology*. pp. 109–125. Upton, NY: Brookhaven National Laboratory.

HILL, J. P. AND LORD, E. M. (1989). Floral development in *Arabidopsis thaliana*: Comparison of the wildtype and the homeotic *pistillata* mutant. *Can. J. Bot.* **67**, 2922–2936.

HOLDER, N. (1979). Positional information and pattern formation in plant morphogenesis and a mechanism for the involvement of plant hormones. *J. theor. Biol.* **77**, 195–212.

KING, P. J. (1988). Plant hormone mutants. *Trends Genet.* **4**, 157–162.

KLEMM, M. (1927). Vergleichende morphologische und entwicklungsgeschiehtliche Untersuchung einer Reihe multipler Allelomorphe bei *Antirrhinum majus. Botanisches Archiv: Zeitschrift für der Gesamte Botanik* Vol. 20, pp. 11 423–11 474, Berlin.

KOMAKI, M. K., OKADA, K., NISHINO, E. AND SHIMURA, Y. (1988). Isolation and characterization of novel mutants of *Arabidopsis thaliana* defective in flower development. *Development* **104**, 195–203.

KOORNNEEF, M., CONE, J. W., KARSSEN, C. M., KENDRICK, R. E., VAN DER VEEN, J. H. AND ZEEVAART, J. A. V. (1985). Plant hormone and photoreceptor mutants in *Arabidopsis* and tomato. *UCLA Symp. Molec. cell. Bio. New Ser.* **35**, 103–114.

KOORNNEEF, M., VAN EDEN, J., HANHART, C. J., STAM, P., BRAAKSMA, F. J. AND FEENNSTA, W. J. (1983). Linkage map of *Arabidopsis thaliana. J. Hered.* **74**, 265–272.

KUNST, L., KLENZ, J. E., MARTINEZ-ZAPATER, J. AND HAUGHN, G. W. (1989). *AP2* gene determines the identity of perianth organs in flowers of *Arabidopsis thaliana. The Plant Cell* **1**, 1195–1208.

LEUTWILER, L. S., HOUGH-EVANS, B. R. AND MEYEROWITZ, E. M. (1984). The DNA of *Arabidopsis thaliana. Molec. gen. Genet.* **194**, 15–23.

LLOYD, A. M., BARNASON, A. R., ROGERS, S. G., BYRNE, M. C., FRALEY, R. T. AND HORSCH, R. B. (1986). Transformation of *Arabidopsis thaliana* with *Agrobacterium tumefaciens. Science* **234**, 464–466.

MASTERS, M. T. (1869). Vegetable teratology: an account of the principle deviations from the usual construction of plants. *Ray Society*, London.

MASTERSON, R. V., FURTEK, D. B., GREVELDING, C. AND SCHELL, J. (1989). A maize Ds transposable element containing a dihydrofolate reductase gene transposes in *Nicotiana tabacum* and *Arabidopsis thaliana. Molec. gen. Genet.* **219**, 461–466.

McHUGHEN, A. (1980). The regulation of tobacco floral organ initiation. *Bot. Gaz.* **141**, 389–395.

MEYER, V. (1966). Flower abnormalities. *Bot. Rev.* **32**, 165–195.

MEYEROWITZ, E. M., SMYTH, D. R. AND BOWMAN, J. L. (1989). Abnormal flowers and pattern formation in floral development. *Development* **106**, 209–217.

MIKSCHE, J. P. AND BROWN, J. A. M. (1965). Development of vegetative and floral meristems of *Arabidopsis thaliana. Amer. J. Bot.* **52**, 533–537.

MÜLLER, A. (1961). Zur Charakterisierung der Blüten und Infloreszenzen von *Arabidopsis thaliana* (L.) Heynh. *Kulturpflanze* **9**, 364–393.

NAM, H.-G., GIRAUDAT, J., DEN BOER, B., MOONAN, F., LOOS, W. D. B., HAUGE, B. M. AND GOODMAN, H. M. (1989). Restriction fragment length polymorphism linkage map of *Arabidopsis thaliana. The Plant Cell* **1**, 699–705.

NORMAN, C., RUNSWICK, M., POLLOCK, R. AND TRIESMAN, R. (1988). Isolation and properties of cDNA clones encoding SRF, a transcription factor that binds to the c-*fos* serum response element. *Cell* **55**, 989–1003.

OKADA, K., KOMAKI, M. K. AND SHIMURA, Y. (1989). Mutational analysis of pistil structure and development of *Arabidopsis thaliana. Cell Diff. Dev.* **28**, 27–38.

PASSMORE, S., MAINE, G. T., ELBLE, R., CHRIST, C. AND TYE, B. K. (1988). A *Saccharomyces cerevisiae* protein involved in plasmid maintenance is necessary for mating of MATa cells. *J. molec. Biol.* **204**, 593–606.

PAYER, J.-B. (1857). Traité d'organogenie comparé de la fleur. *Libraire de Victor Masson*, Paris.

PENZIG, O. (1890–4). *Pflanzen-Teratologie systemattisch geordnet*, vol. 1 (1890) and 2 (1894). A. Ciminago, Genua.

POLOWICK, P. L. AND SAWHNEY, V. K. (1986). A scanning electron microscope study on the initiation and development of floral organs of *Brassica napus* (cv. Westar). *Amer. J. Bot.* **73**, 254–263.

PRUITT, R. E., CHANG, C., PANG, P. P.-Y. AND MEYEROWITZ, E. M. (1987). Molecular genetics and development of *Arabidopsis*. In *Genetic Regulation of Development, 45th Symp. Soc. Devl Biol.* (ed. W. Loomis), pp. 327–338. New York: Liss.

PRUITT, R. E. AND MEYEROWITZ, E. M. (1986). Characterization of the genome of *Arabidopsis thaliana. J. molec. Biol.* **187**, 169–183.

SATTLER, R. (1973). *Organogenesis of flowers*. University of Toronto Press, Toronto.

SCHMIDT, R. AND WILLMITZER, L. (1989). The maize autonomous element Activator (Ac) shows a minimal germinal excision frequency of 0.2 %–0.5 % in transgenic *Arabidopsis thaliana* plants. *Molec. gen. Genet.* **220**, 17–24.

SCHULTZ, E. AND HAUGHN, G. (1990). The role of *FLO10* in specifying the identity of reproductive organs in flowers of *Arabidopsis thaliana*. Abst. *Fourth International Conference On Arabidopsis Research*, Vienna, p. 126.

SCHWARZ-SOMMER, Z., HUIJSER, P., NACKEN, W., SAEDLER, H. AND SOMMER, H. (1990). Genetic control of flower development: homeotic genes in *Antirrhinum majus. Science* **250**, 931–936.

SHANNON, S., JACOBS, C. AND MEEKS-WAGNER, D. R. (1990). Analysis of the shoot apical meristem during the transition to flowering. Abst. *Fourth International Conference on Arabidopsis Research*, Vienna, p. 128.

SMYTH, D. R., BOWMAN, J. L. AND MEYEROWITZ, E. M. (1990). Early flower development in *Arabidopsis. The Plant Cell* **2**, 755–767.

SOMMER, H., BELTRAN, J. P., HUIJSER, P., PAPE, H., LONNIG, W.-E., SAEDLER, H. AND SCHWARZ-SOMMER, Z. (1990). *Deficiens*, a homeotic gene involved in the control of flower morphogenesis in *Antirrhinum majus*: the protein shows homology to transcription factors. *EMBO J.* **9**, 605–613.

STEBBINS, G. L. (1976). Seeds, seedlings, and the origin of Angiosperms. In *Origin and Early Evolution of Angiosperms* (ed. C. B. Beck), pp. 300–311. Columbia University Press: New York.

STUBBE, H. (1966). Genetik und Zytologie von *Antirrhinum* L. *sect. Antirrhinum*, VEB Gustav Fischer Verlag, Jena.

VAN SLUYS, M. A., TEMPE, J. AND FEDEROFF, N. (1987). Studies on the introduction and mobility of the maize activator element in *Arabidopsis thaliana* and *Daucus carota. EMBO J.* **6**, 3881–3889.

VAUGHAN, J. G. (1955). The morphology and growth of the vegetative and reproductive apices of *Arabidopsis thaliana* (L.) heynh., *Capsella bursa-pastoris* (L.) Medic. and *Anagallis arvensis L. J. Linn. Soc. Lond. Bot.* **55**, 279–301.

YANOFSKY, M. F., MA, H., BOWMAN, J. L., DREWS, G. N., FELDMANN, K. A. AND MEYEROWITZ, E. M. (1990). Resemblance of the protein encoded by the *Arabidopsis* homeotic gene *agamous* to transcription factors. *Nature* **346**, 35–39.

Printed in Great Britain © Society for Experimental Biology 1991   117

USING ANTISENSE RNA TO STUDY GENE FUNCTION

WOLFGANG SCHUCH

ICI Seeds, Plant Biotechnology Section, Jealott's Hill Research Station, Bracknell RG12 6EY, UK

Summary

Over the past two years, antisense RNA technology has been developed in plants in several experimental systems. Progress reported in the literature will be reviewed. Special emphasis will be placed on experiments in which antisense RNA has been used to generate novel tomato mutants through the downregulation of polygalacturonase (PG), a major cell wall hydrolase. This work has shown that antisense RNA inhibits PG expression specifically without affecting the expression of other genes. The antisense gene and its phenotypic expression are stably inherited. The inhibitory effect of an antisense gene can be enhanced by increasing the copy number of the gene. The interaction of the antisense gene with its target is only seen when the target gene is expressed.

Antisense RNA technology has been used to clarify the biochemical function of polygalacturonase and its role in tomato fruit softening. Recently, this approach has also been applied to elucidate the biochemical role of another gene from tomato, pTOM13, whose function was previously unknown.

Thus, antisense RNA technology can now be used to generate novel plant mutants that will make possible the definition of the biochemical and biological role of genes whose functions are otherwise unknown.

Introduction

Plant molecular biology has made considerable advances over the past decade. Technical progress in cDNA and gene cloning has contributed to a great abundance of clones now being available in many laboratories. However, only a small number of the clones are being analysed. These clones either encode polypeptides of known function or they have been identified by other criteria such as tissue specific expression, expression after induction by a defined stimulus etc. However, for the majority of clones the biochemical and biological functions are unknown. Until recently only two ways of defining these clones were available: 1. by DNA sequence analysis and database searching. This can in many cases lead to an identification of the biochemical functions of clones if related sequences are already in databases. 2. *In vivo* complementation of mutants has been used to describe the biochemical function of clones. Both methods have only had limited success in the biochemical and biological characterisation of genes.

In other systems, such as yeast, methods of homologous recombination have

Key words: antisense RNA, tomato, polygalacturonase.

been developed that permit new mutants to be generated. These can be used to define the biochemical and biological functions of clones (Scherer and Davis, 1979). These methods have only recently been tested in plants (Paszkowski *et al.* 1988; Lee *et al.* 1990), and it is clear that they are not yet ready for routinely generating new plant mutants for use in the verification of the function of selected genes.

Therefore, considerable effort has been devoted to the development of antisense RNA techniques that promised to provide a general method for the production of novel plant mutants for use in the identification of the biological and biochemical functions of unknown plant genes. In this paper I will review progress which has been made in developing this approach.

Summary of literature reports on the use of antisense RNA experiments in plants

There are now several reports in the literature of gene expression in transgenic plants being inhibited by antisense RNA (Table 1) (for review see Schuch, 1990). This summary excludes any examples on the control of virus infection in transgenic plants. The remarkable observation from Table 1 is that in all but one example (see Delauney *et al.* 1988) a degree of downregulation of the target gene has been achieved.

We have developed antisense RNA techniques for the control of genes expressed during fruit ripening in tomato (Smith *et al.* 1988; Smith *et al.* 1990). This work will be described in detail and the conclusions about the use of antisense RNA will be summarised.

Antisense RNA to modify tomato fruit quality

The tomato fruit ripening system

During tomato fruit ripening several physiological and biochemical changes take place that affect all compartments of fruit cells. On the basis of *in vitro* translation and cDNA cloning experiments it has been suggested that many of these changes are due to the activation of novel genes (Grierson *et al.* 1985; Slater *et al.* 1985). One of the changes that affect the cell wall compartment of the ripening fruit is the induction of the major cell wall hydrolase, polygalacturonase (PG) (Tucker and Grierson, 1982). In the immature and mature green fruit no PG enzyme or mRNA is found. During ripening the level of PG mRNA increases to about 1 % of the mass of tomato fruit mRNA (Dellapenna *et al.* 1986). In the red fruit, PG protein represents approximately 15 % of the total cell wall proteins.

PG has attracted considerable attention as it has been implicated in tomato fruit softening (Hobson, 1964; Hobson, 1965; Crookes and Grierson, 1983; Tigchelaar *et al.* 1978). This work has been founded on several lines of evidence including observations in ripening mutants that lack PG and do not soften normally (Hobson, 1964; Tigchelaar *et al.* 1978). The problem, however, with the available ripening mutants is that they have pleiotropic effects that do not permit the role of PG in fruit softening to be determined (Grierson *et al.* 1987).

Table 1. *Antisense RNA experiments in plants reported in the literature*

	Species	Target gene	Antisense construct		Comments	References
			Promoter	Antisense gene		
Transient assay	Carrot protoplasts	CAT	CaMV35S Nos 'PAL'	Complete gene	Large excess of antisense vector required	Ecker and Davies (1986)
Model systems	Tobacco	Nos	CaMV35S	860 bp 5'	3' End most effective	Rothstein et al. (1987)
		Nos	Cab	Various		Sandler et al. (1988)
		CAT CaMV19S	rbs	Complete	No inhibition	Delauney et al. (1988)
		GUS	CaMV35S	Complete 172 bp 5'	5' End most effective 90 % reduction	Robert et al. (1989)
		GUS	Cab	41 bp surrounding ATG	Up to 100 % reduction	Cannon et al. (1990)
		PAT	CaMV35S TR2	PAT-CAT	90 % reduction transcription and translation affected	Cornelissen and Vandewiele (1989)
Endogenous genes	Tobacco	rbs	CaMV35S	322 bp 5'		Rodermel et al. (1988)
		ME	CaMV35S	Gene construct including intron	Previously reported to be CAD	Schuch et al. (1990)
	Petunia	CHS	CaMV35S CHS	Various	3' End most effective	Van der Krol et al. (1988, 1990)
	Tomato	PG	CaMV35S	730 bp 5'	99 %	Smith et al. (1988, 1990)
		PG	CaMV35S	Complete cDNA	90 %	Sheehy et al. (1988)
		pTOM13	CaMV35S	Various		Hamilton et al. (1990)

Target genes: *Nos*, nopaline synthase; *CAT*, chloramphenicol acetyl transferase; *GUS*, β-glucuronidase; *PAT*, phosphinothricin acetyl transferase. Endogenous genes: *rbs*, ribulose bisphosphate carboxylase; *ME*, malic enzyme; *CAD*, cinnamyl alcohol dehydrogenase; *CHS*, chalcone synthase; *PG*, polygalacturonase.

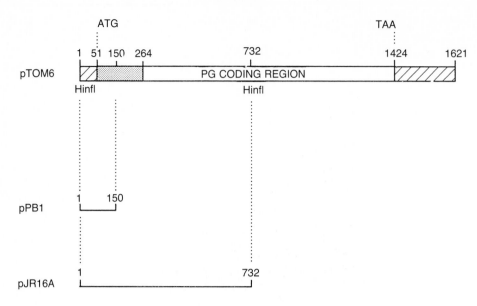

Fig. 1. Polygalacturonase cDNA fragments used to construct antisense RNA vectors.

Thus one of our aims during the development of antisense RNA techniques was the generation of single gene mutants in which only PG gene expression was affected.

The antisense vectors

We have cloned and characterised in detail a PG cDNA (Grierson *et al.* 1986) and the PG gene (Bird *et al.* 1988). We have used the PG cDNA in the construction of antisense vectors. The two vectors described in this study are shown in Fig. 1. We have used a 730 bp fragment covering the 5' untranslated region, the translational initiation region and part of the PG coding region in the construction of pJR16A a binary plant transformation vector based on pBIN19 (Smith *et al.* 1988). We have also constructed another vector in which only 150 bp of the 5' end of the pTOM6 were used in the construction of pPB1 (Schuch *et al.* unpublished). This latter vector covers only the 5'untranslated region, the translational initiation region and part of the region coding for the PG presequence.

PG activity in transgenic tomato plants

These vectors were transferred to tomato (*Lycopersicon esculentum* var. Ailsa Craig) using published techniques (Bird *et al.* 1988). Transgenic plants were selected by their ability to grow on medium containing kanamycin. These plants were regenerated, grown to maturity in the glasshouse and the levels of PG activity determined during fruit ripening.

A total of eight independent lines transformed with pJR16A have been analysed (Smith *et al.* 1988; Smith *et al.* 1990). The levels of PG in the ripe fruit varied

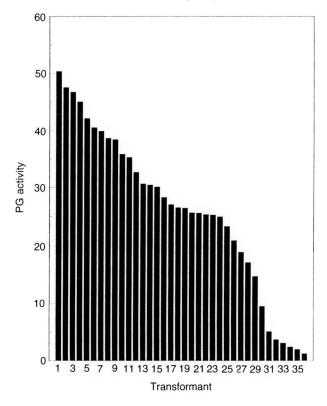

Fig. 2. Plants with the PG antisense gene CaMV 35S-5′150 bp PG antisense-*nos*.

between 51 % and 1 %. The reduction in PG activity is due to a reduction in the level of the PG protein found in the cell wall of these plants (Smith *et al.* 1988). Further characterisation of these plants has indicated that a single antisense gene has been inserted per diploid tomato genome. From the restriction patterns obtained after Southern hybridisations it has also been concluded that these genes are inserted at different chromosomal locations in the independent lines (Smith *et al.* 1990).

Using pPB1 as a transformation vector, 36 independent lines have been regenerated and PG enzyme levels determined in the ripe fruit (Schuch *et al.* unpublished). The level of PG activity varied between 100 % and 3 % (Fig. 2).

Three important conclusions can be drawn from this work: (1) The presence of a single antisense gene in the primary transformant can lead to considerable variation in the level of expression of the target gene. This variation has been observed in all other cases reported in Table 1 and is a general feature of plant transformation. This variation might well prove significant in the biochemical analysis of novel plant mutants generated by antisense RNA, and is already being exploited in the novel flower colour mutants induced through the reduction of CHS (chalcone synthase) in *Petunia*. (2) The maximal level of inhibition of PG in the primary transformant is 99 %. This is remarkable as PG mRNA can account

Table 2. *Antisense gene copy number effect*

Line	Number of alleles	%PG activity
GR16	1	10
GR105	2	0.85
GR43	1	1
GR43.1	2	0.15

The copy number effect of GR16 and GR43, each containing a single antisense gene per diploid genome was doubled by selfing. PG activity was determined in ripe fruit.

for up to 1% of total mRNA in the tomato fruit. This obviously indicates that antisense RNA can be very effective in inhibition of the target gene. This has also been confirmed in experiments by others (see Table 1), who have described 100% inhibition of target genes. (3) Relatively short sequences can be used successfully in antisense vectors. Recently, as few as 41 bases have been used in the construction of antisense vectors to GUS (β-glucuronidase) (Cannon *et al.* 1990).

Analysis of selfed progeny of pJR16A transformants

Two primary transformants, GR16 and GR43, have been chosen for further study (Smith *et al.* 1990; Bird *et al.* unpublished). The difference in the levels of PG activity are approximately 10-fold (GR16:10%; GR43:1%). These plants were selfed, and the S_1 progeny analysed for PG activity and the presence of the antisense gene. Three different classes of plants can be distinguished in the progeny: plants that have normal levels of PG activity and in which the antisense gene is absent; plants that have levels of PG activity comparable to the parent plants and contain only one antisense gene, and plants in which the level of PG activity is reduced beyond that observed in the parent plants and that contain two antisense genes. The levels of PG activity in homozygous progeny of GR16 and GR43 plants are shown in Table 2. This shows clearly that doubling the copy number of the antisense gene through selfing leads to a level of reduction of the expression of the target gene that is considerably greater than that found in the primary transformant. The lowest level of PG activity that we have observed is 0.15% of normal levels (Bird *et al.* unpublished).

Inheritance of the antisense gene and reduction in PG level

The data presented in the previous section also demonstrates that the PG antisense gene is stably inherited. In order to analyse this further we have selfed GR105 and produced a true breeding line, AC105. PG activity in this line has been assessed under commercial glasshouse growing conditions and was found to be 0.64% (Schuch *et al.* 1990b).

The stability of the antisense gene in a backcross experiment has been tested. AC105 was crossed to its untransformed parent line, Ailsa Craig, both as a male and female gamete donor (Boniwell *et al.* unpublished). This has indicated that the

Table 3. *PG activity in selfed progeny and crosses*

	% PG activity
GR16	10
GR105	0.85
AC105	0.64
AC105 progeny	0.58
AC^{++}* AC105	7
AC105*AC^{++}	6

The primary transformant GR16 was identified which contained a single antisense gene insert. GR16 was selfed and in the progeny a homozygous line was identified: GR105. GR105 was selfed and a true breeding line AC105 was obtained. AC105 selfed progeny as well as crosses of AC105 to the parent line Ailsa Craig (AC^{++}) were used for PG analysis in ripe fruit.

antisense gene is stably transferred in this backcross (data not shown) and that the 'potency ' of the antisense gene is maintained: in the backcross plants hemizygous for the antisense gene, PG levels in ripe fruit are comparable to those found in the primary transformant from which it was derived (Table 3). These data clearly demonstrate that antisense genes are stably inherited and that their effect on target genes is maintained.

Specificity of the antisense effect

The specificity of PG antisense RNA on its target has been analysed. Expression of PG antisense RNA does not affect invertase or pectinesterase activities, colour development or ethylene synthesis, total soluble solids, titratable pH, potassium level or dry matter accumulation (Smith *et al.* 1990; Schuch *et al.* 1990*b*).

Experiments on *Petunia* into which CHS antisense RNA was introduced also demonstrate that the expression of the antisense gene specifically inhibits the target gene and has no effect on other physiological parameters (van der Krol *et al.* 1988).

The experiments described so far illustrate clearly that transformation with antisense genes can lead to the successful generation of novel single gene mutants in plants.

RNA analysis

In order to gain some understanding of the possible mechanism by which the introduction of the antisense gene leads to the described effects, RNA was extracted from transgenic antisense plants and analysed. The PG antisense gene is transcribed into stable RNA species that are detectable in leaf tissue, green and red fruit tissues (Smith *et al.* 1988; Grierson *et al.* 1990). In addition to the RNA predicted from the design of the antisense vector, several smaller RNAs were detected. The importance of these RNA species is, at present, not understood.

In orange fruit, PG mRNA is normally expressed at maximal levels. In plants with the antisense gene, the level of PG mRNA is reduced considerably indicating that the reduction in PG enzyme activity is in fact due to a reduction in the level of

PG mRNA. The steady state level of antisense RNA is also reduced in red fruit RNA samples when compared with green fruit samples. These results suggest that an interaction between sense and antisense RNAs has taken place as the levels of both RNA species are reduced.

Although high levels of PG antisense RNA are present in the mature green fruit, an increase in the level of PG mRNA and enzyme activity is observed at the time when PG gene expression normally occurs, during the early phases of fruit ripening (Smith *et al.* 1990). This may mean that the antisense RNA effect can only be realised when a gene is expressed. This observation argues against a direct interaction between the sense and the antisense genes.

In order to analyse the stability of PG mRNA in ripe fruit from transgenic plants with the antisense gene, RNA was extracted and probed with a 'sense' strand-specific probe covering the region of the antisense vector (5' end) and the rest of the cDNA. The results indicate that using the 3' probe only the complete PG mRNA is detectable. Using the 5' probe, the complete PG mRNA can also be detected. In addition several smaller RNA species can be detected with the predominant species corresponding to the size of the antisense RNA (Watson *et al.* personal communication). This result suggests that the PG mRNA is broken down into submolecular RNAs and that as the major RNA species is the size of the antisense RNA, an intermediate is formed, the structure of which is determined by the size of the antisense RNA. The nature of these RNA species are under investigation.

Cannon *et al.* (1990), using a similar hybridisation approach, have recently suggested that the antisense RNA leads to a local instability of the sense RNA in the region that is homologous with the antisense RNA. Further work is required to elucidate the mechanism by which the reduction of the target gene is mediated.

Determination of the biochemical role of PG

The substrate for PG is the pectin molecules in the cell wall of the fruit. In order to elucidate the biochemical role of PG, the relative molecular mass of pectin in the new tomato mutant and controls have been analysed (Smith *et al.* 1990). During ripening of control fruit a reduction in the size of pectins takes place (Seymour and Harding, 1987). In plants with the antisense gene this degradation of pectins is inhibited. The degree of inhibition is dependent on the residual level of PG enzyme activity: in plants heterozygous for the antisense gene (20 % PG activity), the average relative molecular mass of pectin in ripe fruit is intermediate between controls and plants homozygous for the antisense gene with 1 % PG activity. The average relative molecular mass of pectins in the homozygous plants is very similar to that of green fruit. This analysis clearly demonstrates that the biochemical role of PG during fruit ripening is the reduction of pectin chain length. Thus, antisense RNA inhibition of PG has been used to clarify the *in vivo* biochemical role of PG.

Determination of the biological role of PG

PG has been implicated as a key enzyme in tomato fruit softening. The true breeding line, AC105, in which PG levels are reduced to 0.64 %, has been used to assess fruit softening in comparison to control fruit (Schuch *et al.* 1990*b*). For this purpose AC105 and controls were grown under commercial growing conditions and the firmness of fruit determined at different stages of fruit ripening. The results reveal that the rate of softening of fruit from AC105 and controls is indistinguishable. Thus, it must be concluded that PG does not play the key role in tomato fruit softening which had been ascribed to it previously.

This is the first example in which antisense RNA has been used to generate a novel single gene mutant that has been used in the clarification of the biological role of the target enzyme.

Can antisense RNA be used to determine the biochemical function of a gene?

Hamilton *et al.* (1990) have recently taken the development of the antisense technology one step further by generating transgenic plants expressing antisense RNA to a gene of unknown function, pTOM13. This clone had been isolated from a ripe tomato fruit cDNA library (Slater *et al.* 1985), and it has been shown previously that the expression of this gene is correlated with the synthesis of ethylene during fruit ripening and wounding of leaves (Smith *et al.* 1986; Holdsworth *et al.* 1987). In order to test the hypothesis that pTOM13 is involved in ethylene metabolism, Hamilton *et al.* analysed transgenic tomato plants containing one or two antisense genes to pTOM13. Biochemical analysis demonstrated that in these plants ethylene synthesis is reduced in a gene dosage-dependent manner. Plants containing one antisense gene synthesised only 13 % of the normal amount of ethylene, whereas in plants homozygous for the pTOM13 antisense gene the production of ethylene was only 3 % of that found in normal fruits. These results strongly support the hypothesis that pTOM13 is involved in ethylene metabolism. This was tested by measuring the activity of 1-aminocyclopropane 1-carbonylate synthase (ACC) oxidase. The level of ACC oxidase closely correlated with the level of ethylene production, lending support for the hypothesis that pTOM13 encodes ACC oxidase.

This approach clearly demonstrates that antisense RNA techniques can be used to ascribe a biochemical function to genes. These transgenic plants can now be used to elucidate the biological role of the genes under investigation.

Conclusions and outlook

Antisense RNA technology has been developed in several systems including the tomato fruit system. It is clear that antisense RNA vectors can be generated that in general will lead to a successful downregulation of the target gene. Although there does not appear to be a general rule for the construction of the optimal antisense vectors, vectors containing substantial portions of genes can be constructed that in most cases will lead to the down regulation of the target gene. It is interesting to note that there are few examples in the literature in which gene or intron sequences have been used in the construction of antisense vectors. Further work is required to investigate this.

The mechanism of antisense RNA is not fully understood. From the work quoted in this paper it is likely that direct RNA:RNA interactions occur that lead to the rapid degradation of both the antisense and the sense RNA. The PG system offers a good opportunity to study these effects, as, owing to the high level of expression of the PG mRNA, putative degradation intermediates may be detected, particularly in plants in which the down regulation is not as great (e.g. a plant with 20–30 % residual activity might prove the most useful for such a study). In other systems the complete disappearance of both the target and antisense RNA makes this work very difficult.

The reduction in target gene expession using antisense RNA techniques promises to be sufficient for the generation of useful mutants for the discovery of the biochemical and biological function of target genes. Thus, it will prove to be a universally applicable method for the identification of gene function.

References

Bird, C. R., Smith, C. J. S., Ray, J. A., Moureau, P., Bevan, M. J., Bird, A. S., Hughes, S., Morris, P. C., Grierson, D. and Schuch, W. (1988). The tomato polygalacturonase gene and ripening expression in transgenic plants. *Pl. molec. Biol.* **11**, 651–662.

Cannon, M., Platz, J., O'Leary, M., Sookdea, C. and Cannon, F. (1990). Organ-specific modulation of gene expression in transgenic plants using antisense RNA. *Pl. molec. Biol.* **15**, 39–47.

Cornelissen, M. and Vandewiel, M. (1989). Both RNA level and translation efficiency are reduced by antisense RNA in transgenic tobacco. *Nucl. Acids Res.* **17**, 833–843.

Crookes, P. R. and Grierson, D. (1983). Ultrastructure of tomato fruit ripening and the role of polygalacturonase isoenzymes in cell wall degradation. *Pl. Physiol.* **73**, 1088–1093.

Delauney, A. J., Tabaeizadeh, Z. and Verma, D. P. S. (1988). A Stable bifunctional antisense transcript inhibiting gene expression in transgenic plants. *Proc. natn. Acad. Sci. U.S.A.* **85**, 4300–4304.

DellaPenna, D., Alexander, D. C. and Bennet, A. B. (1986). Molecular cloning of tomato fruit polygalacturonase: Analysis of polygalacturonase mRNA levels during ripening. *Proc. natn. Acad. Sci. U.S.A.* **83**, 6420–6424.

Ecker, J. R. and Davies, R. W. (1986). Inhibition of gene expression in plant cells by expression of antisense RNA. *Proc. natn. Acad. Sci. U.S.A.* **83**, 5372–5376.

Grierson, D., Purton, M. E., Knapp, J. E. and Bathgate, B. (1987). Tomato ripening mutants. In *Developmental Mutants in Higher Plants* (ed: H. Thomas and D. Grierson), pp. 73–94. Cambridge University Press, Cambridge.

Grierson, D., Slater, A., Spiers, J. and Tucker, G. A. (1985). The appearance of polygalacturonase mRNA in tomatoes: one of a series of changes in gene expression during development and ripening. *Planta* **163**, 263–271.

Grierson, D., Smith, C. J. S., Watson, C. F., Morris, P. C., Gray, J. E., Davies, K., Picton, S. J., Tucker, G. A., Seymour, G., Schuch, W., Bird, C. R. and Ray, J. (1990). Regulation of expression in transgenic tomato plants by antisense RNA and ripening specific promoters. In *Genetic Engineering of Crop Plants* (ed: G. W. Lycett and D. Grierson) Butterworth, London.

Grierson, D., Tucker, G. A., Keen, J., Bird, C. R. and Schuch, W. (1986). Sequencing and identification of a cDNA clone for tomato polygalacturonase. *Nucl. Acids Res.* **14**, 8595–8603.

Hamilton, A. J., Lycett, G. W. and Grierson, D. (1990). Antisense gene that inhibits synthesis of the hormone ethylene in transgenic plants. *Nature* **346**, 284–287.

HOBSON, G. E. (1964). Polygalacturonase in normal and abnormal tomato fruit. *Biochem. J.* **92**, 342–332.

HOBSON, G. E. (1965). The firmness of tomato fruit in relation to polygalacturonase activity. *J. hort. Sci.* **40**, 66–72.

HOLDSWORTH, M. J., BIRD, C. R., RAY, J., SCHUCH, W. AND GRIERSON, D. (1987). Structure and expression of an ethylene related mRNA from tomato. *Nucl. Acids Res.* **15**, 731–739.

LEE, K. Y., LUND, P., LOWE, K. AND DUNSMUIR, P. (1990). Homologous recombination in plants cells after Agrobacterium-mediated transformation. *The Plant Cell* **2**, 415–425.

PASZKOWSKI, J., BAUR, M., BOGUCKI, A. AND POTRYKUS, I. (1988). Gene targeting in plants. *EMBO J.* **7**, 4021–4026.

ROBERT, L. S., DONALDSON, P. A., LADAIQUE, C., ALTOSAAR, I., ARNISON, P. G. AND FABIJANSKI, S. F. (1989). Antisense RNA inhibition of β-glucuronidase gene expression in transgenic tobacco plants. *Pl. molec. Biol.* **13**, 399–409.

RODERMEL, S. R., ABBOTT, M. S. AND BOGORAD, L. (1988). Nuclear–organelle interactions: nuclear antisense gene inhibits ribulose bisphosphate carboxylase enzyme levels in transformed tobacco plants. *Cell* **55**, 673–681.

ROTHSTEIN, S. J., DIMAIO, J., STRAND, M. AND RICE, D. (1987). Stable and heritable inhibition of the expression of nopaline synthase in tobacco expressing antisense RNA. *Proc. natn. Acad. Sci. U.S.A.* **84**, 8439–8443.

SANDLER, S. J., STAYTON, M., TOWNSEND, J. A., RALSTON, M. L., BEDBROOK, J. R. AND DUNSMUIR, P. (1988). Inhibition of gene expression in transformed plants by antisense RNA. *Pl. molec. Biol.* **11**, 301–310.

SCHERER, S. AND DAVIS, R. W. (1979). Replacement of chromosome segments with altered DNA sequences constructed *in vitro*. *Proc. natn. Acad. Sci. U.S.A.* **76**, 3912–3915.

SCHUCH, W. (1990). The manipulation of plant gene expression using antisense RNA. In: *Plant Gene Research* (ed: E. S. Dennis) Springer Verlag: Vienna.

SCHUCH, W., KANCZLER, J., ROBERTSON, D., HOBSON, G. E., TUCKER, G. A., GRIERSON, D., BRIGHT, S. J. W. AND BIRD, C. R. (1990*b*). Improvement of tomato fruit quality through genetic engineering. Submitted.

SCHUCH, W., KNIGHT, M., BIRD, A., GRIMA-PETTENATI, J. AND BOUDET, A. (1990*a*). Modulation of plant gene expression. In *Genetic Engineering of Crop Plants* (ed: G. W. Lycett and D. Grierson) Butterworth, London.

SEYMOUR, G. B. AND HARDING, S. E. (1987). Molecular size of tomato (Lycopersicon esculentum Mill) fruit polyuronides by gel filtration and low-speed sedimentation equilibrium. *Biochem. J.* **245**, 463–466.

SHEEHY, R. E., KRAMER, M. K. AND HIATT, W. R. (1988). Reduction of polygalacturonase activity in tomato fruit by antisense RNA. *Proc. natn. Acad. Sci. U.S.A.* **85**, 8805–8809.

SLATER, A., MAUNDERS, M. J., EDWARDS, K., SCHUCH, W. AND GRIERSON, D. (1985). Isolation and characterisation of cDNA clones for tomato proteins. *Pl. molec. Biol.* **5**, 137–147.

SMITH, C. J. S., SLATER, A. AND GRIERSON, D. (1986). Rapid appearance of an mRNA correlated with ethylene synthesis encoding a protein of molecular weight 35,000. *Planta* **168**, 94–100.

SMITH, C. J. S., WATSON, C. F., RAY, J., BIRD, C. R., MORRIS, P. C., SCHUCH, W. AND GRIERSON, D. (1988). Antisense RNA inhibition of polygalacturonase gene expression in tomatoes. *Nature* **334**, 724–726.

SMITH, C. J. S., WATSON, C. F., MORRIS, P. C., BIRD, C. R., SEYMOUR, G. B., GRAY, J. E., ARNOLD, C., TUCKER, G. A., SCHUCH, W. AND GRIERSON, D. (1990). Inheritance and effect on ripening of antisense polygalacturonase genes in transgenic tomatoes. *Pl. molec. Biol.* **14**, 369–379.

TIGCHELAAR, E. C., MCGLASSON, W. B. AND BUESCHER, R. W. (1978). Genetic regulation of tomato fruit ripening. *Hort. Sci.* **13**, 508–513.

TUCKER, G. A. AND GRIERSON, D. (1982). Synthesis of polygalacturonase during tomato fruit ripening. *Planta* **155**, 64–67.

VAN DER KROL, A. R., LENTING, P. E., VEENSTRA, J., VAN DER MEER, I. M., KOES, R. E., GERATS, A. G. M., MOL, J. N. M. AND STUITJE, A. R. (1988). An antisense chalcone synthase gene in transgenic plants inhibits flower pigmentation. *Nature* **333**, 866–869.

Printed in Great Britain © *Society for Experimental Biology 1991*   129

AUXIN-BINDING PROTEIN – ANTIBODIES AND GENES

COLIN M. LAZARUS[1], *RICHARD M. NAPIER*[2], *LONG-XI YU*[1],
CAROLINE LYNAS[1] *and MICHAEL A. VENIS*[2]

[1] Department of Botany, University of Bristol, Woodland Road, Bristol BS8 1UG, UK
[2] AFRC Institute of Horticultural Research, East Malling, Maidstone, Kent ME19 6BJ, UK

Summary

Of several auxin-binding systems that have been characterised the auxin-binding protein (ABP) of maize coleoptile membranes is the best candidate for a true auxin receptor. ABP, which exists as a homodimer of $22 \times 10^3 \, M_r$ glycosylated subunits, has been purified, and monoclonal and polyclonal antibodies raised against it. Electrophysiological studies with antibodies indicated the presence of a functional population of auxin receptors on the exterior face of the plasmalemma; electrophysiological experiments with impermeant auxin analogues now reinforce this conclusion.

An epitope mapping kit has been used to identify the major epitopes recognised by antibody preparations. Three major epitopes, bracketing the glycosylation site, have been identified in the polyclonal serum. They are also represented in antisera produced in other laboratories and are conserved in ABP prepared from other plants. One monoclonal antibody recognises an epitope close to the amino terminus of ABP and two others recognise the carboxy terminus. The latter antibodies have been used in a sandwich ELISA to demonstrate that auxin binding induces a conformational change in ABP.

Maize ABP is encoded by a small gene family and cDNA and genomic clones have been isolated. With a single exception, predicted amino acid sequences indicate remarkably little heterogeneity. The exceptional cDNA sequence predicts 87 % amino acid homology with the major class of proteins. Four introns are apparent in the sequence of a complete *ABP* gene; their sequences are very highly conserved in an incompletely-cloned second gene lacking the first exon. The major difference between the two genes lies in the length of the first intron, which has been estimated to exceed 5.2 kb in the incomplete gene. The site of initiation of transcription has not been unambiguously identified in the complete gene, and some evidence suggests that there may be an additional intron.

Homology to maize ABP cDNA has been detected in the genomes of *Arabidopsis*, spinach and strawberry but not in that of tobacco. A sequence located within the 3'-half of the maize cDNA is highly repeated in the strawberry genome, from which clones with homology to both halves of the maize cDNA (i.e. putative ABP genes) have been isolated.

Key words: auxin receptor, auxin-binding-protein epitopes, auxin-binding-protein genes, impermeant auxins.

Introduction

Auxins are one of the groups of naturally occurring compounds that are usually considered to be plant hormones. Their involvement in a large number of physiological and developmental responses has been well documented (see Davies, 1987 and Trewavas, 1981, for reviews). Implicit in the characterisation of any biological compound such as a hormone is the expectation that there must be a cognate receptor somewhere in the cell, and a particularly appropriate site for hormone perception to occur is the cell surface (plasmalemma in plant cells). However, auxins are lipophilic weak acids that are freely able to permeate the plasmalemma, so a response cascade could originate from auxin perception within the cell. Not all responses need emanate from the same hormone – receptor interaction, and receptors initiating diverse responses could have different subcellular locations and/or different structures.

Progress towards the characterisation of auxin receptors began with the identification of membrane binding sites for which auxin binding affinities and structural specificities were found to be consistent with the biological activities of the individual auxins (Ray et al. 1977). The identification of natural inhibitors of binding (Venis and Watson, 1978) and the establishment of a correlation between the amount of auxin binding and the capacity of a tissue to respond to auxin (Walton and Ray, 1981) provided further circumstantial evidence that binding sites were (or contained) auxin receptors.

Membrane-associated auxin-binding sites were first identified in maize (Hertel et al. 1972), but only relatively recently has it been possible to purify an auxin-binding protein (ABP) that could be implicated in mediating auxin-induced responses (Löbler and Klämbt, 1985; Shimomura et al. 1986; Napier et al. 1988). On release from microsomal membranes, by acetone or butanol treatment, the native protein behaves as a soluble homodimer of $22 \times 10^3 M_r$ glycoprotein subunits. Sophisticated experiments have been carried out with the purified protein and antibodies raised against it, and faith in its identity as an auxin receptor has grown. The most compelling data have been obtained from electrophysiological experiments involving auxin-induced hyperpolarisation of the plasmalemma of tobacco leaf mesophyll protoplasts. Barbier-Brygoo and coworkers (Barbier-Brygoo et al. 1989, 1990a, 1990b) demonstrated that the sensitivity of the protoplasts to auxin could be decreased by incubation with anti-maize-ABP serum or increased by the addition of purified maize ABP. As well as suggesting very strongly that ABP is indeed an auxin receptor these results demonstrate that a functional ABP population is accessible on the exterior face of the plasma membrane. However, the latter is a relatively minor site of ABP localisation, and most of the protein is found in the endoplasmic reticulum (ER).

Genetic studies now complement the biochemical and physiological characterisation of ABP. cDNA clones for the major (ER) form of ABP have been isolated and the full protein sequence deduced (Hesse et al. 1989; Inohara et al. 1989; Tillmann et al. 1989; our unpublished results). The 201-codon reading frame encodes a mature protein of $18.4 \times 10^3 M_r$ (163 amino acids) whose subcellular location is explained by a preceding signal peptide (38 amino acids) and a carboxy-terminal KDEL ER-lumen-retention signal (Munro and Pelham, 1987). Secondary structure predictions indicate that ABP is neither an integral nor peripheral membrane protein.

Antibodies and cloned genes are currently at the forefront of investigations aimed at gaining a deeper understanding of the mode of action of ABP and its physiological significance. In this paper we describe recent findings on the characterisation of ABP epitopes recognised by our monoclonal antibodies and polyclonal serum, and we present further data that support the notion of ABP as an auxin receptor. We also present information on an ABP isoform identified by cDNA cloning, the structure of ABP genes in maize and the identification of homologous ABP genes in dicotyledonous plants.

Monoclonal antibodies (mAbs) detect auxin-induced conformational change

A sandwich ELISA has been developed that enables a competitive reaction to be established between mAbs and auxin for binding to polyclonal-immobilised ABP. In this assay, binding of two mAbs, MAC 256 and MAC 259 (Napier *et al.* 1988), is reduced by auxins in a concentration-dependent manner. As an example, the effect of some phenoxyacetic acids, and benzoic acids is shown in Fig. 1. The

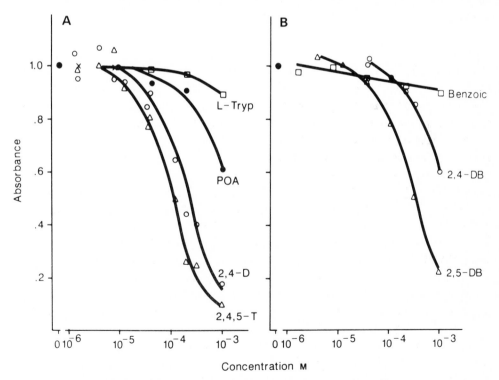

Fig. 1. Effect of phenoxyacetic acid (POA) analogues and L-tryptophan (A) and of benzoic acid analogues (B) on binding of MAC 256 in the sandwich ELISA.

potent auxins 2,4,5-T and 2,4-D are both active in displacing MAC 256, although the parent compound phenoxyacetic acid (with no auxin activity) has very little effect. Likewise, the auxin 2,5-dichlorobenzoic acid competes, but not the inactive benzoic acid, and the 2, 4-dichloro-analogue, an auxin antagonist, is intermediate. A wide range of auxins and inactive analogues has now been examined and the effect appears to be auxin-specific (Napier and Venis, 1990).

This competition between auxin and mAbs for binding to ABP might be explained in various ways. The mAb might bind to the auxin-binding site, competing directly; it might obscure the auxin-binding site, or might bind to an epitope elsewhere on the protein that is conformationally active. Auxin-binding activity is inhibited by a polyclonal antiserum (Venis and Napier, 1990), but we have been unable to inhibit auxin binding with MAC 256, which indicates that the mAb does not bind directly to the auxin-binding site. If the binding site were just obscured by the mAbs, the structural specificity observed (Fig. 1) would not be expected. Instead, therefore, we propose that MAC 256 and MAC 259 bind to an epitope that is conformationally active, i.e. that on binding to ABP, the auxin induces a conformational change that masks the antibody epitope. Since both these mAbs have been mapped to epitopes within $1 \times 10^3 M_r$ of the C terminus (Napier and Venis, 1990), this region would appear to be conformationally active. This region also includes, right at the C terminus, the KDEL tetrapeptide for protein retention within the lumen of the endoplasmic reticulum (Inohara *et al.* 1989).

Epitope mapping

There are three major epitopes for our original ABP polyclonal antiserum (Napier *et al.* 1988), bracketing the glycosylation site. Two of these (residues 85–90) are overlapping epitopes. This appears to be a highly antigenic region, since ABP antisera from other laboratories (Palme, Köln; Ordowski, Bonn) also map to two of the same three epitopes.

We are also using this technique to examine whether any of these epitopes are conserved in ABPs of different species, starting with barnyard grass (*Echinocloa crusgalli*) that contains homologous $24 \times 10^3 M_r$ subunit ABP at an abundance approx. 7- to 8-fold lower than $22 \times 10^3 M_r$ ABP of maize (Venis and Napier, 1990). The method is to separate partly-purified *Echinocloa* ABP by preparative SDS–PAGE, immunoblot against polyclonal anti-maize ABP, dissociate off at low pH the antibody fraction binding to the $24 \times 10^3 M_r$ band and then map this fraction on the hexapeptide pins. This reveals clearly that one of the three main epitopes is conserved between *Zea* and *Echinocloa*. We are currently applying the method to a dicot species.

We have also used the kit with all five of our mAbs (Napier *et al.* 1988). We already had evidence that MAC 257 recognises an epitope within $7 \times 10^3 M_r$ of the N terminus (Venis and Napier, 1990) and with the kit this has been confirmed and pinpointed. Interestingly, neither MAC 256 nor MAC 259 maps using this system.

Nevertheless, there is excellent evidence that these mAbs do in fact recognise a region near the C terminus (Napier and Venis, 1990) and this has been reinforced by finding that both antibodies strongly recognise protein disulphide isomerase, (PDI) another KDEL protein, kindly supplied by Dr Freedman, University of Kent, (unpublished results). Indeed, since there is no other homology between PDI and ABP, the epitope of both mAbs must be (or include) KDEL itself. The fact that neither antibody is detected by the epitope mapping kit is almost certainly because the hexapeptides are synthesised on activated pins, from the C-terminal ends. Therefore, the final leucine carboxyl group will be blocked in the synthetic hexapeptide (but not in ABP). It can be concluded that a free carboxyl group in the terminal leucine residue is an absolute requirement in the epitope of MAC 256 and MAC 259.

Impermeant auxins

Elegant electrophysiological studies of Barbier-Brygoo *et al.* (1990*a,b*) have shown that the optimum concentration for auxin-induced hyperpolarisation of isolated tobacco protoplasts can be shifted in both directions, spanning several orders of magnitude, by adding either anti-maize ABP antibodies or maize ABP. These findings suggest strongly that plasmalemma-located ABP or isoforms of ABP are functional in auxin perception and action and that the site of perception is accessible at the outer face of the plasma membrane. Further support for the latter point comes from experiments with non-permeant auxin analogues.

Protein conjugates with 5-substituted NAA derivatives are able to stimulate growth of excised sections provided that the protein is small enough (<approx. $100 \times 10^3\, M_r$) to traverse the cell wall (Venis *et al.* 1990). In order to see whether such conjugates were active in more rapid assays they were evaluated for effects on tobacco protoplast E_m (transmembrane potential difference), where the full response is seen within 1 min of auxin addition. The activity of the free analogues was in overall agreement with their behaviour in elongation tests (Venis and Thomas, 1990), including the lack of activity shown by 5-amino-NAA (Fig. 2A). After diazotisation and azo-coupling to protein, the molecule will more closely resemble the growth-active 5-azido-NAA, and both bovine serum albumin (BSA, M_r 68 000) and keyhole limpet haemocyanin (KLH, $M_r > 10^6$) conjugates induce protoplast hyperpolarisation (Fig. 2B). In the absence of a cell wall, protein size is no longer a limitation on activity. The conjugate between 5-azido-NAA and BSA is also active, eliciting the bell-shaped dose-response curve seen with free auxins (Fig. 2B). These results provide further evidence that functional auxin receptors are present at the exterior face of the plasma membrane.

Multiple genes for ABP in maize

Previous reports of ABP sequences have indicated the existence of a multigene family encoding this protein. Inohara *et al.* (1989) described minor sequence

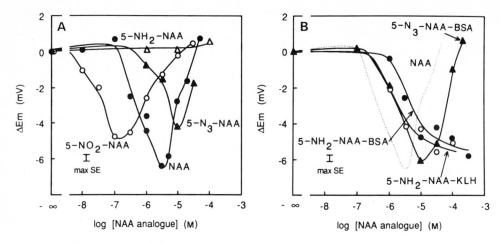

Fig. 2. Effect of NAA and 5-substituted derivatives (A) and of NAA derivative-protein conjugates on transmembrane potential difference (ΔE_m) of tobacco mesophyll protoplasts. (From Venis *et al.* 1990).

variation among the set of cDNA clones they obtained and Hesse *et al.* (1989) detected differences between the N-terminal amino acid sequences of major and minor isoforms of ABP. Our evidence, at the level of genomic Southern blotting (see below) and gene isolation and sequencing, confirms that ABP is encoded by a small multigene family in maize.

Sequences of the maize genes

From cDNA clones

cDNA clones were isolated from two libraries that were both constructed in the vector λgt11. A single clone, designated cAux1, was isolated from the first library by expression screening with polyclonal anti-ABP serum. Partly due to ineffective *Eco*RI methylation of cDNA prior to linker addition and *Eco*RI cleavage, this cDNA was very short (254 bp, covering only an internal part of the coding region); it was used as a probe in screening the second (amplified) library by hybridisation. Several cDNA clones were obtained, of which the longest (designated cAux4) extended from the poly[A] tail to 32 bp upstream of the coding region. The sequence of this cDNA is given in Fig. 3. cAux4 is almost identical in sequence to the clones previously described (Inohara *et al.* 1989; Hesse *et al.* 1989; Tillmann *et al.* 1989); with respect to the predicted amino acid sequence it is novel only in encoding a glycine residue at position -26 (the 13th residue of the signal peptide) where previously described clones encode alanine or proline.

Sequence comparisons between clones cAux1 and cAux4 revealed several differences, many of which result in amino acid substitutions. Attempts to isolate a full-length version of cAux1 from the amplified library were unsuccessful, although the existence of such full-length clones was indicated by the polymerase chain reaction (PCR). To obtain more sequence information on cAux1-type

```
                              M  A  P  D  L  S  E  L  A  A  A  A  G   -26
         CGACATTCACGTGCAGCTGTCGGGAGCAGGCAATGGCGCCGGATCTAAGCGAACTCGCCGCCGCCGCCGGA


 A  R  G  A  Y  L  A  G  V  G  V  A  V  L  L  A  A  S  F  L  P  V  A  E  S   -1
GCCCGTGGCGCCTACCTCGCCGGCGTCGGTGTCGCGGTCCTCCTCGCTGCCTCCTTCCTCCCAGTAGCCGAGTCG


 S  C  V  R  D  N  S  L  V  R  D  I  S  Q  M  P  Q  S  S  Y  G  I  E  G  L   25
TCCTGCGTGCGAGATAACTCATTGGTGAGAGACATAAGCCAAATGCCGCAAAGCAGCTATGGGATTGAAGGATTG

          .   .   .   T   .   .   A   .   T   .   .   .   .   .   .   .   .   .   F   G   .
 S  H  I  T  V  A  G  A  L  N  H  G  M  K  E  V  E  V  W  L  Q  T  I  S  P   50
TCACATATAACAGTTGCTGGTGCGCTCAATCATGGGATGAAGGAGGTGGAAGTGTGGCTTCAGACAATAAGTCCA
........T..CA.A.....T..TGC...C....C...................A..A...T.TG.....


 .   .   .   .   .   .   .   .   .   .   .   I   .   .   .   .   .   .   .
 G  Q  R  T  P  I  H  R  H  S  C  E  E  V  F  T  V  L  K  G  K  G  T  L  L   75
GGTCAAAGGACGCCAATCCACAGGCATTCCTGTGAAGAAGTTTTCACTGTCCTCAAAGGGAAGGGTACGCTCTTG
..........C................T.........G.......T.........G.....A..C........A


 L   .   .   .   .   .   .   .   .   .   V   .   V   .   .   .   .   .   .
 M  G  S  S  S  L  K  Y  P  G  Q  P  Q  E  I  P  F  F  Q  N  T  T  F  S  I   100
ATGGGATCAAGCTCACTAAAGTACCCAGGGCAGCCACAGGAAATTCCTTTCTTTCAGAATACCACATTTTCAATC
C.C..G..G.....G..G..........A.....G......G....CG...............G...


 .   .   .   .   .   .   .   .   .   N   .   .   .   .   .   .   .   .
 P  V  N  D  P  H  Q  V  W  N  S  D  E  H  E  D  L  Q  V  L  V  I  I  S  R   125
CCTGTAAATGATCCACACCAGGTTTGGAATTCTGACGAGCACGAAGATTTGCAAGTTCTTGTGATCATATCGAGA
................T.....C.........A.T.........................AC.C


 .   .   V   .   .   I   .   .   .   .   .   .   .   .   K   .   .   .   .   Y   F
 P  P  A  K  I  F  L  Y  D  D  W  S  M  P  H  T  A  A  V  L  K  F  P  F  V   150
CCGCCTGCTAAGATATTTTTTATATGATGATTGGAGCATGCCTCATACAGCCGCGGTACTGAAATTCCCCTTCGTC
..A..G.TC.........A.................T...............T..AAAG.............A.T..


 .   .   .   .   L   P   .   P   .   .   .   .   .
 W  D  E  D  C  F  E  A  A  K  D  E  L ***                                163
TGGGATGAGGACTGCTTCGAAGCAGCAAAAGACGAACTCTAGGTCACAAGTGTTTCCTGCAATTTATCTGCTTCA
................GCC....C.......T...........T--.....G.............AA....G


T------CCATGATCCTGCTGGTGCTGGACTACTACAATTCTCAGCACTAGTTGTAATAAAGCCA---GTGCGCT
TGTTCGT...CA....---.....---......---G....A..T.--.CAA..-TG...A...TTA...TA..


TTTCATGTATAATTCTGTATTGTGGCTCGCGAAAATAAAATTTGGCAACGGTTTATGA(43)C
C.......G..G..T...G.....A...TG.G....CC.TA...A....-.......-A(4)T
```

Fig. 3. Nucleotide and deduced amino acid sequences of auxin-binding-protein-encoding cDNAs. The complete DNA sequence of cAux4 is presented with the protein sequence in single letter code above it. Positive numbering refers to the mature protein (+1 to +163), negative numbering to the signal peptide (−38 to −1); the amino acid sequence of the latter is in italics. The glycosylation site is marked in bold type and the KDEL in bold italics. The nucleotide and amino acid sequences of cAux1 are presented below and above those of cAux4 respectively, with identical positions represented by dots. In the 3′ untranslated regions of both clones gaps (hyphens) have been introduced into each nucleotide sequence to maximise alignment. The original cAux1 clone contained the region including amino acids 27 to 111; the remainder of the sequence (to the 3′ end) was determined directly from DNA amplified from the cDNA library.

mRNA and protein, products amplified from the library (using one internal primer based on the sequence of cAux1 and one external primer adjacent to the cloning site in λgt11) were directly sequenced by a PCR method (Murray, 1989). Although it has not proved possible to extend the sequence of cAux1 toward the 5′ end of the mRNA the known sequence of this message now extends from codon 27 (of the cAux4 sequence) through the end of the coding region to the 3′ end (or very close to it).

Comparison of the cAux4 and (extended) cAux1 coding sequences (Fig. 3) revealed 85.5 % and 87 % homology at the nucleotide and predicted amino acid levels respectively (60 bases differ out of the total of 414, causing 18 amino acid substitutions from the total of 138 codons). The 3′ untranslated regions of the two messages have diverged quite widely, although the introduction of a limited number of gaps into the sequences permits a good alignment along their whole lengths. Many (though certainly not all) predicted amino acid differences between the two protein types are conservative, and none affect the single glycosylation site (Asn Thr Thr) nor the KDEL (Lys Asp Glu Leu Stop). It seems unlikely, therefore, that the cAux1 and cAux4 classes of ABP proteins differ significantly in function or intracellular location. A feature of the extended sequence of cAux1 is that no polyadenylation signal (AAUAAA) nor extensive poly[A] tract was apparent. Prior to amplification the cDNA library must have contained at least two independent clones of the cAux1 type, since PCR products were obtained (and sequenced) using priming sites in both vector arms (i.e. cDNAs inserted in both possible orientations). Sequences obtained for the two orientations differ only in that one is 4 bp shorter at the 3′ end than that shown in Fig. 3. The possibility that cAux1-type mRNAs are not polyadenylated is remote since cDNA synthesis was primed with oligo dT; the absence of extensive 3′-terminal poly[A] tracts in cAux1-type clones more likely results from incomplete second strand synthesis or cloning artefacts. Such an artefact is apparent in the sequence of cAux4, in which a single C residue is present between the poly[A] tract and the adapter sequence used in cloning; the terminal T residue of the cAux1 sequence could likewise be artefactual and the immediately preceding run of 4 A's could mark the start of the poly[A] tail. This seems particularly likely in view of the alignment of the cAux1 and cAux4 sequences shown in Fig. 3.

Northern blots of coleoptile poly[A]$^+$ RNA probed with oligonucleotides that distinguish between the cAux1 and cAux4 cDNAs demonstrated that cAux1-class mRNA is present in much lower abundance than the cAux4-class message (results not shown). This may simply reflect relative gene number, since lack of heterogeneity in the extended cAux1 DNA sequence indicates that all such clones in the cDNA library emanate from a single gene.

From genomic clones

Genomic clones containing ABP genes have not yet been described in any detail, although Hesse et al. (1989) referred to a clone indicating considerable intron content. Our results bear out this conclusion.

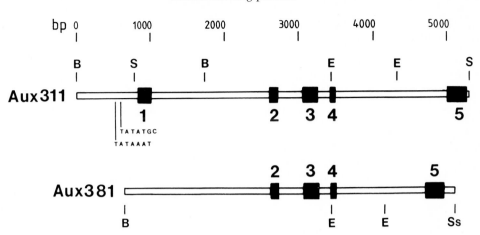

Fig. 4. Structures of two maize genes encoding auxin binding protein. Exons (black boxes) are numbered 1 to 5 in the direction of transcription (*Aux381* is incomplete, lacking the 5′ exon). Introns and non-transcribed sequences are represented by open bars. Putative TATA box sequences are indicated upstream of *Aux311* exon 1. The scale is in base pairs (bp) and restriction sites are represented by single letters: B, *Bam*HI; E, *Eco*RI; S, *Sal*I; Ss, *Sst*I.

We isolated two clones by constructing a maize genomic library in λEMBL3 and screening it with cAux4 DNA. Subcloning and sequencing the hybridising regions has shown one clone, designated λAux311, to contain a complete gene, whereas the other, designated λAux381, is incomplete. Comparison of genomic with cDNA sequences splits the known transcribed region into five exons separated by two short and two long introns. The structures of the two genes *Aux311* and *Aux381* are shown in Fig. 4.

The sequences of the two genes are very highly conserved and they are both of the cAux4 type. Only two single nucleotide changes, 2 bp apart in a highly repetitive sequence, distinguish *Aux311* from cAux4; one change is silent but the other converts the glycine residue at position −26 to alanine (see above). Since even the untranslated regions of cAux4 are completely identical to the cognate sequences in *Aux311* it seems likely that the sequence differences arise from cloning or sequencing artefacts and that the mRNA from which cAux4 was derived was a transcript of the *Aux311* gene. The major single difference between the *Aux311* and *Aux381* genes is the length of the first intron, which we estimate to be in excess of 5.2 kb in *Aux381* (see below). Intron 4 also differs significantly in length between the two genes due to a number of discrete sequence insertions in *Aux311* relative to *Aux381*; intron sequences are otherwise very highly conserved. Some of these data are summarised in Table 1.

Searching for the promoter of *Aux311*

Subcloning of λAux311 restriction fragments for sequencing was initially restricted to those hybridising to cAux4 DNA. As can be seen in Fig. 3 all five

Table 1. *Comparison of DNA sequences for ABP genes Aux311 and Aux381 (excluding exon 1 and intron 1)*

Region	Exon 2	Intron 2	Exon 3	Intron 3	Exon 4	Intron 4	Exon 5 Coding	Exon 5 Non-coding
Length *Aux311* (bp)	107	340	201	171	68	1522	103	156
Length *Aux381* (bp)	107	343	201	171	6	1209	103	157
Single nucleotide substitutions	2	9	3	4	0	27	0	4
Substitutions causing amino acid changes	2	–	0	–	0	–	0	–
Single nucleotide insertions (+) and deletions (−)*	0	+4 −1	0	0	0	+8 −13	0	+1
Multiple nucleotide insertions (+) and deletions (−)*	0	0	0	0	0	−308	0	−6

*The total number of base pairs involved in insertions/deletions is quoted for *Aux381* relative to *Aux311*.

exons are contained within a 4.5 kb *Sal*I fragment, but the *Sal*I site lies only 74 bp upstream of the cAux4 sequence start and only 60 bp upstream of the 5' end of the longest cDNA reported (Inohara *et al.* 1989). To extend the sequence in the 5' direction the 1.7 kb *Bam*HI fragment shown in Fig. 3 was subcloned and sequenced.

The surprising outcome of this work was that no TATA box indicating the promoter region could be identified in the immediate vicinity of sequences present in cDNA clones considered to be 'full length' (Inohara *et al.* 1989; Hesse *et al.* 1989; Tillman *et al.* 1989). The closest upstream sequences that are significantly related to the TATA box consensus sequence are TATAAAT at −310 (with respect to the translation initiation codon) and TATATGC at −247. The position of these sequences is indicated in Fig. 4. Attempts (data not presented) to determine the transcription initiation site by conventional means have yielded inconsistent results which may, however, indicate the presence of an additional intron (and exon). S_1 mapping with a DNA fragment labelled within Exon 1 produced a protected fragment extending just upstream of the 5' end of the longest known transcribed sequence (Inohara *et al.* 1989). Immediately upstream of this sequence is an AG dinucleotide that could represent the 3' border of an intron. Several GT dinucleotides further upstream could represent the 5' border of such an intron. Primer extension from sites within Exon 1 has yielded a range of products indicating that the 5' untranslated region of the mRNA may be up to 35 bp longer than indicated by analysis of cDNA clones. We are currently attempting to resolve this problem by amplification and sequencing of reverse transcription products from the 5' end of ABP mRNA (Frohman, 1990).

Length of intron 1 in *Aux381*

Hybridisation of cAux4 DNA to λAux381 was confined to the 4.5 kb *Bam*HI–*Sst*I restriction fragment shown in Figs 4 and 5B, but exon 1 was not located within the sequence of this fragment. Exon 1 was therefore not present in λAux381 DNA even though a further 3.7 kb of genomic DNA was present between the *Bam*HI site and the vector arm (Fig. 5B), and the first intron of *Aux381* appeared to be very much longer than the cognate intron in *Aux311*. A trivial explanation of this result was that genomic sequences contiguous with exons 2–5 ceased somewhere upstream of exon 2 at the ligation point of two separate *Sau*3A partial digestion products derived from non-adjacent regions of the genome. A Southern blotting experiment comparing the hybridisation of a cAux4 probe to restriction digests of λAux381 and maize genomic DNA was used to eliminate this latter possibility, and the results are shown in Fig. 5A. Since ABP is encoded by a small multigene family restriction fragments with a range of sizes hybridised to the cDNA probe; only single bands hybridised in the λDNA digests, but in each case they were of identical mobility to bands in the respective genomic digests (7.7 kb in *Sst*I digests and 4.5 kb in *Bam*HI+*Sst*I digests). Clearly the 7.7 kb *Sst*I restriction fragment of λAux381 is an authentic genomic fragment and the

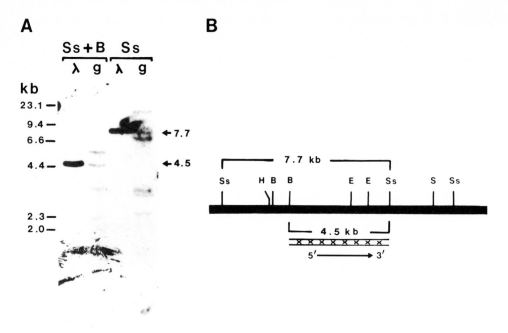

Fig. 5. ABP genes in the maize genome and λAux381. (A) DNA blot hybridisation. Tracks contained 10 μg maize genomic DNA (g) or 100 pg λAux381 DNA (λ), digested with *Sst*I (Ss) or *Sst*I and *Bam*HI (Ss+B). The probe was cAux4 DNA (complete cDNA insert), and hybridisation was carried out at 65 °C. The scale is in kilobase pairs and hybridising bands of 7.7 and 4.5 kb are indicated by arrows. (B) Restriction map of the genomic insert in λAux381 showing the region hybridising to cAux4 DNA (open bar with crosses), with the arrow indicating the direction of transcription. Restriction sites: B, *Bam*HI; E, *Eco*RI; H, *Hin*dIII; S, *Sal*I; Ss, *Sst*I.

minimum size of intron 1 of Aux381 is defined by the distance between the 5′ border of exon 2 and the upstream *Sst*I site i.e. 5.2 kb; at least three times the length of intron 1 of *Aux311*. As far as the determined sequence is concerned this difference in length is explained by extensive insertions with respect to the *Aux311* intron 1 sequence; the entire sequence of *Aux311* intron 1 from nucleotide 1819 to its 3′ border at nucleotide 2625 (see Fig. 4) is contained between the *Bam*HI site and the 3′ border of *Aux381* intron 1, split into four blocks of very highly conserved sequence.

Homologous ABP genes in monocot and dicot plants

Southern blot hybridisations have been carried out with genomic DNA isolated from maize and a number of dicot plants. The results of some of these experiments are shown in Figs 5A, 6 and 7. Consistent with the discovery of sequence heterogeneity among cDNA and genomic clones for ABP genes is the observation of multiple hybridising bands in all restriction digests of maize DNA probed either with the full-length cAux4 DNA (Fig. 5A) or with probes covering only the 5′- or

3'-halves of it (Fig. 6). It is not possible to equate band numbers with gene numbers since the blotting efficiency of bands of different sizes varies (leading to differences in band intensities) and some restriction enzymes cleave some of the genes into two or more hybridising fragments. We estimate that there may be in the region of 6–10 ABP genes per genome, at least in the F_1 hybrid variety used in our experiments.

Fig. 6 shows that maize cDNA probes identify homologous sequences in Southern blots of DNA from the dicotyledonous plants spinach and strawberry, and hybridisation has also been observed to *Arabidopsis* genomic DNA (data not shown). No significant hybridisation was observed to tobacco DNA, a result that has been observed consistently for two varieties of this plant and using the full-length cAux4 probe. This is a surprising result in view of the fact that protein immunologically related to maize ABP is clearly present in this species (Barbier-Brygoo *et al.* 1989, 1990*a,b*). It would appear that presumed conservation of epitope structure is not reflected by conservation of gene sequence in tobacco.

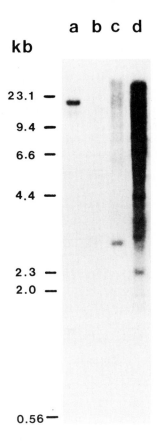

Fig. 6. ABP genes in the genomes of several plants. Tracks contained 5–10 μg total DNA from strawberry (a), tobacco (b), spinach (c) and maize (d) digested with *Bam*HI. The probe was the 5' half of cAux4 DNA (5' untranslated region to codon 111), and hybridisation was carried out at 60 °C.

A single locus for ABP in strawberry, but related sequences are repetitive

Initial Southern blotting experiments to look for maize ABP-related sequences in dicots utilised the full-length cAux4 DNA as probe, rather than the 5'-half probe used in the experiments shown in Fig. 6. The results for strawberry, shown in Fig. 7A, indicated that ABP could be encoded by a sizeable multigene family in this plant, with many of the members having a similar structure. The latter suggestion arose because, even in the incomplete *Bam*HI digest shown in Fig. 7A, strong hybridisation was observed to a low relative molecular mass fragment (<0.56 kb) that could have emanated from conserved *Bam*HI sites within each gene.

To clone a strawberry ABP gene a genomic library was constructed in λGEM-11 and approximately 100 000 independent recombinant plaques were blotted onto duplicate filters; hybridisations were carried out with non-overlapping maize ABP probes covering the 5'- and 3'-halves of cAux4. This approach was adopted to enable us to distinguish between plaques with a high probability of containing a

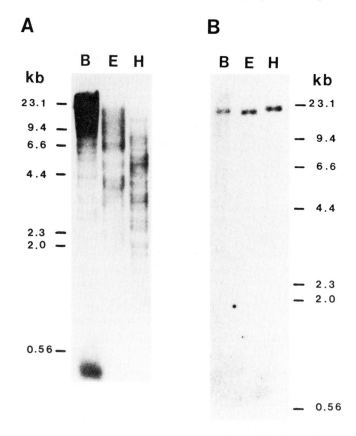

Fig. 7. Homology to maize ABP sequences in the strawberry genome.
6 μg total strawberry DNA was digested with *Bam*HI (B), *Eco*RI (E) or *Hind*III (H). Hybridisation was at 60 °C to the complete cDNA insert of cAux4 (panel A) or to the 5' half (5' untranslated region to codon 111) of cAux4 DNA (panel B).

complete ABP gene (signals at identical positions on the duplicate autoradio-graphs) and artefacts caused by non-specific binding of probe DNA to the filters (signals present only on one or other of the duplicate autoradiographs). Plaques containing grossly incomplete genes with homology to only one half of the maize cDNA could be discarded as 'artefacts'.

Hybridisations with the two probes gave vastly different results: only two relatively weak signals were obtained with the 5'-half probe, whereas the 3'-half probe yielded a very large number of strong plaque-like signals (>1000), together with weak signals in identical positions to those obtained with the 5'-half probe. Both weakly-hybridising plaques were purified and their hybridisation with equal intensity to each half of the cAux4 DNA confirmed. Additionally 4 plaques were chosen arbitrarily and purified; they hybridise only to the 3'-half of cAux4.

The results of the cloning exercise suggested that ABP genes may indeed not be in particularly high copy number in strawberry, but rather that a sequence contained within the 3'-half of maize ABP mRNA is highly repeated within the strawberry genome. To investigate this, Southern blot hybridisation to strawberry DNA was repeated using just the 5' half of cAux4 as probe; the results are shown in Fig. 6B. Single hybridising fragments can be seen in each of the three restriction digests leading to the conclusion that there is only a single gene for ABP per haploid strawberry genome or that if there are more than one gene they are tightly linked at a single locus. A new interpretation of the small strongly-hybridising *Bam*HI fragment is that it represents the monomeric repeat unit of a tandemly-repeated sequence containing a single *Bam*HI site and a region (of undefined length) with homology to maize ABP mRNA. Analysis of the four repeat sequence clones should provide definitive information.

Structure of a strawberry ABP gene

Preliminary analysis of one of the strawberry clones hybridising to both halves of cAux4 has been carried out; a crude restriction map has been constructed and the hybridising region located. The data (not shown) are consistent with an authentic ABP gene since the regions hybridising to the two half-cDNA probes are contiguous.

Discussion

Strong evidence supporting the rôle of ABP as an auxin receptor has accumulated since the protein was purified and antibodies raised against it became available for physiological studies. The detailed characterisation of our polyclonal serum has shown the most antigenic region of ABP to be located roughly in the middle of the primary sequence, flanking the glycosylation site. The three epitopes identified are also represented in antisera raised against the maize protein in other laboratories, although one is conserved in the ABP of the monocot, barnyard grass and another is conserved in the dicot, mung bean. It is perhaps surprising that none of these major epitopes features among our collection of five monoclonal antibodies, which have all been mapped to alternative locations. The epitopes recognised by MAC 256 and MAC 259 may be identical, since both map within $1 \times 10^3 M_r$ of the carboxy terminus but neither could be pinpointed with the kit, implying involvement of the free carboxylic acid group on the C-terminal leucine residue. The carboxy terminus, as defined by MAC 256 and 259, is conformationally active, apparently becoming inaccessible to these antibodies when ligands bind to ABP. Conformational change is certainly a feature that may be expected of a receptor on ligand binding, but our results cannot be taken to imply that the carboxy terminus plays any functional rôle in the interaction.

Although evidence accumulates that a functional population of auxin receptors is exposed on the outer face of the plasmalemma, the presence of approximately 95 % of the ABP within the ER of maize can perhaps be explained by the KDEL motif (Hesse et al. 1989; Inohara et al. 1989; Tillmann et al. 1989) but its function there remains unexplained. The relationship between plasmalemma and ER forms of ABP also needs to be clarified. One possibility is that the ER acts as a store of ABP molecules that are destined to find their way to the plasmalemma by a mechanism involving 'leakage' or specific proteolytic cleavage of the KDEL. Alternatively there may be distinct plasmalemma and ER isoforms of ABP, in which case the plasmalemma form would be expected to lack the KDEL, but no KDEL-lacking form is predicted from the several cDNA and genomic clone sequences now available. Indeed in all but cAux1 the variation is extremely restricted: Inohara et al. (1989) reported serine or asparagine as alternatives at residue 103, and we predict glycine at residue 11 and glutamine at residue 16 in the protein encoded by *Aux381* in place of the aspartic acid and proline residues predicted at these positions in all the other clones. The protein encoded by cAux1 is very different from all the others and we feel justified in assigning it to a separate class. Nevertheless the extended sequence of this clone still predicts a KDEL, although whether it could function as an ER-retention signal is not known since we have not been able to extend the sequence of this cDNA to the 5′ end and consequently do not know whether a signal peptide is encoded upstream of the mature protein. Even a very short extension in the 5′ direction would also have allowed us to compare the predicted amino-terminal sequence of the protein with that of isoform ABP2 determined by Hesse et al. (1989), for which homology with the major class sequence lies in the range 76–83 % over the first 29 amino acids.

In previous experiments with protein conjugates of NAA derivatives we concluded that long-term (growth) effects could be induced by impermeant auxin analogues, provided that their size did not preclude their passage through the cell wall (Venis and Napier, 1990). Electrophysiological data now confirm this conclusion (Venis et al. 1990) and extend it to rapid responses (less than one minute). Auxin-induced membrane hyperpolarisation operates *via* the plasma-lemma H^+-pumping ATPase (Barbier-Brygoo et al. 1989), which, according to the 'acid growth theory' (Hager et al. 1971) would also be responsible for the cell wall acidification and loosening necessary for cell elongation. An auxin receptor location in the plasmalemma would seem to be consistent with these results, but auxin responses extend beyond the directly physiological, and a range of genes is known to be regulated by this hormone (see Key, 1989 and Lazarus, 1991 for reviews). No information is yet available linking auxin perception with genetic responses, so the location of the receptor involved remains an open question. A further subject of contemporary speculation is the mechanism by which the information of auxin binding at an external or internal membrane site is transduced to the responsive genes.

The isolation of clones encoding ABP has yielded important information on the primary structure of the protein, but genetic analysis has not yet proceeded far beyond the purely descriptive. The small multigene family in maize defined by genomic Southern blots and sequencing of clones appears, except in the case of cAux1, to be extremely highly conserved with minimal variation in the predicted protein sequences. We do not know if the protein predicted by cAux1 has any different functional rôle or subcellular location from the major class of ABP proteins but it should be possible to raise class-specific antibodies from peptides synthesised to maximise this variation. Such antibodies should at least be of use in studies involving denatured forms of the proteins (such as western blots), although they may be ineffective in studies of the native protein especially *in situ* (Laver *et al.* 1990).

Both of the genomic clones we have analysed encode proteins of the major class of ABP, and the only significant variation detected between them is located in the length of two of the four introns. This length variation arises exclusively from discrete insertion/deletion events that have occurred within very highly conserved sequences. Whether either of these features (conservation of sequence and variation in length) are of any functional significance for the expression of the genes is not known, and comparisons with other plant genes are not helpful. For some plant genes, such as those encoding actin, the positions of the introns are conserved but the sequences are not (Shah *et al.* 1983). In contrast, almost complete similarity was observed in the first intron sequences of two pairs of wheat α-amylase class 2 genes; some localised similarities were detected between the pairs and also with a fifth member of this multigene family (Huttly *et al.* 1988). Plant introns have also been shown to exert an influence on the level of gene expression in transiently and stably transformed plants. Callis *et al.* (1987) obtained enhanced levels of expression of several genes combined with various promoters by inserting maize *Adh1* intron 1 into their constructs, however removal of introns from a *Petunia rbcS* gene expressed in transgenic tobacco resulted in a reduction of the steady-state mRNA level compared to the intact (intron-containing) gene (Dean *et al.* 1989).

Introns have often been found to split coding sequences roughly into functional domains, and this is certainly true of the first intron of the maize ABP genes, which interrupts the fifth codon of the mature protein. Such inexact coding separation of the signal peptide from the mature protein is reminiscent of wheat and barley α-amylase genes (Baulcombe *et al.* 1987; Knox *et al.* 1987; Whittier *et al.* 1987), and also of the separation of mature protein from the transit peptide of ribulose 1,5-bisphosphate carboxylase small subunit genes (Broglie *et al.* 1983).

An unusual feature of the *Aux311* gene is the absence of any TATA box sequence immediately upstream of the 5' mRNA sequences determined from cDNA clones. We are currently investigating the possibility that an additional intron interrupts the region of the gene encoding the 5' untranslated leader. Such an intron position would not be novel in plant genes; precedents include phytochrome genes from oat (Hershey *et al.* 1987) and pea (Sato, 1988), and soybean actin genes (Pearson and Meagher, 1990).

Southern blot hybridisations with maize ABP cDNA probes has demonstrated the likely presence of homologous genes in some dicotyledonous plants including *Arabidopsis*, spinach and strawberry, but tobacco proved to be exceptional in that no hybridisation was observed. This is surprising in view of the fact that all the electrophysiological work of the type reported in this paper has been carried out with tobacco protoplasts. Conservation of ABP epitopes between maize and tobacco is implicit in the work of Barbier-Brygoo *et al.* (1989, 1990*a,b*) with maize polyclonal antibodies, so conserved epitopes may be encoded by short homologous regions in otherwise widely diverged genes or the genes are totally dissimilar and related epitopes are a feature of the three-dimensional structures of proteins with unrelated primary sequences. Clearly maize cDNAs are not suitable as probes for tobacco ABP genes.

Manipulation of auxin sensitivity of cells (protoplasts) with purified ABP and anti-ABP serum has been achieved *in vitro* (Barbier-Brygoo *et al.* 1990*a,b*), and we would like to emulate these experiments *in vivo* by over- and under-expressing ABP genes in transgenic plants. Tobacco is an obvious candidate for over-expression using maize genes, but the apparent lack of homology between the incoming and endogenous genes precludes an antisense approach to under-expression. Strawberry presents an attractive alternative to tobacco for such studies since it can be transformed and regenerated (James *et al.* 1990), and easily observed functions such as fruit ripening are regulated by auxin (Given *et al.* 1988). The observation that there may be only a single ABP gene in strawberry could be of advantage, and we have made progress towards carrying out these experiments by isolating strawberry genomic clones with homology to the maize ABP genes.

Support for this work was received from the Agricultural and Food Research Council (research grants to C.M.L.) and under the Biotechnology Action Programme of the European Economic Community (to M.A.V.).

References

BARBIER-BRYGOO, H., EPHRITIKHINE, G., KLÄMBT, D., GHISLAIN, M. AND GUERN, J. (1989). Functional evidence for an auxin receptor at the plasmalemma of tobacco mesophyll protoplasts. *Proc. natn. Acad. Sci. U.S.A.* **86**, 891–895.

BARBIER-BRYGOO, H., EPHRITIKHINE, G., SHEN, W. H., DELBARRE, A., KLÄMBT, D. AND GUERN, J. (1990*a*). Characterization and modulation of the sensitivity of plant protoplasts to auxin. In

Transducing Pathways: Activation and Desensitization (ed. T. H. Konijn, M. D. Houslay and P. J. M. van Haastert), pp. 231–244. New York: Springer Verlag.

BARBIER-BRYGOO, H., GUERN, J., EPHRITIKHINE, G., SHEN, W. H., MAUREL, C. AND KLÄMBT, D. (1990*b*). The sensitivity of plant protoplasts to auxins: modulation of receptors at the plasmalemma. In *Plant Gene Transfer* (ed. C. Lamb and R. Beachy), pp. 165–173. New York: Alan R. Liss.

BAULCOMBE, D. C., HUTTLY, A. K., MARTIENSSEN, R. A., BARKER, R. F. AND JARVIS, M. G. (1987). A novel wheat α-amylase gene (α-*Amy3*). *Molec. gen. Genet.* **209**, 33–40.

BROGLIE, R., CORUZZI, G., LAMPPA, G., KEITH, B. AND CHUA, N.-H. (1983). Structural analysis of nuclear genes coding for the precursor to the small subunit of wheat ribulose-1, -5-bisphosphate carboxylase. *Biotechnology* **1**, 55–61.

CALLIS, J., FROMM, M. AND WALBOT, V. (1987). Introns increase gene expression in cultured maize cells. *Genes and Dev.* **1**, 1183–1200.

DAVIES, P. J. (1987). *Plant Hormones and their Role in Plant Growth and Development.* Dordrecht: Martinus Nijhoff, Kluwer Academie.

DEAN, C., FAVREAU, M., BOND-NUTTER, D., BEDBROOK, J. AND DUNSMUIR, P. (1989). Sequences downstream of translation start regulate quantitative expression of two petunia *rbcS* genes. *Pl. Cell* **1**, 201–208.

FROHMAN, M. A. (1990). RACE: Rapid amplification of cDNA ends. In *PCR Protocols: A guide to methods and applications* (ed. M. A. Innes, D. H. Gelfand, J. J. Sninsky and T. J. White), pp. 28–38. London: Academic Press.

GIVEN, N. K., VENIS, M. A. AND GRIERSON, D. (1988). Hormonal regulation of ripening in the strawberry, a non-climacteric fruit. *Planta* **174**, 402–406.

HAGER, A., MENZEL, H. AND KRAUSS, A. (1971). Versuche und Hypothese zur Primärwirkung des Auxins beim Streckungswachstum. *Planta* **100**, 47–75.

HERSHEY, H. P., BARKER, R. F., IDLER, K. B., MURRAY, M. G. AND QUAIL, P. H. (1987). Nucleotide sequence and characterization of a gene encoding the phytochrome polypeptide from *Avena*. *Gene* **61**, 339–348.

HERTEL, R., THOMPSON, K.-ST. AND RUSSO, V. E. A. (1972). *In vitro* auxin binding to particulate cell fractions from corn coleoptiles. *Planta* **107**, 325–340.

HESSE, T., FELDWISCH, J., BALSHÜSEMANN, D., BAUW, G., PUYPE, M., VANDEKERCKHOVE, J., LÖBLER, M., KLÄMBT, D., SCHELL, J. AND PALME, K. (1989). Molecular cloning and structural analysis of a gene from *Zea mays* (L.) coding for a putative receptor for the plant hormone auxin. *EMBO J.* **8**, 2453–2461.

HUTTLY, A. K., MARTIENSSEN, R. A. AND BAULCOMBE, D. C. (1988). Sequence heterogeneity and differential expression of the α-*Amy2* gene family in wheat. *Molec. gen. Genet.* **214**, 232–240.

INOHARA, N., SHIMOMURA, S., FUKUI, T. AND FUTAI, M. (1989). Auxin-binding protein located in the endoplasmic reticulum of maize shoots: molecular cloning and complete primary structure. *Proc. natn. Acad. Sci. U.S.A.* **86**, 3564–3568.

JAMES, D. J., PASSEY, A. J. AND BARBARA, D. J. (1990). *Agrobacterium*-mediated transformation of the cultivated strawberry (*Fragaria×Anannassa* Duch.) using disarmed binary vectors. *Plant Sci.* **69**, 79–94.

KEY, J. L. (1989). Modulation of gene expression by auxin. *Bioessays* **11**, 52–58.

KNOX, C. A. P., SONTHAYANON, B., CHANDRA, G. R. AND MUTHUKRISHNAN, S. (1987). Structure and organisation of two divergent α-amylase genes from barley. *Pl. molec. Biol.* **9**, 3–17.

LAVER, W. G., AIR, G. M., WEBSTER, R. G. AND SMITH-GILL, S. J. (1990). Epitopes on protein antigens: misconceptions and realities. *Cell* **61**, 553–556.

LAZARUS, C. M. (1991). Hormonal regulation of plant gene expression. In *Developmental Regulation of Plant Gene Expression* (ed. D. Grierson), pp. 42–74. Glasgow: Blackie.

LÖBLER, M. AND KLÄMBT, D. (1985). Auxin-binding protein from coleoptile membranes of corn (*Zea mays* L.). *J. biol. Chem.* **260**, 9848–9853.

MUNRO, S. AND PELHAM, H. R. B. (1987). A C-terminal signal prevents secretion of luminal ER proteins. *Cell* **48**, 899–907.

MURRAY, V. (1989). Improved double-stranded DNA sequencing using the linear polymerase chain reaction. *Nucl. Acids Res.* **17**, 8889.

NAPIER, R. M. AND VENIS, M. A. (1990). Monoclonal antibodies detect an auxin-induced conformational change in the maize auxin-binding protein. *Planta* **182**, 313–318.

NAPIER, R. M., VENIS, M. A., BOLTON, M. A., RICHARDSON, L. I. AND BUTCHER, G. W. (1988). Preparation and characterisation of monoclonal and polyclonal antibodies to maize membrane auxin-binding protein. *Planta* **176**, 519–526.

PEARSON, L. AND MEAGHER, R. B. (1990). Diverse soybean transcripts contain a large intron in the 5' untranslated leader: structural similarity to vertebrate muscle actin genes. *Pl. molec. Biol.* **14**, 513–526.

RAY, P. M., DOHRMANN, U. AND HERTEL, R. (1977). Specificity of auxin-binding sites on maize coleoptile membranes as possible receptors for auxin. *Pl. Physiol.* **60**, 585–591.

SATO, N. (1988). Nucleotide sequence and expression of the phytochrome gene in *Pisum sativum*: Differential regulation by light of multiple transcripts. *Pl. molec. Biol.* **11**, 697–710.

SHAH, D. M., HIGHTOWER, R. C. AND MEAGHER, R. B. (1983). Genes encoding actin in higher plants: intron positions are highly conserved but the coding sequences are not. *J. molec. appl. Genet.* **2**, 111–126.

SHIMOMURA, S., SOTOBAYASHI, T., FUTAI, M. AND FUKUI, T. (1986). Purification and properties of an auxin-binding protein from maize shoot membranes. *J. Biochem.* **99**, 1513–1524.

TILLMANN, U., VIOLA, G., KAYSER, B., SIEMEISTER, G., HESSE, T., PALME, K., LÖBLER, M. AND KLÄMBT, D. (1989). cDNA clones of the auxin-binding protein from corn coleoptiles (*Zea mays* L.): isolation and characterization by immunological methods. *EMBO J.* **8**, 2463–2467.

TREWAVAS, A. J. (1981). How do plant growth substances work? *Pl. Cell Environ.* **4**, 203–228.

VENIS, M. A. AND NAPIER, R. M. (1990). Characterisation of auxin receptors. In *Hormone perception and signal transduction in plants* (ed. J. Roberts, C. Kirk and M. Venis), pp. 55–65. Cambridge: Company of Biologists.

VENIS, M. A. AND THOMAS, E. W. (1990). Synthesis and auxin activity of 5-substituted 1-naphthaleneacetic acids. *Phytochemistry* **29**, 381–383.

VENIS, M. A., THOMAS, E. W., BARBIER-BRYGOO, H., EPHRITIKHINE, G. AND GUERN, J. (1990). Impermeant auxin analogues have auxin activity. *Planta* **182**, 232–235.

VENIS, M. A. AND WATSON, P. J. (1978). Naturally occurring modifiers of auxin-receptor interaction in corn: identification as benzoxazolinones. *Planta* **142**, 103–107.

WALTON, J. D. AND RAY, P. M. (1981). Evidence for receptor function of auxin binding sites in maize. Red light inhibition of mesocotyl elongation and auxin binding. *Pl. Physiol.* **68**, 1334–1338.

WHITTIER, R. F., DEAN, D. A. AND ROGERS, J. C. (1987). Nucleotide sequence analysis of α-amylase and thiol protease genes that are hormonally regulated in barley aleurone cells. *Nucl. Acids Res.* **15**, 2515–2535.

Printed in Great Britain © *Society for Experimental Biology 1991*

GENETIC ANALYSIS OF ETHYLENE RESPONSES IN *ARABIDOPSIS THALIANA*

ANTHONY B. BLEECKER

Department of Botany, University of Wisconsin-Madison, 132 Birge Hall, 430 Lincoln Drive, Madison, Wisconsin 53706, USA

Summary

The plant hormone ethylene mediates a number of developmental processes and responses to environmental stress in higher plants. Our research efforts over the last three years have been focused on developing an understanding of the molecular basis of ethylene action in plants. To this end, we have isolated mutants in *Arabidopsis thaliana* with altered responses to ethylene. One such mutant, designated *etr*, shows no measurable responses to ethylene and has reduced ethylene binding in leaf tissue indicating that the mutation may directly affect the ethylene receptor. We have genetically mapped the *etr* mutation and by chromosome walking have isolated an 18 kb fragment of genomic DNA that contains the mutant gene. Sequence analysis of cDNAs which map to the 18 kb fragment has produced a candidate for the ETR gene which codes for a putative transmembrane protein kinase. Sequence analysis indicates a domain composed of 9 copies of a 23 amino acid leucine-rich repeat unit and a domain containing a serine/threonine type protein kinase. These two domains are separated by a single 24 amino acid hydrophobic domain. A model is presented that describes the possible mechanism of action of the protein kinase with respect to ethylene-mediated responses in plants.

Introduction

The plant hormone ethylene is an endogenous regulator of growth and development in higher plants. Increases in the level of ethylene influence many developmental processes, ranging from seed germination and seedling growth (Abeles, 1973; Abeles, 1986; Eisinger, 1983), to such terminal events as abscission (Abeles, 1973; Lloyd and Seagull, 1985), fruit ripening (Burg and Burg, 1967; Grierson *et al.* 1986; McGlasson *et al.* 1978) and senescence (Abeles, 1973; Davies and Grierson, 1989). A number of environmental stresses including oxygen deficiency, wounding and pathogen invasion stimulate ethylene synthesis (Yang and Hoffman, 1984). Stress-induced ethylene elicits adaptive changes in plant development. For example, wounding and pathogen invasion may result in ethylene-mediated acceleration of senescence and abscission of infected organs

Key words: ethylene, chromosome walking, RFLP mapping, *Arabidopsis* and in the induction of specific defense proteins such as chitinase, glucanases and hydroxyproline rich proteins (Boller *et al.* 1983).

A number of advances have occurred over the last decade in our understanding of the molecular processes involved in the regulation of ethylene synthesis and ethylene-mediated responses. The biosynthetic pathway (see Fig. 1) in all higher plants examined begins with the conversion of S-adenosylmethionine (SAM) to 1-aminocyclopropane-1-carboxylic acid (ACC) catalyzed by the enzyme ACC synthase (Bleecker *et al.* 1986; Yang and Hoffman, 1984). The ACC is subsequently oxidized to ethylene by the ethylene forming enzyme (EFE) system (Yang and Hoffman, 1984). ACC synthase appears to be the rate limiting enzyme in ethylene biosynthesis in most systems that have been studied (Yang and Hoffman, 1984). This enzyme has been purified to homogeneity (Bleecker *et al.* 1986; Sato and Theologis, 1989; Tsai *et al.* 1988) and the cloning of the gene has recently been reported (Sato and Theologis, 1989). Studies at the protein and nucleic acid levels indicate that developmental and environmental induction of ethylene synthesis involves transcriptional activation of the ACC synthase gene (Bleecker *et al.* 1988*b*; Sato and Theologis, 1989).

The molecular mechanisms that are involved in the ultimate responses to ethylene have also been the subject of a number of recent studies. Regulation of cell expansion appears to involve an ethylene-mediated 90° shift in the orientation of the cortical microtubule array that results in subsequent reorientation of cellulose microfibril deposition into the cell wall (Eisinger, 1983; Lloyd and Seagull, 1985). This process causes a change in the direction of cell expansion from a primarily longitudinal to a more lateral one resulting in an inhibition of elongation of developing shoots and roots. Other responses to ethylene clearly involve transcriptional activation of specific genes. For example, fruit ripening in many species is the result of ethylene-mediated changes in the pattern of gene expression within the fruit (Grierson *et al.* 1986). In mature green fruit, applied ethylene can activate ripening-specific genes within 15 min (Lincoln *et al.* 1987). Ethylene-induced changes in gene expression also occur in developing seedlings (Broglie *et al.* 1986; Zurfluh *et al.* 1982), in abscission zones (Sexton *et al.* 1985), and in leaves (Broglie *et al.* 1986; Davies and Grierson, 1989). The *cis* and *trans* acting genetic elements that are involved in ethylene-mediated transcriptional activation of specific genes are being studied by a number of groups (Broglie *et al.* 1989; Cordes *et al.* 1989; Holdsworth and Laties, 1989).

Although both the synthesis of ethylene and the molecular basis for some of the ultimate responses to ethylene are becoming much better understood, we still know very little about the events that occur in between. That is, how do plant tissues perceive the level of ethylene and how is this signal quantitatively transduced into the variety of well characterized responses? Responses to ethylene are usually related to the log ethylene concentration over a range of two to three orders of magnitude, suggesting a simple ligand–receptor interaction analogous to many animal hormone responses (Goth, 1981; Kende and Gardener, 1976). The requirements for high affinity and specificity for an ethylene receptor have led to the suggestion that ethylene–receptor interactions involve binding to a transition metal complex (Burg and Burg, 1967). In support of this idea, various agonists and antagonists of ethylene action form complexes with silver ions in the same order they mimic or inhibit ethylene action. High affinity binding sites for ethylene have been detected in plant tissues, both *in vivo* (Sisler, 1979) and *in vitro* (Sisler, 1980; Thomas *et al.* 1985). Although the K_ds for binding were consistent with the concentration of ethylene required for half saturation of the response, no direct connection has ever been established between the detectable binding and any physiological response.

Direct biochemical approaches to unravelling the mechanisms of ethylene action have not been successful to date for several reasons. Given the low concentration of ethylene required to elicit the responses, it is reasonable to assume that the initial events in ethylene responses occur at very low levels relative to the well studied major biochemical pathways in plants. The earliest measurable responses to ethylene occur only after 15 min exposure to the compound (Abeles, 1973; Lincoln *et al.* 1987). In this amount of time, any number of undetected biochemical and/or biophysical events could have occurred. Without any clue as to what these early events might be, we cannot bring the powerful tools of biochemistry to bear on this problem.

As an alternative to the direct biochemical approach, the genetic approach to investigating complicated cellular processes has been very successful, particularly in understanding the complex regulatory circuits involved in the regulation of gene expression in microorganisms (Miller and Reznikoff, 1980; Oshima, 1982). This approach involves generating and analyzing mutations affecting the processes under study. The genetic approach had not previously been utilized to any degree in the study of ethylene action. One spontaneous tomato mutant, *diageotropica*, which requires ethylene for normal growth, has been described (Zobel and Roberts, 1978). However, recent reports suggest that the genetic lesion in *diageotropica* may not be directly related to the ethylene regulatory system (Kelly and Bradford, 1986).

Recent advances in molecular genetics now make it possible to clone mutant genes based on the phenotypic segregation of the mutation. Cloning of such genes can provide a number of tools for studying the biochemical nature of the gene product. Primary protein sequence data derived from the clone may indicate both the physical properties and possible functional domains of the gene product. The DNA clones and antibodies raised against the gene product can be used as probes to localize the normal gene product at both the tissue and intracellular level.

Transformation of plants with the mutant and normal genes could provide valuable insight into the function of the gene product. One of the most promising plant systems being developed for cloning genes based on mutant phenotype is the cruciferous plant, *Arabidopsis thaliana*. Having a small genome with little repetitive DNA (Leutwiler *et al.* 1984), and a good genetic map (Koornneef *et al.* 1983), *Arabidopsis* is currently being used to develop several cloning strategies, including RFLP (restriction fragment length polymorphism) mapping/chromosome walking, shotgun cloning, and transposon mutagenesis. Two separate RFLP linkage maps, comprising a total of over 200 markers, have recently been constructed (Chang *et al.* 1988; Nam *et al.* 1989).

Isolation of ethylene response mutants

We have recently undertaken a systematic genetic approach to studying ethylene action in *Arabidopsis*. The screen for ethylene response mutants in *Arabidopsis* involves the 'triple response' of developing seedlings, first described for pea epicotyls by Neljubov (see Abeles, 1973). The triple response in pea includes an ethylene-mediated decrease in epicotyl elongation, a concomitant increase in radial growth, and an alteration in the geotropic orientation of growth of the developing epicotyl (Abeles, 1973). The biological significance of the triple response was elegantly demonstrated by Goeschel *et al.* (1966). Mechanical resistance to seedling growth induces ethylene biosynthesis. The resulting alterations in growth habit mediated by ethylene allow the seedling to grow through soils having varying degrees of compactness.

The ethylene-mediated inhibition of hypocotyl elongation provides a very effective screen for mutants with altered responses to ethylene. In preliminary experiments with *Arabidopsis*, hypocotyl elongation was inhibited by up to 70 % by $1 \mu l \, l^{-1}$ ethylene after 4 days of growth in the dark (Bleecker *et al.* 1988a). Thus, mutants lacking or having reduced responsiveness to ethylene can be readily identified in a large population of mutagenized seed. Populations of the M_2 generation from EMS (ethyl methanesulphonate) mutagenized seed were plated on agar in 15 cm Petri dishes (5000 seeds per plate) and placed in a chamber through which $5 \mu l \, l^{-1}$ ethylene was circulated. Seedlings with hypocotyls showing greater than 1 cm of growth after 4 days were selected as potential ethylene response mutants. These seedlings were transplanted in soil, allowed to develop, and M_3 seed collected. Mutant lines that showed inheritance of the ethylene-resistant phenotype were analyzed for alterations in other ethylene responses such as accelerated leaf senescence. One class of dominant mutations, which map to a locus designated *ETR*, has been extensively characterized. These mutants and a second recessive complementation group are affected in a range of ethylene responses throughout the plant indicating that the mutations occur early in the signal transduction chain. A third group of mutants has been identified that is affected in hypocotyl elongation but not in ethylene-mediated acceleration of senescence indicating that the lesion in these mutants occurs later in the transduction chain specific to ethylene effects on axis elongation. Using a different screen, a fourth group of mutations has been identified that show normal ethylene responses in the hypocotyl but lack the senescence response. The schematic representation of ethylene response pathways in *Arabidopsis* with the putative locations of the four types of mutations is shown in Fig. 1.

The original *etr* mutant line was isolated in a screen of 75000 M_2 EMS mutagenized seed. Initial genetic analysis indicated that *etr* was a dominant

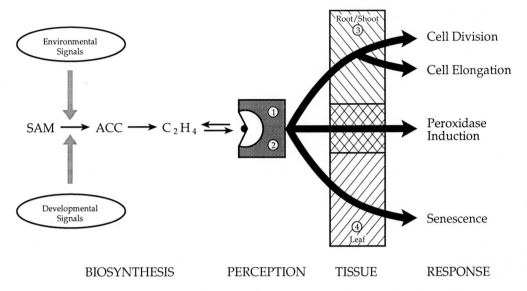

BIOSYNTHESIS   PERCEPTION   TISSUE   RESPONSE

Fig. 1. Schematic representation of ethylene response pathways in higher plants. Numbers 1 through 4 indicate positions in the pathway where mutations have been identified in *Arabidopsis*. SAM, S-adenosylmethionine; ACC, 1-aminocyclopropane-1-carboxylic acid.

mutation that mapped near the flower mutation *ap-1* on chromosome 1 of the *Arabidopsis* genome (Bleecker *et al.* 1988*a*). More recently, a number of additional dominant ethylene-insensitive mutants have been isolated in our laboratory and by other researchers (Guzman and Ecker, 1989; I. Raskin, personal communication). At least some of these mutants map near *ap-1* and are probably alleles of *etr*. To date, no recessive alleles of the *ETR* locus have been isolated, although a second complementation group of recessive ethylene-insensitive mutants has been mapped to chromosome 4 (Guzman and Ecker, 1989).

The *etr* mutant seedlings show no inhibition of hypocotyl or root elongation even at very high ethylene concentrations (Bleecker *et al.* 1988*a*). Other ethylene responses, which can be measured in different tissues and at other stages in the life cycle of the wild-type plant, were also absent in the *etr* mutant plant. These include promotion of seed germination, acceleration of leaf senescence, enhancement of peroxidase activity, and feedback inhibition of ethylene synthesis.

The fact that a variety of ethylene responses occurring in different tissues are affected by the single *etr* mutation is consistent with the idea that a single receptor system is responsible for all of the above responses. It can also be argued that the *etr* mutation must occur quite early in the signal transduction pathway since the biochemical bases for the responses examined are quite different. To examine whether the ethylene receptor is directly affected by the mutation, the capacity of wild-type and mutant plants to bind [^{14}C]ethylene *in vivo* was determined by the isotope dilution assay of Sisler (1979). The results indicate that saturable ethylene binding in *etr* mutant leaves is 5-fold lower than that in wild-type leaves (Bleecker *et al.* 1988*a*). These data are consistent with the possibility that receptor function in the mutant is impaired.

Cloning of the *ETR* locus
RFLP mapping of the ETR *locus*

In order to gain insight into the biochemical function of the *ETR* gene product, we are in the process of cloning the *etr* mutant gene. Genetic fine mapping of the *ETR* locus was accomplished through two genetic crosses. A line (Columbia background) that is homozygous for the *etr* mutation was crossed to either a Landsberg *erecta* line carrying the recessive *ap-1* flower mutation or carrying the recessive *clv-2* flower mutation. Linkage analysis indicated that the *etr* mutation mapped 3 cM distal to the *clv-2* mutation and 5 cM proximal to the *ap-1* mutation. Lines in the F_2 generations that carry recombinations between the *etr* locus and either *ap-1* or *clv-2* were selected by phenotype. These recombinant lines provided a map resolution of about 0.25 cM. Assuming that the average nucleotide distance per cM is 140 kb in *Arabidopsis* and that the recombinations occur randomly within the region, the recombinant lines provide a map resolution of 40 to 50 kb in the region of the genome containing the *ETR* locus. Since the genetic crosses were between Columbia and Landsberg *erecta* ecotypes, the Southern blots of recombinant lines were probed with genomic clones that show RFLPs for these lines and map near the *ETR* locus (Chang *et al.* 1988). One RFLP fragment, designated alpha, segregated with the *ETR* genotype in all recombinant lines, indicating that this polymorphic fragment lay within 50 kb of the *ETR* locus.

Cloning of the ETR *locus using chromosome walking*

The alpha polymorphic fragment was used as a probe to initiate a chromosome walk in the region. Overlapping cosmid and yeast artificial chromosome (YAC) clones were isolated from genomic libraries. The cosmid library was constructed using DNA from *etr* mutant plants. Since the cosmid vector used contained sequences that allowed homologous recombination of clones directly into the Ti plasmid of *Agrobacterium*, the cosmid clones could be transferred into *Arabidopsis*. Transformation of wild-type plants with a cosmid carrying the dominant *etr* gene was expected to confer ethylene insensitivity to the transgenic plant. Two cosmids from the walk which share 18 kb of overlapping sequence both confer at least partial insensitivity to ethylene in transgenic lines, indicating the presence of the *etr* mutant gene. In order to identify genes that are transcribed from this region, the cosmid DNA was used as a probe to screen *Arabidopsis* cDNA libraries. cDNA clones representing two different genes have been identified.

Analysis of cDNA clones for transcribed regions covering the ETR *locus*

One of the cDNA clones, designated LC5-1, detects a 3.2 kb transcript on northern blots. The nucleotide sequence of the LC5-1 clone has recently been obtained (unpublished). Translational analysis of the sequence revealed a single large open reading frame. An homology search of the PIR protein sequence data base using the PROSCAN program (DNASTAR) revealed three functional domains. The amino terminus of the derived polypeptide is composed of a stretch of hydrophobic residues indicating a membrane targeting function (von Heijne, 1990). This is followed by nine imperfect copies of a 22 amino acid leucine-rich repeating unit that shows homology to similar repeat units found in proteins from yeast (Kataoka *et al.* 1985), *Drosophila* (Hashimoto *et al.* 1988), and human (Lopez *et al.* 1988). In each of these cases the leucine-rich repeat is thought to be involved in protein/protein interactions. Adjacent to the leucine repeats, there is a stretch of 24 hydrophobic amino acids indicating a transmembrane domain (McCrea *et al.* 1988). The carboxy end of the derived polypeptide contains 11 subdomains that are common to all eukaryotic protein kinases (Hanks *et al.* 1988).

From the deduced amino acid sequence, the LC5 gene product thus appears to be a transmembrane protein with a protein kinase activity on the cytoplasmic (carboxy-terminal) side and a domain involved in protein/protein interaction on the extracytoplasmic side of the membrane. This molecular topology is analogous to that of a well studied class of growth factor receptors found in higher animals, the receptor tyrosine kinases (Ullrich and Schlessinger, 1990). Members of this class of receptors also contain a single transmembrane domain, a cytoplasmic protein-kinase domain, and an extracytoplasmic domain that is involved in both ligand binding and protein/protein interactions. This tyrosine kinase class of cell surface receptors has been divided into five subclasses based on structural features of the extracytoplasmic domains (Ullrich and Schlessinger, 1990). One feature that all members of this class share is that the ligand for the receptor appears to activate the protein kinase activity by causing dimerization of receptor subunits (Schlessinger, 1988). The dimerization is induced by protein/protein interactions between receptor subunits on the extracellular side of the membrane as a result of ligand binding. This interaction on the extracellular side of the membrane brings about the association of protein kinase domains on the cytoplasmic side, resulting in their mutual activation. This dimerization mechanism is the only known mechanism by which the ligand binding signal is transduced across the membrane in transmembrane kinases with this type of topology.

While the LC5 gene is genetically linked to the *etr* mutation, there is as yet no direct evidence that the LC5 gene is involved in mediating ethylene responses. Efforts are currently underway to determine if the LC5 gene derived from DNA of the *etr* mutant is capable of conferring ethylene insensitivity in transgenic plants. In addition, the genomic sequences of the LC5 gene from wild-type and *etr* mutant DNA are being compared to determine whether a specific genetic lesion is present in the LC5 sequence from the mutant.

Outlook

Whether or not the LC5 gene is responsible for the *etr* phenotype, it is gratifying to discover that transmembrane signaling mechanisms that are so well characterized in animal systems may also occur in plants. In this regard, it is of interest to note that the cloning of a putative transmembrane kinase from maize has recently been reported (Walker and Zhang, 1990).

Special thanks to my collaborators Hans Kende and Elliot Meyerowitz and to Postdoctoral Fellow Caren Chang for the characterization of the LC5-1 clone. This work is supported by grants from the DOE (DE-FG03-88ER13873) and the NSF (DMB-9005164).

References

ABELES, F. B. (1973). *Ethylene in Plant Biology*. Academic Press; N.Y.

ABELES, F. B. (1986). Role of ethylene in *Lactuca sativa* cv Grand-rapids seed germination. *Pl. Physiol.* **81**, 780–787.

BLEECKER, A. B., ESTELLE, M. E., SOMERVILLE, C. AND KENDE, H. (1988a). Insensitivity to ethylene conferred by a dominant mutation in *Arabidopsis thaliana*. *Science* **241**, 1086–1089.

BLEECKER, A. B., KENYON, W. H., SOMERVILLE, S. C. AND KENDE, H. (1986). Use of monoclonal antibodies in the purification and characterization of 1-aminocyclopropane-1-carboxylate synthase, an enzyme in ethylene biosynthesis. *Proc. natn. Acad. Sci. U.S.A.* **83**, 7755–7759.

BLEECKER, A. B., ROBINSON, G. AND KENDE, H. (1988b). Studies on the regulation of 1-aminocyclopropane-1-carboxylate synthase in tomato using monoclonal antibodies. *Planta* **173**, 385–390.

BOLLER, T., GEHRI, A., MAUCH, F. AND VOGELI, U. (1983). Chitinase in bean *Phaseolus vulgaris* leaves induction by ethylene purification properties and possible function. *Planta* **157**, 22–34.

BROGLIE, K. E., BIDDLE, P., CRESSMAN, R. AND BROGLIE, R. (1989). Functional analysis of DNA sequences responsible for ethylene regulation of a bean chitinase gene in transgenic tobacco. *Pl. Cell* **1**, 599–607.

BROGLIE, K. E., GAYNOR, J. J. AND BROGLIE, R. M. (1986). Ethylene-regulated gene expression – molecular cloning of the genes encoding an endochitinase from *Phaseolus vulgaris*. *Proc. natn. Acad. Sci. U.S.A.* **83**, 6820–6824.

BURG, S. P. AND BURG, E. A. (1967). Molecular requirements for the biological activity of ethylene. *Pl. Physiol.* **42**, 144–152.

CHANG, C., BOWMAN, J. L., DEJOHN, A. W., LANDER, E. S. AND MEYEROWITZ, E. M. (1988). Restriction fragment length polymorphism linkage map for *Arabidopsis thaliana*. *Proc. natn. Acad. Sci. U.S.A.* **85**, 6856–6860.

CORDES, S., DEIKMAN, J., MARGOSSIAN, L. J. AND FISCHER, R. L. (1989). Interaction of a developmentally regulated DNA-binding factor with sites flanking 2 different fruit-ripening genes from tomato. *Plant Cell* **1**, 1025–1034.

DAVIES, K. M. AND GRIERSON, D. (1989). Identification of cDNA clones for tomato (*Lycopersicon esculentum* Mill) messenger-RNAs that accumulate during fruit ripening and leaf senescence in response to ethylene. *Planta* **179**, 73–80.

EISINGER, W. (1983). Regulation of pea internode expansion by ethylene. *A. Rev. Pl. Physiol.* **34**, 225–240.

GOESCHL, J. D., RAPPAPORT, L. AND PRATT, H. K. (1966). Ethylene as a factor regulating growth of pea epicotyls subjected to physical stress. *Pl. Physiol.* **41**, 877–884.

GOTH, A. (1981).Medical Pharmacology, C. V. Mosby: St. Louis.

GRIERSON, D., MAUNDERS, M. J., SLATER, A., RAY, J., BIRD, C. R., SCHUCH, W., HOLDSWORTH, M. J., TUCKER, G. A. AND KNAPP, J. E. (1986). Gene expression during tomato ripening. *Phil. Trans. R. Soc. Lond.* **B314**, 399–440.

GUZMAN, P. AND ECKER, J. R. (1989). Exploiting the triple response of *Arabidopsis* to identify ethylene-related mutants. *Pl. Cell* **2**, 513–523.

HANKS, S. K., QUINN, A. M. AND HUNTER, T. (1988). The protein-kinase family – conserved features and deduced phylogeny of the catalytic domains. *Science* **241**, 42–52.

HASHIMOTO, C., HUDSON, K. L. AND ANDERSON, K. V. (1988). The toll gene of *Drosophila*, required for dorsal–ventral embryonic polarity, appears to encode a transmembrane protein. *Cell* **52**, 269–279.

HOLDSWORTH, M. J. AND LATIES, G. G. (1989). Site-specific binding of a nuclear factor to the carrot extensin gene is influenced by both ethylene and wounding. *Planta* **179**, 17–23.

JERIE, P. H., SHAARI, A. R. AND HALL, M. A. (1979). The compartmentation of ethylene in developing cotyledons of *Phaseolus vulgaris*. *Planta* **144**, 503–507.

KATAOKA, T., BROEK, D., WIGLER, M. (1985). DNA sequence and characterization of the *Saccharomyces cerevisiae* gene encoding adenylate cyclase. *Cell* **43**, 493–505.

KELLY, M. O. AND BRADFORD, K. J. (1986). Insensitivity of the *Diageotropica* tomato mutant to auxin. *Pl. Physiol.* **82**, 713–717.

KENDE, H. AND GARDENER, G. (1976). Hormone binding in plants. *A. Rev. Pl. Physiol.* **27**, 267–290.

KOORNNEEF, M., VAN EDEN, J., HANHART, C. J., STAM, P., BRAAKSMA, F. J. AND FEENSTRA, W. J. (1983). Linkage map of *Arabidopsis thaliana*. *J. Hered.* **74**, 265–272.

LEUTWILER, L. S., HOUGH-EVANS, B. R. AND MEYEROWITZ, E. M. (1984). The DNA of *Arabidopsis thaliana*. *Molec. gen. Genet.* **194**, 15–23.

LINCOLN, J. E., CORDES, S., READ, E. AND FISCHER, R. L. (1987). Regulation of gene expression by ethylene during *Lycopersicon esculentum* (tomato) fruit-development. *Proc. natn. Acad. Sci. U.S.A.* **84**, 2793–2797.

LLOYD, C. W. AND SEAGULL, R. W. (1985). A new spring for plant cell biology – microtubules as dynamic helices. *Trends in Biochem. Sci.* **10**, 476–478.

LOPEZ, J. A., CHUNG, D. W., FUJIKAWA, K., HAGEN, F. S., DAVIE, E. W. AND ROTH, G. J. (1988). The alpha and beta chains of human platelet glycoprotein-IB are both transmembrane proteins containing a leucine-rich amino acid sequence. *Proc. natn. Acad. Sci. U.S.A.* **85**, 2135–2139.

MCCREA, P. D., ENGLEMAN, D. M. AND POPOT, J. L. (1988). Topography of integral membrane proteins – hydrophobicity analysis vs immunolocalization. *Trends in Biochem. Sci.* **13**, 289–290.

MCGLASSON, W. B., WADE, D. L. AND ADATO, I. (1978). Phytohormones and fruit ripening. In *Phytohormones and Related Compounds: a Comprehensive Treatise* (eds Letham, D. S., Goodwin, P. B., Higgins, T. J. V.). Elsevier: Holland.

MILLER, J. H., REZNIKOFF, W. S., eds (1980). *The Operon*, Cold Springs Harbor Lab, N.Y.

NAM, H. G., GIRAUDAT, J., BOER, B., MOONAN, F., LOOS, W. D. B., HAUGE, B. M. AND GOODMAN, H. M. (1989). Restriction fragment length polymorphism linkage map of *Arabidopsis thaliana*. *Pl. Cell* **1**, 699–705.

OSHIMA, Y. (1982). Regulatory circuits for gene expression: the metabolism of galactose and phosphate. In *The Molecular Biology of the Yeast Saccaromyces. Metabolism and Gene Expression* (eds: Strathern, J. N., Jones, E. W. Broach, J. R.). Cold Springs Harbor, N.Y.

SATO, T. AND THEOLOGIS, A. (1989). Cloning the messenger-RNA encoding 1-aminocyclopropane-1-carboxylate synthase, the key enzyme for ethylene biosynthesis in plants. *Proc. natn. Acad. Sci. U.S.A.* **86**, 6621–6625.

SCHLESSINGER, J. (1988). Signal transduction by allosteric receptor oligomerization. *Trends in Biochem. Sci.* **13**, 443–447.

SEXTON, R., LEWIS, L. N., TRAWAVAS, A. J. AND KELLY, P. (1985). Ethylene and abscission. In *Ethylene and Plant Development* (eds: Roberts, J. A., Tucker, G. A.). Butterworths, London.

SISLER, E. C. (1979). Measurement of ethylene binding in plant tissue. *Pl. Physiol.* **64**, 538–542.

SISLER, E. C. (1980). Partial purification of an ethylene binding component from plant tissue. *Pl. Physiol.* **66**, 404–406.

THOMAS, C. J. R., SMITH, A. R. AND HALL, M. A. (1985). Partial purification of an ethylene-binding site from *Phaseolus vulgaris* cotyledons. *Planta* **164**, 272–277.

TSAI, D. S., ARTEC, R. N., BACHMAN, J. M. AND PHILLIPS, A. T. (1988). Purification and characterization of 1-aminocyclopropane-1-carboxylate synthase from etiolated mung bean hypocotyls. *Arch. Biochem. Biophys.* **264**, 632–640.

ULLRICH, A. AND SCHLESSINGER, J. (1990). Signal transduction by receptors with tyrosine kinase activity. *Cell* **61**, 203–212.

VON HEIJNE, (1990). The signal peptide. *J. Membr. Biol.* **115**, 195–201.

WALKER, J. C. AND ZHANG, R. (1990). Relationship of a putative receptor protein-kinase from maize to the S-locus glycoproteins of *Brassica*. *Nature* **345**, 743–746.

YANG, S. F. AND HOFFMAN, N. J. (1984). Ethylene biosynthesis and its regulation in higher plants. *A. Rev. Pl. Physiol.* **35**, 155–190.

ZOBEL, R. W. AND ROBERTS, L. W. (1978). Effects of low concentrations of ethylene on cell division and cyto differentiation in lettuce pith explants. *Can. J. Bot.* **56**, 987–990.

ZURFLUH, L. L. AND GUILFOYLE, T. J. (1982). Auxin induced and ethylene induced changes in the population of translatable messenger RNA in basal sections and intact soybean *Glycine max* cultivar Wayne hypocotyl. *Pl. Physiol.* **69**, 338–340.

Note added in proof

We have recently obtained evidence which indicates that the LC5 gene, while genetically linked to the *ETR* locus, does not represent the *ETR* gene.

Printed in Great Britain © *Society for Experimental Biology 1991*   159

INOSITOL PHOSPHOLIPIDS AS PLANT SECOND MESSENGERS

QIUYUN CHEN, IRENA BRGLEZ and WENDY F. BOSS

Botany Department, North Carolina State University, Raleigh, NC 27695-7612, USA

Summary

Two plasma membrane lipids, phosphatidylinositol monophosphate (PtdIns4P) and phosphatidylinositol bisphosphate (PtdIns(4,5)P_2), have been shown to be key intermediates in stimulus response pathways in many animal cells. PtdIns4P and PtdIns(4,5)P_2 act as sources of second messengers and they directly alter the activity of membrane enzymes. These lipids and the enzymes involved in their metabolism are found in the plasma membranes of plant cells. A clear role for the inositol phospholipids in plant signal transduction, however, has not emerged. In this chapter we present some of the questions raised by current work in the area and propose an alternative focus for phosphoinositide metabolism in plants: a role for PtdIns4P and PtdIns(4,5)P_2 as direct effectors of membrane structure and function.

Introduction

The plasma membrane lipids, phosphatidylinositol-4-monophosphate (PtdIns4P) and phosphatidylinositol-4,5-bisphosphate (PtdIns(4,5)P_2), have been shown to be key intermediates in stimulus response pathways in many animal cells. In addition to acting as sources of second messengers, PtdIns4P and PtdIns(4,5)P_2 can act directly on membrane enzymes. These lipids and the enzymes that synthesize them have been shown to be present in the plasma membranes of higher plants (Sandelius and Sommarin, 1986; Wheeler and Boss, 1987). The precise role(s) of these lipids in plants signal transduction, however, remains elusive. Recently, there have been several reviews concerning inositol phospholipid metabolism and signal transduction in plants (Boss, 1989; Lehle and Ettlinger, 1990; Morse *et al.* 1990; Sandelius and Sommarin, 1990; Rincón and Boss, 1990; Einspahr and Thompson, 1990). Our intent in this chapter is not to review the field but rather to add some perspective with regard to the current literature and to identify areas of focus for future research.

As a brief background, consider the two major roles for inositol phospholipids in animal cells. The inositol polyphospholipids, PtdIns4P and PtdIns(4,5)P_2, can act as direct membrane effectors (Schäfer *et al.* 1987; Chauhan and Brockerhoff, 1988) and as a source of the second messengers, inositol-1,4,5-trisphosphate (Ins(1,4,5)P_3) and diacylglycerol (DAG) (for review see Majerus *et al.* 1986;

Key words: phosphoinositide, inositol phospholipid kinase, phospholipase A$_2$, signal transduction.

Michell, 1986; Berridge, 1987a,b; Berridge and Irvine, 1989) in animal cells. The ratio of PtdIns4P to PtdIns(4,5)P_2 is about 1:2 in the rabbit iris muscle (Akhtar and Abdel-Latif, 1980). In these cells and others, such as brain and blowfly salivary gland, which have a relatively high percentage of PtdIns(4,5)P_2, PtdIns(4,5)P_2 is hydrolyzed by phospholipase C in response to external stimuli and produces the second messengers, Ins(1,4,5)P_3 and DAG. Once formed, DAG activates a calcium-phospholipid-dependent protein kinase (protein kinase C) (Nishizuka, 1984) and Ins(1,4,5)P_3 releases calcium from non-mitochondrial intracellular stores (Streb et al. 1983) thus activating calcium-dependent enzymes and altering cell physiology.

In contrast, it was the almost total lack of PtdIns(4,5)P_2 formation by skeletal muscle sarcoplasmic reticulum that led Schäfer et al. (1987) to propose that PtdIns4P might act as a membrane effector rather than as an intermediate in Ins(1,4,5)P_3 formation. PtdIns4P and PtdIns(4,5)P_2 can directly affect the activity of membrane enzymes. PtdIns4P increases the specific activity of E_1-E_2-type ATPases found in animal cells such as the Ca^{2+}-transport ATPase of skeletal muscle sarcoplasmic reticulum (Varsanyi et al. 1983; Schäfer et al. 1987) and the nuclear envelope ATPase of hepatocytes (Smith and Wells, 1983). PtdIns4P and PtdIns(4,5)P_2 enhance the activity of the plasma membrane Ca^{2+}-transport ATPase of erythrocytes (Choquette et al. 1984) and hepatocytes (Lin and Fain, 1985). Similarly, inositol phospholipids increased the activity of the plasma membrane vanadate-sensitive ATPase of plant cells (Memon et al. 1989a). In addition, PtdIns(4,5)P_2 increases protein kinase C activity (Chauhan and Brockerhoff, 1988; Chauhan et al. 1989) and can stimulate the release of calcium from skeletal muscle sarcoplasmic reticulum (Kobayashi et al. 1989). With these potential roles in mind, let us consider the function(s) of PtdIns4P and PtdIns(4,5)P_2 in plants.

A perspective of plant phosphoinositide metabolism

There is no question that the inositol phospholipids, PtdIns4P and PtdIns(4,5)P_2, are present in plant cells (Boss and Massel, 1985; Helsper et al. 1986; Irvine et al. 1989; van Breemen et al. 1990; for review see Boss, 1989). The distribution of these lipids, however, is quite different from that found in reversibly responsive animal cells such as the rabbit iris muscle. For example, the ratio of PtdIns4P:PtdIns(4,5)P_2 ranges from 10:1 to 20:1 in most plant cells and PtdIns(4,5)P_2 ranges from 0.0 to 0.5 % of the total inositol phospholipid (Drøbak et al. 1988; Heim and Wagner, 1989; Boss, 1989). It has been estimated in animal cells that only 10 to 20 % of the PtdIns(4,5)P_2 pool turns over in response to stimuli. If PtdIns(4,5)P_2 is the source of second messengers in plants, then one has to argue that the turnover rate must be very high since the pool is relatively small. Alternatively, in specialized cells (e.g. the flexor and extensor cells of pulvini or the guard cells of the stomata) there may be relatively large amounts of PtdIns(4,5)P_2. Analysis of inositol phospholipids of these specialized cells needs to

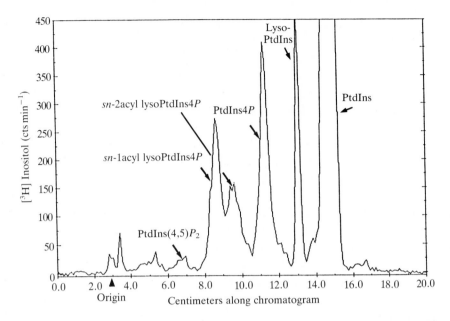

Fig. 1. The distribution of [³H]inositol lipids extracted from wild carrot cells grown in suspension culture. Cells were labeled for 18 h with *myo*[2-³H]inositol (185 kBq/0.4 g fresh wt). The lipids were extracted and chromatographed as previously described (Wheeler and Boss, 1987) except that the solvent system used was CHCl₃:MeOH:NH₄OH:H₂O (86:76:6:18). The cells were used in the mid-log phase of growth. The relative distribution of the lipids shown in the above chromatogram was PtdIns, 79 %; lysoPtdIns, 3.4 %; PtdIns4*P*, 6.1 %; *sn*-2acyl lysoPtdIns4*P*, 3.3 %; *sn*-1acyl lysoPtdIns4*P*, 4.6 %; PtdIns(4,5)*P*₂, 0.5 % of a total of 1356 cts min⁻¹. Reprinted with permission from Bioscan, Inc. (Cho and Boss, unpublished results).

be done. As for the higher plants studied thus far, the inositol phospholipid profile is more typical of those animal systems where the inositol phospholipids are thought to act as direct membrane effectors. An example of the profile of [³H]inositol lipids extracted from wild carrot cells grown in suspension culture is given in Fig. 1. In addition to phosphatidylinositol (PtdIns), PtdIns4*P*, and PtdIns(4,5)*P*₂, lysolipids (Wheeler and Boss, 1987, 1990*a,b*; Coté *et al.* 1990) and inositol glycolipids (Carter and Kisic, 1969; Low and Saltiel, 1988) are prevalent in plant cells. These inositol containing lipids may not only be important in regulating cell metabolism, but they also complicate the analysis of PtdIns4*P* and PtdIns(4,5)*P*₂ (Boss, 1989; Cho and Boss, unpublished results).

Differences in the inositol phospholipids in plants and animals are not only evident from *in vivo* inositol labeling studies, but also from the *in vitro* phosphorylation studies. [³²P]PtdIns4*P* and [³²P]PtdIns(4,5)*P*₂ biosynthesis and catabolism by membranes from several plant and animal cell types are compared in Tables 1 and 2 of Sandelius and Sommarin (1990). Of interest is the fact that although plants hydrolyze PtdIns(4,5)*P*₂ at about the same rate as the animal cells and PtdIns4*P* at approximately a 100-fold faster rate, the lipid kinase activity in the *in vitro* assay is about 10-fold less in higher plants than in animals. The differences

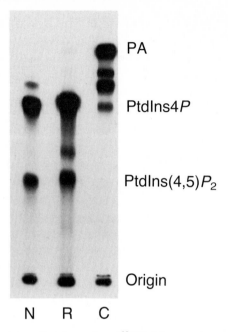

PA

PtdIns4P

PtdIns(4,5)P_2

Origin

N   R   C

Fig. 2. An autoradiogram showing the [^{32}P]lipids produced after *in vitro* phosphorylation using 20 μg membrane protein from *Neurospora crassa* (**N**, provided by Dr. John Lenard, Rutgers University) and rat liver (**R**, provided by Dr. D. James Morré, Purdue University) and carrot culture cells (**C**). After 15 min phosphorylation with [γ^{32}P]ATP, the lipids were extracted and chromatographed as previously described (Memon and Boss, 1990). Standards migrated as noted.

in the lipid kinase activities are demonstrated in Fig. 2 where we have compared plasma membrane from *Neurospora*, rat liver, and carrot. Each assay contained the same amount of membrane protein and the same amount of [^{32}P]ATP. The *in vitro* data are consistent with the low levels of PtdIns(4,5)P_2 found from *in vivo* labeling studies of the carrot cells.

In addition, other differences in plant and animal lipid metabolism have become evident from the *in vitro* studies. For example, PtdIns(4,5)P_2 phospholipase C activity is primarily associated with the membrane or particulate fraction in plants; whereas, in rat brain the activity is found primarily in the soluble fraction (McMurray and Irvine, 1988; Melin *et al.* 1987; Sandelius and Sommarin, 1990). Furthermore, there is no evidence for GTP stimulation of the PtdIns4P- or PtdIns(4,5)P_2-specific phospholipase C of plasma membranes isolated from higher plants (Melin *et al.* 1987; McMurray and Irvine, 1988; Tate *et al.* 1989). In contrast, Einspahr *et al.* (1989) reported that GTP stimulated PtdIns(4,5)P_2 hydrolysis in *Dunaliella salina* plasma membranes, suggesting that algal cells are more like animal cells in this regard.

Another intriguing and often overlooked aspect of plant lipid phosphorylation is that phosphatidic acid (PA) is the primary product in the *in vitro* assay representing about 50 % of the [^{32}P]lipid recovered with endogenous substrate (Fig. 2). The [^{32}P]lipids migrating between PA and PtdIns4P in the carrot extract are probably lysoPA and PA pyrophosphate; however, their identities have yet to be confirmed. The formation of PA indicates the presence of DAG in large quantities. Whatever the source of DAG, if there were a DAG-calcium-phospholipid-regulated protein kinase (i.e. protein kinase C) in plants, which is at best controversial (Harmon, 1990), it should be continuously 'on' and potentially insensitive to the production of DAG by external stimuli (Chiarugi *et al.* 1989).

What about the presence of the other potential second messenger in the polyphosphoinositide pathway, $Ins(1,4,5)P_3$? $Ins(1,4,5)P_3$ has been found in plants at levels at or below the limits of detection of conventional analytical techniques, which has made it difficult to monitor (Morse *et al.* 1987; Rincón *et al.* 1989; Heim and Wagner, 1989; Coté *et al.* 1990). Initial studies of $Ins(1,4,5)P_3$ responses included adding $Ins(1,4,5)P_3$ to protoplasts (Rincón and Boss, 1987), vacuoles (Ranjeva *et al.* 1988) and isolated tonoplast vesicles (Schumaker and Sze, 1987), and measuring calcium transport. From these studies it was concluded that plant cells did respond to $Ins(1,4,5)P_3$ by releasing calcium and the vacuole was the $Ins(1,4,5)P_3$ responsive organelle in plants. Recent data from Brosnan and Sanders (1990) showing that the $Ins(1,4,5)P_3$ response in beet microsomes is heparin-sensitive suggests that there may be an $Ins(1,4,5)P_3$ receptor similar to the heparin-binding receptor isolated from animal cells (Supattopone *et al.* 1988). In contrast, in *Neurospora crassa* the $Ins(1,4,5)P_3$-stimulated release of calcium from vacuoles was not inhibited by heparin (Cornelius *et al.* 1989).

Another approach to studying $Ins(1,4,5)P_3$ responses has been to microinject 'caged'-$Ins(1,4,5)P_3$, i.e. an inactive, photolabile form of $Ins(1,4,5)P_3$ that releases $Ins(1,4,5)P_3$ upon photolysis of the chemical cage. These data are discussed in detail in another chapter in this volume (Trewavas *et al.*). The studies have confirmed that $Ins(1,4,5)P_3$ can increase intracellular calcium from the vacuole and suggest that $Ins(1,4,5)P_3$ may be involved in stimulus-induced closure of stomata. However, if $Ins(1,4,5)P_3$ and DAG are both produced in the guard cell in response to stimuli, one has somewhat of a paradox since Lee and Assmann (1990) have shown that DAG causes stomata to open. It may be that the guard cell response is biphasic (Berridge, 1986), and the $Ins(1,4,5)P_3$ produced causes a rapid closure of the stomata as a result of the transient increase in calcium. DAG would then be involved in a slower recovery (i.e. opening). Many questions need to be addressed in this system e.g., what happens when $InsP_2$ or Pi, the potential products of $Ins(1,4,5)P_3$ metabolism, are microinjected? Microinjecting $InsP_2$ was more effective than $Ins(1,4,5)P_3$ in stopping cell-to-cell transport in *Setcreasea* stamen hairs (Tucker, 1990). $InsP_2$ was also found to stop particle movement and cytoplasmic streaming in *Acetabularia* (Allen and O'Connor, 1990). Most importantly, changes in DAG, $Ins(1,4,5)P_3$, and $PtdIns(4,5)P_2$ must be monitored during stimulus-induced closing and opening of the stomata.

In animal cell studies of $Ins(1,4,5)P_3$ metabolism, the 5'phosphatase is membrane associated and yields $Ins(1,4)P_2$ (for review see Shears, 1989). When $Ins(1,4,5)P_3$ was added to plant extracts both $Ins(4,5)P_2$ and $Ins(1,4)P_2$ were formed rapidly (Joseph *et al.* 1989). The formation of $Ins(4,5)P_2$ was stimulated by less than $1\,\mu\text{M}\ \text{Ca}^{2+}$. In contrast, using soluble extracts from rat brain under the same assay conditions, calcium inhibited inositol phosphate metabolism (Joseph *et al.* 1989). In addition, neither the inositol-1-phosphate phosphatase (Gumber *et al.* 1984) nor Ins(1,4,5)P_3 phosphatase (Memon *et al.* 1989*b*) was inhibited by 10 mM Li$^+$. On the other hand, molybdate, a nonspecific phosphatase inhibitor, did inhibit dephosphorylation of Ins(1,4,5)P_3 by microsomal membranes (Memon *et al.* 1989*b*).

Although distinct differences have been demonstrated in the metabolism of Ins(1,4,5)P_3 by extracts from plants and animals, more extensive studies are necessary. With all the studies of Ins(1,4,5)P_3 metabolism by soluble enzymes from plants, the hydrolytic enzymes of the vacuole were included. What is needed in order to delineate the metabolic pathway is a complete study of Ins(1,4,5)P_3 metabolism by the plasma membrane, cytosol (excluding the vacuolar contents), and tonoplast. In addition, an Ins(1,4,5)P_3 receptor must be isolated and *in situ* localization studies done.

Thus, we do not have definitive evidence from combined biochemical, physiological, and cell biological studies of one system for the role of Ins(1,4,5)P_3 and DAG in the stimulus–response mechanisms in plants. This can be attributed in part to the complexity of plant tissues, which contain many cell types, and in part, to the fact that attempts have been made to 'fit' plant metabolism to the paradigm for rabbit iris muscle and brain cells without recognizing the complexity of the various animal pathways much less the difference in plant and animal cell response mechanisms. In specialized cells such as the extensor or flexor cells of the pulvini or in stomatal guard cells, it may be eventually proven that the PtdIns(4,5)P_2–Ins(1,4,5)P_3 pathway plays a critical role in transducing external signals to bring about physiological responses. Let us consider, however, possible roles for the inositol phospholipids in plant responses other than the reversible, turgor-driven responses of these specialized cells.

An alternative focus

Data accumulated thus far would suggest that in many plant cells, the concentration of PtdIns(4,5)P_2 may be too low to be a significant source of second messengers. As mentioned, PtdIns4P and PtdIns(4,5)P_2 are negatively charged and can directly affect the activity of membrane enzymes. Chauhan and Brockerhoff (1988) have shown that PtdIns4P and PtdIns(4,5)P_2 can regulate protein phosphorylation. They proposed that in some cells, a PtdIns–PtdIns4P–PtdIns(4,5)P_2 shuttle rather than the metabolism of the inositol phospholipids to DAG by phospholipase C was important for regulating protein kinases. Whether signal transduction in plants involves a direct effect of the inositol phospholipids on the plasma membrane or the production of Ins(1,4,5)P_3, the responsive state of the cell depends on the rapid turnover of the inositol phospholipids and thus PtdIns and PtdIns4P kinase activity.

Interestingly, PtdIns kinase activity may reflect the physiological state of a cell.

For example, PtdIns kinase was closely associated with the tyrosine kinase of platelet-derived growth factor (PDGF) receptor (Whitman *et al.* 1987) and PtdIns kinase activity increased upon PDGF-induced transformation (Kaplan *et al.* 1987). Recently, Auger *et al.* (1989) reported that PtdIns3*P* kinase appeared to be associated with cell proliferation and PDGF stimulation and suggested that the products, PtdIns3*P*, PtdIns(3,4)P_2 and PtdIns(3,4,5)P_3 or their metabolites might be important mediators of the mitogenic response. Our working hypothesis is that inositol phospholipid kinase activity correlates with growth responses in higher plant cells.

*Rapid changes in PtdIns and PtdIns4*P *kinase activity*

We have observed rapid (within seconds) changes in PtdIns and PtdIns4*P* kinase activity and corresponding changes in the plasma membrane ATPase activity in two systems with three types of stimuli. The stimuli were given *in vivo* and the response was monitored in the isolated plasma membranes. In each instance, under conditions where growth would be inhibited, the PtdIns and PtdIns4*P* kinase activity and ATPase activity decreased. When wall loosening occurred, which would allow for growth, the lipid kinase and ATPase activities increased. These data are summarized briefly in the paragraphs that follow.

The first stimulus used was light. There was a decrease in the production of PtdIns4*P* and PtdIns(4,5)P_2 by plasma membranes isolated from etiolated sunflower hypocotyls exposed to 10 s of light compared with dark grown controls (Memon and Boss, 1990). When comparing membranes from dark and light irradiated plants, we found no change in the production of Ins(1,4,5)P_3 from endogenous substrate indicating that the decreased production of PtdIns4*P* and PtdIns(4,5)P_2 was not the result of increased product degradation by phospholipase C. The decrease in PtdIns4*P* and PtdIns(4,5)P_2 also was not caused by limited substrate since when exogenous substrate was added similar differences in dark- and light-irradiated plants were observed.

In addition to changes in the plasma membrane PtdIns4*P* kinase activity in response to light, rapid changes were observed in the plasma membrane vanadate-sensitive ATPase activity (Memon and Boss, 1990). As with the lipid kinases, the ATPase activity decreased in response to light treatment. Others have reported changes in the plasma membranes of etiolated seedlings after short-term exposure to light. Light-induced changes in membrane potential preceded light-induced inhibition of growth in etiolated cucumber (Spalding and Cosgrove, 1989). Rapid changes in protein phosphorylation were observed in plasma membranes from etiolated pea stems (Gallagher *et al.* 1988; Short and Briggs, 1989). Whether light-induced changes in protein phosphorylation, lipid phosphorylation, membrane potential, or ATPase activity are part of the same or different signalling pathways remains to be determined. Isolation of the signalling peptides and antibodies to these peptides will allow a delineation of the sequence of events in the near future.

The light-induced effects on the PtdIns4*P* kinase and ATPase activity were observed in a system where light inhibited growth i.e. etiolated seedlings. Cell wall

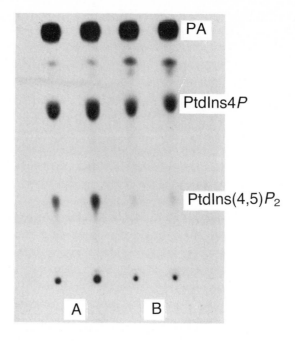

Fig. 3. An autoradiogram showing the effects of *in vivo* treatment with Driselase on plasma membrane inositol phospholipid kinase activities. The cells were treated for 10 min with 2 % Driselase in 0.4 osmolal sorbitol with 2 mM MES buffer (pH 4.8) (A), or in the sorbitol buffer alone (B). The plasma membranes were isolated by aqueous two-phase partitioning and 15 μg membrane protein plus 20 μg PIP were used per assay. Standards migrated as noted. (Reprinted from Chen and Boss, 1990.)

loosening is essential to allow growth to occur. As a result of our studies of the effects of cell wall digestion on the plasma membranes of cultured cells, we observed a rapid increase in the PtdIns4P kinase activity (Fig. 3) and vanadate-sensitive ATPase activity (Chen and Boss, 1990). The response was rapid (5 s) and was seen with purified hemicellulase as well as with a crude mixture of wall degrading enzymes, Driselase. The stimulus was heat labile and the response was only seen if whole cells were treated. Adding the wall degrading enzymes to the purified plasma membranes either had no effect or inhibited lipid kinase activity. These data suggested that either the wall degrading enzymes were not acting directly on the membrane or that some component of the cytoskeleton, cytosol, or cell wall was necessary to activate the lipid kinases. From these studies and others (Strasser *et al.* 1986) there was no evidence to indicate that carbohydrate elicitors increase inositol phospholipid kinase activity.

Curiously, treating cells with purified cellulase did not activate the plasma membrane enzymes, and in fact, at the highest concentration used (1 %, w/v), cellulase inhibited the enzyme activities. This was consistent with the fact that treating cells with cellulase has been shown to produce xyloglucans that are thought to be able to act as feedback regulators to control auxin-induced growth (York *et al.* 1984; McDougall and Fry, 1989). Also consistent with this observation

Table 1. *Correlation between the changes in PtdIns4P kinase activity and the vanadate-sensitive ATPase activity*

Plant material	Stimulus	PtdIns4P kinase activity	Vanadate-sensitive ATPase activity
Etiolated sunflower hypocotyl	Light	Decreased 50%	Decreased 47%
Carrot culture cells	Driselase	Increased 200%	Increased 125%
	Hemicellulase	Increased 70%	Increased 63%
	Pectinase	Increased 68%	Increased 100%
	0.5% Cellulase	No Change	No Change
	1.0% Cellulase	Decreased 40%	Decreased 50%
	Hypertonic stress	Decreased 40%	Decreased 50%

is the fact that when isolated xyloglucans were added to the cells there was no effect or a slight inhibition of the PtdIns4P kinase activity (Chen and Boss, unpublished results).

In the course of studying the effects of cell wall degradation, we also noticed a transient effect of osmotic stress on the lipid kinases and vanadate-sensitive ATPase activity. Further studies of osmotic stress are currently underway in our laboratory and indicate that there is a transient change in the PtdIns and PtdIns4P kinase and ATPase activities when a hypertonic stress of $0.2 \, \text{osmol kg}^{-1}$ is imposed upon the carrot culture cells (Cho and Boss, unpublished results).

With each of the stimuli used, when there was a change in the lipid kinase activity there was a concomitant change in the vanadate-sensitive ATPase activity. These data are summarized in Table 1. The fact that adding PtdIns4P and PtdIns(4,5)P_2 to the isolated plasma membranes increased the ATPase activity (Memon *et al.* 1989a) and that the lipid kinases responded to the wall degrading enzymes at concentrations where there was no significant change in the ATPase activity, lead us to hypothesize that the lipid kinases were affected first, and that changes in the amount of PtdIns4P and PtdIns(4,5)P_2 resulted in a change in the ATPase activity (Chen and Boss, 1990). The effect of PtdIns4P and PtdIns(4,5)P_2 on the ATPase was inhibited by adding the positively charged aminoglycoside, neomycin, indicating a charged interaction and not a nonspecific detergent effect (Lipsky and Lietman, 1980; Chen and Boss, 1991). While these data suggest a cause and effect relationship between the lipid kinases and the ATPase, they do not indicate that this is the only pathway affected by the stimuli. In fact, our current studies have begun to reveal the complexity of the transduction pathways.

Activation of phospholipase A_2

At least one other plasma membrane enzyme, phospholipase A_2, responds

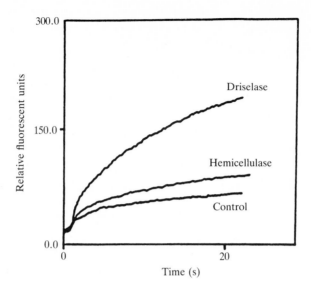

Fig. 4. Phospholipase A_2 activity was assayed spectrofluorometrically monitoring the release of NBD-caproic acid from sn-2-NBD-caproyl-phosphatidylcholine.
The change in relative fluorescence is monitored over time. Only the initial slope was used to calculate the rates given in Table 2.

when cells are treated with hemicellulase and Driselase. The phospholipase A_2 assay was done with an exogenous substrate 1-acyl-2-(N-4-nitrobenzo-2-oxa-1,3-diazole) aminocaproyl phosphatidylcholine (NBD-PC), monitoring the release of NBD-caproic acid as described by Wittenauer *et al.* (1984). The fluorescence of NBD-caproic acid increases 50-fold when it is hydrolyzed from the phospholipid. Representative fluorescence tracings are shown in Fig. 4. The data are summarized in Table 2. In comparing Tables 1 and 2 one can see that when a stimulus increased the PtdIns4P kinase activity and the vanadate-sensitive ATPase, the phospholipase A_2 activity also increased. We have yet to separate the

Table 2. *Phospholipase A_2 activity of plasma membranes*

Treatment	nmoles NBD-caproic acid $min^{-1}mg$ protein^{-1}
Sorbitol	115 ± 8
0.5 % Cellulase	135 ± 15
15.8 % Hemicellulase	210 ± 10
0.005 % Driselase	187 ± 27
2 % Driselase	453 ± 63
Desalted 2 % Driselase	609 ± 4

Plasma membranes were isolated from carrot cells and treated for 5 min with either 0.4 molal sorbitol or with cell wall degrading enzymes in 0.4 molal sorbitol. The desalted Driselase was eluted through a G-25 column to remove low relative molecular mass compounds. Phospholipase A_2 activity was assayed spectrofluorometrically monitoring the release of NBD-caproic acid from sn-2-NBD-caproyl-phosphatidylcholine.

responses of the three enzymes temporally; however, our preliminary studies indicate that the phospholipase A_2 is the most sensitive enzyme responding to the lowest concentrations of Driselase (0.005%) (Brglez and Boss, unpublished results).

What effects might activating the plasma membrane phospholipase A_2 have on cell physiology? The products produced, fatty acids and lysolipids, have been shown to activate protein kinases isolated from zucchini, wheat germ and beet leaves (Scherer *et al.* 1988; Lucantoni and Polya, 1987; Klucis and Polya, 1987). Lysolipids also have been shown to increase the plasma membrane ATPase (Palmgren *et al.* 1988), H^+ pump (Palmgren and Sommarin, 1989), and NADH oxidase activity (Brightman *et al.* 1990). Lysolipids inhibit PtdIns and PtdIns4P kinase activity (Wheeler and Boss, 1990a) and they also affect 1,3β glucan synthase activity (Kauss and Jeblick, 1986; Sloan and Wasserman, 1989).

There is one report of auxin stimulating phospholipase A_2 activity (Scherer and André, 1989); however, this appears not to be a plasma membrane associated enzyme (Wilkinson *et al.* 1990). Lesham (1987) proposed that phospholipase A_2 is an important enzyme in regulating plant growth and senescence, and the potential roles of lysolipids in signal transduction were recently reviewed (Morré, 1990). As more data accumulate, it will be interesting to learn what roles these potentially toxic lysolipids play in regulating plant growth and development.

A perspective

Some might argue that all cells are genetically similar and, therefore, have the same signal transduction pathways. And yet, it is well established that while some aspects of metabolism are the same, key pathways are often regulated differently. For example, the regulation of the biosynthesis of fructose-2,6-bisphosphate (F-2,6-BP), which stimulates glycolysis in both plant and animal cells, is different in the two systems. Biosynthesis of F-2,6-BP is under metabolite control in the former and hormonal control in the latter. In addition, although F-2,6-BP stimulates glycolysis in both, in plants a pyrophosphate-dependent fructose-6-phosphate phosphotransferase is the site of regulation and in animals it is an ATP-dependent phosphofructokinase (for review see Huber, 1986). Thus, it is not illogical to suppose that the regulation of inositol phospholipid and inositol phosphate metabolism might also differ in plants and animals.

The divergence of plant responses from the animal paradigms is in fact reassuring. One could ask, should signal transduction in a stationary organism that undergoes irreversible growth in response to stimuli be the same as one that is mobile and/or that undergoes a reversible response? In many animal cells and in pulvini or stomata where a reversible response is involved, the transient production of $Ins(1,4,5)P_3$ or other transiently produced second messengers would be advantageous. In elongating cells or tissue undergoing irreversible growth (cell differentiation or transformation) a transient change may not be adequate (Berridge, 1987b). Either continued stimuli or a change in the physiological state of the cell – e.g. an increase or decrease in inositol phospholipid biosynthesis may be involved.

Cross-talk

We have only mentioned a few of the enzymes in the plasma membrane involved in stimulus–response pathways. These are shown in Fig. 5. Not depicted are NADH oxidase, which has been shown to be stimulated by auxin and lysolipids (Brightman *et al.* 1988; Morré, 1989; Brightman *et al.* 1990; Morré, 1990) or membrane receptors such as the auxin receptor or fusicoccin receptor. Also not shown is the potential for phosphatidylcholine to play an important role in signal transduction (Pelech and Vance, 1989). Stimulus-induced hydrolysis of phosphatidylcholine by phospholipases also leads to the production of DAG, fatty acids, and lysolipids.

Assuredly as we ask more questions the stimulus–response pathways will become even more complex. This complexity contributes to the variety of responses by different plant tissues to the same stimuli. 'Cross-talk' between the signalling systems and the physiological state of the cell will be key factors in determining the magnitude and duration of the physiological response. Mechanisms of recovery or for controlling cell growth should not be overlooked (Fig. 5). For example, if phospholipase A_2 is important for growth responses then lysolipid lipases and fatty acyltransferases will be key components in bringing about controlled growth or recovery from a stimulus. Protein phosphatases

MECHANISMS OF RESPONSE

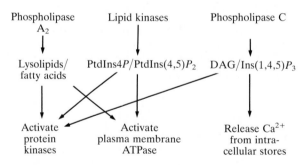

MECHANISMS FOR RECOVERY OR GROWTH CONTROL

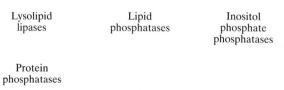

Fig. 5. Proposed mechanisms for signal transduction and cross-talk *via* changes in phospholipid metabolism in plasma membranes of higher plants.

(Hardie, 1989; Witters, 1990), lipid phosphatases, or inositol phosphate phosphatases play similar roles in maintaining homeostasis and thus controlling cell growth. These and many other avenues of plant signal transduction are only beginning to be investigated. These are exciting times for plant physiologists.

References

AKHTAR, A. AND ABDEL-LATIF, A. (1980). Requirement for calcium ions in acetylcholine-stimulated phosphodiesteratic cleavage of phosphatidyl-myo-inositol 4,5-bisphosphate in rabbit iris smooth muscle. *J. Biochem.* **192**, 783–791.

ALLEN, N. S. AND O'CONNOR, S. A. (1990). Inhibition of intracellular particle motions in *Acetabularia acetabulum* L. by phosphoinositides. In *Inositol Metabolism in Plants* (ed. D. J. Morré, W. F. Boss and F. A. Loewus) pp. 301–310. Wiley-Liss, New York: U.S.A.

AUGER, K. R., SERUNIAN, L. A., SOLTOFF, S. P., LIBBY, P. AND CANTLEY, L. C. (1989). PDGF-dependent tyrosine phosphorylation stimulated production of novel polyphosphoinositides in intact cells. *Cell*. **57**, 167–175.

BERRIDGE, M. J. (1986). Second messenger dualism in neuromodulation and memory. *Nature* **323**, 294–295.

BERRIDGE, M. J. (1987*a*). Inositol trisphosphate and diacylglycerol: Two interacting second messengers. *A. Rev. Biochem.* **56**, 159–193.

BERRIDGE, M. J. (1987*b*). Inositol lipids and cell proliferation. *Biochim. biophys. Acta* **907**, 33–45.

BERRIDGE, M. J. AND IRVINE, R. F. (1989). Inositol phosphates and cell signaling. *Nature*. **341**, 197–205.

BOSS, W. F. (1989). Phosphoinositide metabolism: Its relation to signal transduction in plants. In *Second Messengers in Plant Growth and Development*. (ed. W. F. Boss and D. J. Morré) pp. 29–56. Alan R. Liss, New York: U.S.A.

BOSS, W. F. AND MASSEL, M. O. (1985). Polyphosphoinositides are present in plant tissue culture cells. *Biochim. biophys. Res. Commun.* **132**, 1018–1023.

BRIGHTMAN, A. O., BARR, R., CRANE, F. L. AND MORRÉ, D. J. (1988). Auxin- stimulated NADH oxidase purified from plasma membrane of soybean. *Pl. Physiol.* **86**, 1264–1269.

BRIGHTMAN, A. O., ZHU, X. Z. AND MORRÉ, D. J. (1990). Activation of NADH oxidase of soybean plasma membrane by products of phospholipase A activity. *Pl. Physiol.* **93**, 293S.

BROSNAN, J. M. AND SANDERS, D. (1990). Inositol trisphosphate-mediated Ca^{2+} release in beet microsomes inhibited by heparin. *FEBS Letts*. **260**, 70–72.

CARTER, H. E. AND KISIC, A. (1969). Counter current distribution of inositol lipids of plant seeds. *J. Lipid Res.* **10**, 356–362.

CHAUHAN, A., CHAUHAN, V. P. S., DESHMUKH, D. S. AND BROCKERHOFF, H. (1989). Phosphatidylinositol 4,5-bisphosphate completely inhibits phorbol ester binding to protein kinase C. *Biochemistry* **28**, 4952–4956.

CHAUHAN, V. P. S. AND BROCKERHOFF, H. (1988). Phosphatidylinositol-4,5-bisphosphate may antecede diacylglycerol as activation of protein kinase C. *Biochem. biophys. Res. Commun.* **155**, 18–23.

CHEN, Q. AND BOSS, W. F. (1990). Short-term treatment with cell wall degrading enzymes increases the inositol phospholipid kinase and vanadate-sensitive ATPase activity. *Pl. Physiol.* **94**, 1820–1829.

CHEN, Q. AND BOSS, W. F. (1991). Neomycin inhibits the phosphatidylinositol monophosphate and phosphatidylinositol bisphosphate stimulation of plasma membrane ATPase activity. *Pl. Physiol.* (in press).

CHIARUGI, V., BRUNI, P., PASQUALI, F., MAGNELLI, L., BASI, G., RUGGIERO, M. AND FARNARARO, M. (1989). Synthesis of diacylglyceride *de novo* is responsible for permanent activation and down-regulation of protein kinase C in transformed cells. *Biochem. biophys. Res. Commun.* **164**, 816–823.

CHOQUETTE, D., HAKIM, E., FILOTEO, A. E., PLISHKER, E. A., BOSTWICK, J. R. AND PENNISTON,

J. T. (1984). Regulation of plasma membrane Ca^{2+} ATPases by lipids of the phosphatidylinositol cycle. *Biochem. biophys. Res. Commun.* **125**, 908–915.

CORNELIUS, G., GEBAUER, G. AND TECHEL, D. (1989). Inositol trisphosphate calcium release from *Neurospora crassa* vacuoles. *Biochem. biophys. Res. Commun.* **162**, 852–856.

COTÉ, G. G., QUARMBY, L. M., SATTER, R. L., MORSE, M. J. AND CRAIN, R. L. (1990). Extraction, separation, and characterization of the metabolites of the inositol phospholipid cycle. In *Inositol Metabolism in Plants* (ed. D. J. Morré, W. F. Boss and F. A. Loewus) pp. 113–137. Wiley-Liss, New York: U.S.A.

DRØBAK, B. K., FERGUSON, I. B., DAWSON, A. P. AND IRVINE, R. F. (1988). Inositol containing lipids in suspension cultured plant cells. An isotopic study. *Pl. Physiol.* **87**, 217–222.

EINSPAHR, K. J., PEELER, T. C. AND THOMPSON, G. A., JR (1989). Phosphatidylinositol 4,5-bisphosphate phospholipase C and phosphomonoesterase in *Dunaliella salina* membranes. *Pl. Physiol.* **90**, 1115–1120.

EINSPAHR, K. J. AND THOMPSON, G. A., JR (1990). Transmembrane signalling *via* phosphatidylinositol 4,5-bisphosphate hydrolysis in plants. *Pl. Physiol.* **93**, 361–366.

GALLAGHER, S., SHORT, T. W., RAY, P. M., PRATT, L. H. AND BRIGGS, W. R. (1988). Light-mediated changes in two proteins associated with plasma membrane fractions from pea stem sections. *Proc. natn. Acad. Sci. U.S.A.* **85**, 8003–8007.

GUMBER, S. C., LOEWUS, M. W. AND LOEWUS, F. A. (1984). Further studies on *myo*-inositol-1-phosphatase from the pollen of *Lillium longiflorum* Thunb. *Pl. Physiol.* **76**, 40–41.

HARDIE, D. G. (1989). Protein phosphorylation and dephosphorylation. *Current Opinion in Cell Biology* **1**, 220–226.

HARMON, A. C. (1990). Lipid activated protein kinases. In *Inositol Metabolism in Plants* (ed. D. J. Morré, W. F. Boss and F. A. Loewus), pp. 319–334. Wiley-Liss: New York.

HEIM, S. AND WAGNER, K. G. (1989). Inositol phosphates in the growth cycle of suspension plant cells. *Pl. Science.* **63**, 159–165.

HELSPER, J. P. F. G., DE GROOT, P. F. M., LINSKENS, H. F. AND JACKSON, J. F. (1986). Phosphatidylinositol monophosphate in *Lillium* pollen and turnover of phospholipid during pollen tube extension. *Phytochem.* **25**, 2193–2199.

HUBER, S. C. (1986). Fructose 2,6-bisphosphate as a regulatory metabolite in plants. *A. Rev. Pl. Physiol.* **37**, 233–246.

IRVINE, R. F., LETCHER, A. J., LANDER, D. J., DROBAK, B. K., DAWSON, A. P. AND MUSGRAVE, A. (1989). Phosphatidylinositol(4,5)bisphosphate and phosphatidylinositol(4)phosphate in plant tissues. *Pl. Physiol.* **89**, 888–892.

JOSEPH, K. S., ESCH, T. AND BONNER, W. D., JR (1989). Hydrolysis of inositol phosphates by plant cell extracts. *J. Biochem.* **264**, 851–856.

KAPLAN, D. R., WHITMAN, M., SCHAFFHAUSEN, B., PALLAS, D. C., WHITE, M., CAUTLEY, L. AND ROBERTS, T. M. (1987). Common elements in growth factor stimulation and oncogenic transformation: 85 kDa phosphoprotein and phosphatidylinositol kinase activity. *Cell* **50**, 1021–1029.

KAUSS, H. AND JEBLICK, W. (1986). The influence of free fatty acids, lysophosphatidylcholine, platelet-activating factor, acylcarnitine, and echinocandin B on 1,3-B-D-glucan synthase and callose synthesis. *Pl. Physiol.* **80**, 7–13.

KLUCIS, E. AND POLYA, G. M. (1987). Calcium-dependent activation of two plant leaf calcium-regulated protein kinases by unsaturated fatty acids. *Biochem. biophys. Res. Commun.* **147**, 1041–1047.

KOBAYASHI, M., MUROYAMA, A. AND OHIZUMI, Y. (1989). Phosphatidylinositol 4,5-bisphosphate enhances calcium release from sarcoplasmic reticulum of skeletal muscle. *Biochem. biophys. Res. Commun.* **163**, 1487–1491.

LEE, Y. AND ASSMANN, S. (1990). Diacylglycerol induces both ion pumping in patch-clamped guard cell protoplasts and stomatal opening. *Pl. Physiol.* **93**, 95S.

LEHLE, L. AND ETTLINGER, C. (1990). Phosphoinositides and cell cycle control. In *Inositol Metabolism in Plants* (ed. D. J. Morré, W. F. Boss and F. A. Loewus) pp. 217–226. Wiley-Liss, New York: U.S.A.

LESHAM, Y. Y. (1987). Membrane phospholipid catabolism and Ca^{2+} activity in control of senescence. *Physiol. Plant.* **69**, 551–559.

LIN, S. H. AND FAIN, J. N. (1985). Calcium-magnesium ATPase in rat hepatocyte plasma membranes: inhibition by vasopressin and purification of the enzyme. *Prog. clin. biol. Res.* **168**, 25–30.

LIPSKY, J. J. AND LIETMAN, P. S. (1980). Neomycin inhibition of adenosine triphosphatase: Evidence for a neomycin phospholipid interaction. *Antimicrobial Agents and Chemotherapy.* **18**, 532–535.

LOW, M. G. AND SALTIEL, A. R. (1988). Structural and functional roles of glycosyl-phosphatidylinositol in membranes. *Science* **234**, 268–275.

LUCANTONI, A. AND POLYA, G. M. (1987). Activation of wheat embryo calcium-regulated protein kinase by unsaturated fatty acids in the presence and absence of calcium. *FEBS Lett* **221**, 33–36.

MAJERUS, P. W., CONNOLLY, T. M., DECKMYN, H., ROSS, T. S., BROSS, T. E., ISHII, H., BANSAL, V. S. AND WILSON, D. B. (1986). The metabolism of phosphoinositide-derived messenger molecules. *Science* **234**, 1519–1526.

MCDOUGALL, G. J. AND FRY, S. C. (1989). Structure-activity relationships for xyloglucan oligosaccharides with antiauxin activity. *Pl. Physiol.* **89**, 883–887.

MCMURRAY, W. C. AND IRVINE, R. F. (1988). Phosphatidylinositol 4,5-bisphosphate phosphodiesterase in higher plants. *Biochem. J.* **249**, 877–881.

MELIN, P. M., SOMMARIN, M., SANDELIUS, A. S. AND JERGIL, B. (1987). Identification of Ca^{2+}-stimulated polyphosphoinositide phospholipase C in isolated plant plasma membranes. *FEBS Lett.* **223**, 87–91.

MEMON, A., CHEN, Q. AND BOSS, W. F. (1989*a*). Inositol phospholipids activate plasma membrane ATPase in plants. *Biochem. biophys. Res. Commun.* **162**, 1295–1301.

MEMON, A., RINCÓN, M. AND BOSS, W. F. (1989*b*). Inositol trisphosphate metabolism in carrot (*Daucus carota* L.) cells. *Pl. Physiol.* **91**, 477–480.

MEMON, A. AND BOSS, W. F. (1990). Rapid light-induced changes in phosphoinositide kinases and ATPase activity in sunflower hypocotyls. *J. biol. Chem.* **245**, 14817–14821.

MICHELL, R. H. (1986). Inositol lipids and their role in receptor function: History and general principles. In *Phosphoinositides and Receptor Mechanisms* (ed. J. W. Putney, Jr.) pp. 1–24. Allan R. Liss, New York: U.S.A.

MORRÉ, D. J. (1989). Stimulus-response coupling in auxin regulation of plant cell elongation. In *Second Messengers in Plant Growth and Development* (ed. W. F. Boss and D. J. Morré) pp. 81–114. Alan R. Liss, New York: U.S.A.

MORRÉ, D. J. (1990). Activation of phospholipase A: An alternative mechanism for signal transduction. In *Inositol Metabolism in Plants* (ed. D. J. Morré, W. F. Boss and F. A. Loewus) pp. 227–257. Wiley-Liss, New York: U.S.A.

MORSE, M. J., CRAIN, R. C., COTÉ, G. G. AND SATTER, R. L. (1990). Light-signal transduction *via* accelerated inositol phospholipid turnover in *Samanea* pulvini. In *Inositol Metabolism in Plants* (ed. D. J. Morré, W. F. Boss and F. A. Loewus) pp. 201–215. Wiley-Liss, New York: U.S.A.

MORSE, M. J., CRAIN, R. C. AND SATTER, R. L. (1987). Light-stimulated inositol phospholipid turnover in *Samanea saman* leaf pulvini. *Proc. natn. Acad. Sci. U.S.A.* **84**, 7075–7078.

NISHIZUKA, Y. (1984). Turnover of inositol phospholipids and signal transduction. *Science.* **225**, 1365–1370.

PALMGREN, M. G. AND SOMMARIN, M. (1989). Lysophosphatidylcholine stimulates ATP-dependent proton accumulation in isolated oat root plasma membrane vesicles. *Pl. Physiol.* **90**, 1009–1014.

PALMGREN, M. G., SOMMARIN, M., ULVSKOV, P. AND JORGENSEN, P. L. (1988). Modulation of plasma membrane H^+-ATPase from oat roots by lysophosphatidylcholine, free fatty acids and phospholipase A_2. *Physiologia Pl.* **74**, 11–19.

PELECH, S. L. AND VANCE, D. E. (1989). Signal transduction *via* phosphatidylcholine cycles. *Trends in Biochem. Sci.* **14**, 28–30.

RANJEVA, R., CARRASCO, A. AND BOUDET, A. M. (1988). Inositol trisphosphate stimulates the release of calcium from intact vacuoles isolated from *Acer* cells. *FEBS Lett.* **230**, 137–141.

RINCÓN, M. AND BOSS, W. F. (1987). *Myo*-inositol trisphosphate mobilizes calcium from fusogenic carrot (*Daucus carota* L.) protoplasts. *Pl. Physiol.* **83**, 389–398.

RINCÓN, M. AND BOSS, W. F. (1990). The second messenger role of phosphoinositides. In

Inositol Metabolism in Plants (ed. D. J. Morré, W. F. Boss and F. A. Loewus) pp. 173–200. Wiley-Liss, New York: U.S.A.

RINCÓN, M., CHEN, Q. AND BOSS, W. F. (1989). Characterization of inositol phosphates in carrot (*Daucus carota* L.) cells. *Pl. Physiol.* **89**, 126–132.

SANDELIUS, A. S. AND SOMMARIN, M. (1986). Phosphorylation of phosphatidylinositol in isolated plant membranes. *FEBS Lett.* **201**, 282–286.

SANDELIUS, A. S. AND SOMMARIN, M. (1990). Membrane-localized reactions involved in polyphosphoinositide turnover. In *Inositol Metabolism in Plants* (ed. D. J. Morré, W. F. Boss and F. A. Loewus) pp. 139–161. Wiley-Liss, New York: U.S.A.

SCHÄFER, M., BEHLE, G., VARSANYI, M. AND HEILMEYER, L. M. G., JR (1987). Ca^{2+} regulation of 1-(3-sn-phosphatidyl)-1D-myo-inositol 4-phosphate formation and hydrolysis on sarcoplasmic-reticular Ca^{2+}-transport ATPase: A new principle of phospholipid turnover regulation. *Biochem J.* **247**, 579–587.

SCHERER, G. F. E. AND ANDRÉ, B. (1989). Auxin stimulates phospholipase A_2 *in vivo* and *in vitro*. *Biochem. biophys. Res. Commun.* **163**, 111–117.

SCHERER, G. F. E., MARTINY-BARON, G. AND STOFFEL, B. (1988). A new set of regulatory molecules in plants: A plant phospholipid similar to platelet activating factor stimulates protein kinase and proton-translocating ATPase in membrane vesicles. *Planta.* **175**, 241–253.

SCHUMAKER, K. S. AND SZE, H. (1987). Inositol 1,4,5-trisphosphate releases Ca^{2+} from vacuolar membrane vesicles of oat roots. *J. biol. Chem.* **262**, 3944–3946.

SHEARS, S. B. (1989). Metabolism of the inositol phosphates produced upon receptor activation. *Biochem. J.* **260**, 313–324.

SHORT, T. W. AND BRIGGS, W. R. (1989). Characterization of a rapid, blue light-mediated change in detectable phosphorylation of a plasma membrane protein from etiolated pea (*Pisum sativum* L.) seedlings. *Pl. Physiol.* **92**, 179–185.

SLOAN, M. E. AND WASSERMAN, B. P. (1989). Susceptibility of UDP-glucose:(1,3)-β-glucan synthase to inactivation by phospholipases and trypsin. *Pl. Physiol.* **89**, 1341–1344.

SMITH, C. D. AND WELLS, W. W. (1983). Phosphorylation of rat liver envelopes. II. Characterization of *in vitro* lipid phosphorylation. *J. biol. Chem.* **258**, 9368–9373.

SPALDING, E. P. AND COSGROVE, D. J. (1989). Large plasma-membrane depolarization precedes rapid blue-light-induced growth inhibition in cucumber. *Planta.* **178**, 407–410.

STRASSER, H., HOFFMAN, C., GRISEBACH, H. AND MATERN, U. (1986). Are polyphosphoinositides involved in signal transduction of elicitor-induced phytoalexin synthesis in cultured plant cells? *Z. Naturforsch.* **41c**, 717–724.

STREB, H., IRVINE, R. F., BERRIDGE, M. J. AND SCHULTZ, I. (1983). Release of Ca^{2+} from a nonmitochondrial intracellular store in pancreatic acinar cells by inositol-1,4,5-trisphosphate. *Nature* **306**, 67–68.

SUPATTAPONE, S., WORLEY, P. F., BARABAN, J. M. AND SNYDER, S. H. (1988). Solubilization, purification, and characterization of an inositol trisphosphate receptor. *J. biol. Chem.* **263**, 1530–1534.

TATE, B. F., SCHALLER, G. E., SUSSMAN, M. R. AND CRAIN, R. C. (1989). Characterization of polyphosphoinositide phospholipase C from the plasma membrane of *Avena saliva*. *Pl. Physiol.* **91**, 1275–1279.

TUCKER, E. B. (1990). Inositol phosphates and diacylglycerol inhibit cell-to-cell transport. In *Inositol Metabolism in Plants* (ed. D. J. Morré, W. F. Boss and F. A. Loewus) pp. 311–318. Wiley-Liss, New York: U.S.A.

VAN BREEMEN, R., WHEELER, J. J. AND BOSS, W. F. (1990). Negative ion fast atom bombardment mass spectrometry of plant phosphoinositides. *Lipids* **25**, 328–334.

VARSANYI, M., TOLLE, H. G., HEILMEYER, L. M. G., JR, DAWSON, R. M. C. AND IRVINE, R. F. (1983). Activation of sarcoplasmic reticular Ca^{2+} transport ATPase by phosphorylation of an associated phosphatidylinositol. *EMBO J.* **2**, 1543–1548.

WHEELER, J. J. AND BOSS, W. F. (1987). Polyphosphoinositides are present in plasma membrane from fusogenic carrot cells. *Pl. Physiol.* **85**, 389–392.

WHEELER, J. J. AND BOSS, W. F. (1990a). Inositol lysophospholipids. In *Inositol Metabolism in Plants* (ed. D. J. Morré, W. F. Boss and F. A. Loewus) pp. 163–172. Wiley-Liss, New York: U.S.A.

WHEELER, J. J. AND BOSS, W. F. (1990b). The presence of sn-1-palmitoyllysophosphatidylinositol monophosphate correlates positively with the fusion permissive state of the plasma membrane of fusogenic carrot cells grown in suspension culture. *Biochim. biophys. Acta.* **984**, 33–40.

WHITMAN, M., KAPLAN, D., ROBERTS, T. AND CANTLEY, L. (1987). Evidence for two distinct phosphatidylinositol kinases in fibroblasts. *Biochem. J.* **247**, 165–174.

WILKINSON, F. E., BRIGHTMAN, A. O., ZHU, X. Z. AND MORRÉ, D. J. (1990). Differential *in vitro* response of microsomal subfractions to auxin. *Pl. Physiol.* **93**, 273S.

WITTENAUER, L. A., SHIRAI, K., JACKSON, R. L. AND JOHNSON, J. D. (1984). Hydrolysis of a fluorescent phospholipid substrate by phospholipase A$_2$ and lipoprotein lipase. *Biochem. biophys. Res. Commun.* **118**, 894–901.

WITTERS, L. A. (1990). Protein phosphorylation and dephosphorylation. *Current Opinion in Cell Biology* **2**, 212–220.

YORK, W. S., DARVILL, A. G. AND ALBERSHEIM, P. (1984). Inhibition of 2,4-Dichlorophenoxyacetic acid-stimulated elongation of pea stem segments by a xyloglucan oligosaccharide. *Pl. Physiol.* **75**, 295–297.

Printed in Great Britain © *Society for Experimental Biology 1991*   177

VISUALISATION AND MEASUREMENT OF THE CALCIUM MESSAGE IN GUARD CELLS

M. D. FRICKER[1]*, *S. GILROY*[2]†, *N. D. READ*
and A. J. TREWAVAS[2]

[1] Department of Plant Sciences, University of Oxford, South Parks Road, Oxford, OX1 3RB, UK

[2] Institute of Cell and Molecular Biology, University of Edinburgh, The King's Buildings, Mayfield Road, Edinburgh, EH9 3JH, UK

Summary

We have applied several novel technologies to investigate the role of cytosolic free calcium $[Ca^{2+}]_i$ in signal transduction in guard cells of *Commelina communis* L. Fluorescence ratio imaging and photometry together with the fluorescent Ca^{2+} indicator Indo-1 were used to directly visualise and measure dynamic spatial and temporal changes in $[Ca^{2+}]_i$ in response to various exogenous stimuli. More subtle manipulation of the Ca^{2+} signal transduction pathway was achieved through the use of photoactivateable, caged Ca^{2+} and caged inositol-1,4,5-triphosphate ($InsP_3$) released directly into the cytoplasm of the guard cell after microinjection. In these experiments, changes in $[Ca^{2+}]_i$ were simultaneously monitored with the fluorescent Ca^{2+} indicator, Fluo-3. Resting levels of $[Ca^{2+}]_i$ (100–200 nM) increased in response to elevated $[Ca^{2+}]_e$, lowering $[K^+]_e$, application of the ionophore A-23187 or cytosolic release of either Ca^{2+} or $InsP_3$ from their caged forms. Stomatal closure was triggered if $[Ca^{2+}]_i$ increased above a threshold of about 600 nM. Abscisic acid (ABA) had little effect on $[Ca^{2+}]_i$ in the majority of cells studied, being elevated in only a minority of cells investigated. However, stomatal closure occurred in all cases after ABA application. This suggests that ABA acts through both Ca^{2+}-independent and Ca^{2+}-dependent pathways. The imaging data revealed a substantial heterogeneity in $[Ca^{2+}]_i$ within the guard cell. Cytoplasmic regions, particularly near the nucleus, often showed marked elevations and sometimes oscillations. The origin and kinetics of the Ca^{2+} fluxes leading to the dynamic spatial patterns is discussed along with several new approaches directed towards identification of the source of the Ca^{2+}. These methods include optical sectioning and 3-D reconstruction of both the

Abbreviations: ABA, abscisic acid; $[Ca^{2+}]_e$, extracellular free calcium concentration; $[Ca^{2+}]_i$, cytosolic free calcium concentration; caged-$InsP_3$, trisodium myo-inositol 1,4,5-triphosphate, $P^{4(5)}$-1-(2-nitrophenyl)ethyl ester; EGTA, Ethyleneglycol-bis-(o-amino-ethyl)tetraacetic acid; $(K^+)_e$, extracellular potassium concentration; Mes, 2-(N-morpholino)-ethanesulphonic acid.

*Author for correspondence.

† Current address: Department of Plant Biology, University of California, Berkeley, California 94720, USA.

Key words: calcium, *Commelina communis*, signal transduction, stomatal guard cells.

endomembrane system and $[Ca^{2+}]_i$ in living guard cells using confocal microscopy. Overall, our data is consistent with multiple sources for $[Ca^{2+}]_i$, including uptake across the plasma membrane and $InsP_3$- or Ca^{2+}-induced Ca^{2+} release from internal stores.

Introduction

Stomatal pores in the leaf epidermis regulate both water loss *via* transpiration and CO_2 uptake for photosynthesis through changes in the turgor of the surrounding guard cells. The signal transduction pathway(s) initiating stomatal movements in response to several closing stimuli are thought to involve changes in cytosolic free calcium levels ($[Ca^{2+}]_i$) (Fricker *et al.* 1990). However, most evidence supporting this hypothesis is indirect from experiments in which external Ca^{2+} levels ($[Ca^{2+}]_e$) have been modulated, Ca^{2+}-channel inhibitors and Ca^{2+}-ionophores have been applied, or measurements have been made of $^{45}Ca^{2+}$ fluxes (De Silva *et al.* 1985a; De Silva *et al.* 1985b; Schwartz, 1985; Inoue and Katoh, 1987; Schwartz *et al.* 1988; Smith and Willmer, 1988; MacRobbie, 1989).

Direct demonstration of signal-response coupling *via* Ca^{2+} requires measurement of changes in $[Ca^{2+}]_i$ during stomatal responses *in vivo*. This has recently become possible with the introduction of a family of fluorescent dyes (Quin-2, Indo-1, Fura-2 and Fluo-3) that exhibit a high affinity for Ca^{2+} (Grynkiewicz *et al.* 1985). Of these, Indo-1 and Fura-2 exhibit a marked spectral shift on binding Ca^{2+}, allowing them to be used for fluorescence ratio imaging or photometry. The ratio method involves obtaining a ratio of the fluorescence intensity at two wavelengths, of which at least one is influenced by $[Ca^{2+}]_i$. Photometric measurements are averaged over a defined area of the specimen, whilst ratio images are calculated pixel by pixel from images acquired separately at each wavelength. The ratio is proportional to the ion concentration but crucially is independent of the amount of dye and therefore effectively removes many optical artifacts (eg. uneven cytoplasmic thickness or loss of signal from leakage or photobleaching, Cobbold and Rink, 1987).

A second powerful experimental approach for studies on the role of $[Ca^{2+}]$ *in vivo* has recently been developed. Caged compounds can be introduced directly into the cytoplasm and then released in their active form by brief flashes of ultraviolet (uv) light. Of particular interest are photoactivateable forms of Ca^{2+} and inositol-1,4,5-triphosphate ($InsP_3$). The latter is a second messenger which is now well characterised as mobilising the internal stores of Ca^{2+} in animal cells (Berridge and Irvine, 1989). Flashing with UV light normally precludes the use of the ratio dyes Indo-1 and Fura-2, which are usually excited at 340–350 nm. However $[Ca^{2+}]_i$ can still be monitored with the non-ratio Ca^{2+} indicator Fluo-3 (excitation wavelength 490 nm) in cells loaded with caged probes.

A third recent development is the application of confocal laser scanning microscopy (CLSM) to obtain high resolution, blur-free serial optical sections of living cells (Shotton, 1989). The increased spatial resolution and 3-D sampling capability facilitate a more precise description of the internal membrane morphology *in vivo* and the sites of intracellular Ca^{2+} release. In this chapter we report the use of these novel experimental approaches to investigate the spatial and temporal patterns in $[Ca^{2+}]_i$ during responses of single living guard cells to a variety of stimuli.

Methodology

Abaxial epidermis from *Commelina communis* L. was mounted in a microscope chamber and perfused with CO_2-free media (50 mM KCl, 1 mM MES, pH 6.1 with KOH). 1–10 μM solutions of the dual emission dye Indo-1 were then introduced into the guard cells iontophoretically (Gilroy *et al.* 1991). Guard cells with dye localised in the cytoplasm were viewed by epifluorescence microscopy using a Nikon Diaphot inverted microscope. Fluorescence intensity was measured at wavelengths of 405 nm and 480 nm with excitation at 350 nm using either a dual emission photomultiplier system (Newcastle Photonic Systems, Newcastle, UK) or an Extended ISIS CCD camera (Photonic Science, Cambridge, UK) fitted with a computer-controlled filter wheel (Newcastle Photonic Systems). Ratio images were produced using a Dell 310 personal computer fitted with a Synapse framestore and Semper-6-plus image processing language (Synoptics Ltd; Cambridge, UK) and maths co-processor. Dark current and average auto fluorescence over the whole cell (for photometry) or cytosol (for imaging) were subtracted on line from the data at each wavelength and ratios calculated.

$[Ca^{2+}]_i$ and InsP_3 were artificially increased by UV photolysis of their caged forms (Nitr-5 and InsP_3-nitrophenylethyl ester, respectively) with Ca^{2+} continuously monitored by photometry of Fluo-3 (excitation wavelength 405 nm: emission wavelength 530 nm). UV photolysis of these caged probes was performed using 350 nm wavelength light from the 75 W microscope illuminator with a manually controlled shutter (Gilroy *et al.* 1990*a,b*).

Confocal microscopy and 3-D reconstruction were performed using a BioRad MRC500/Nikon Diaphot combination operated in fluorescence mode with excitation at 488 nm (Fricker and White, 1990). Endoplasmic reticulum was stained with DiOC$_6$ (1 μM, 10 min) and the vacuole stained with acridine orange (1 μM, 10 min).

Fluorescent dyes and caged probes were obtained from Molecular Probes Inc. (Eugene, Oregon, USA) and Calbiochem Corporation (San Diego, California, USA). Unless otherwise stated, all other chemicals were of analytical grade or higher and supplied either by Sigma (Poole, Dorset, UK) or BDH (Poole, Dorset, UK).

Visualisation and measurement of $[Ca^{2+}]_i$ transients

Typical results illustrating measurement of $[Ca^{2+}]_i$ using fluorescence ratio photometry of single microinjected guard cells are shown in Figs 1, 11 and 12. In

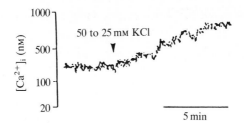

Fig. 1. Fluorescence ratio photometric trace of $[Ca^{2+}]_i$ in guard cells microinjected with Indo-1 after which $[K^+]_e$ was lowered from 50 to 25 mM at the point indicated (data from Gilroy *et al.* 1990*a*, with permission).

the absence of stimulation a constant resting level of 100–200 nM was maintained. Fig. 1 shows one example of the effects of stimulation. In this case $[Ca^{2+}]_i$ increased to >500 nM within 10 min of lowering the external K^+ ($[K^+]_e$) from 50 to 25 mM. Two time points from a set of comparable imaging data are presented in Figs 2 and 3. Although the photometry gives an overall impression of the rise in $[Ca^{2+}]_i$, the imaging data contains far more spatial information and illustrates the heterogeneity in Ca^{2+} response in different parts of the cytoplasm. Transient increases occurred in the cytoplasm adjacent to the nucleus and vacuole whilst $[Ca^{2+}]_i$ in the nucleoplasm remained at a constant low level. Similar changes in $[Ca^{2+}]_i$ have been observed in most, but not all, of the stomatal closing responses examined (see below; Gilroy *et al.* 1991).

The main results from a series of such experiments using both photometry and imaging are summarised below (data from Fricker *et al.* 1990; Gilroy *et al.* 1990*a,b*; Gilroy *et al.* 1991):

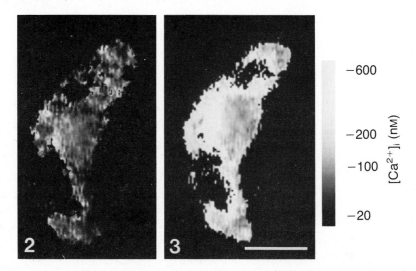

Figs 2, 3. Fluorescence ratio changes in $[Ca^{2+}]_i$ in a guard cell microinjected with Indo-1 before (Fig. 2) and after (Fig. 3) lowering $[K^+]_e$ from 50 to 25 mM. Scale bar 10 μm (data from Gilroy *et al.* 1990*a*, with permission).

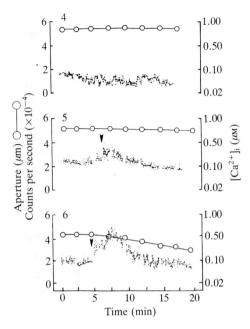

Figs 4–6. Photometric traces showing the effects of caged Ca^{2+} on $[Ca^{2+}]_i$ and stomatal aperture (circles) from guard cells previously microinjected with Fuo-3. Fig. 4. No photolysis (control). Fig. 5. UV photolysis (arrow) for 5 s. Fig. 6. UV photolysis (arrow) for 10 s. Note that stomatal closure only seems to be triggered after a threshold $[Ca^{2+}]_i$ is reached. Also note that Ca^{2+} elevation apparently occurs in two stages, which possibly indicates Ca^{2+} induced Ca^{2+} release from intracellular stores (data from Gilroy *et al.* 1990*b*, with permission).

(1) Resting $[Ca^{2+}]_i$ levels were maintained between 100 and 200 nM ($n=75$) (Figs 1, 11 and 12).

(2) $[Ca^{2+}]_i$ increased in response to several treatments, including: raising extracellular Ca^{2+} concentration $[Ca^{2+}]_e$, lowering extracellular K^+ concentration $[K^+]_e$ (Figs 1–3), application of the non-fluorescent Ca^{2+}-ionophore Br-A23187, and release of Ca^{2+} or $InsP_3$ from their caged forms directly into the cytoplasm (Figs 5, 6, 8 and 9).

(3) An increase in $[Ca^{2+}]_i$ caused stomatal closure if a threshold of about 600 nM was exceeded, irrespective of the origin of the stimulus (Figs 6, 8 and 9).

(4) Chelation of $[Ca^{2+}]_e$ with EGTA reduced $[Ca^{2+}]_i$ but did not promote stomatal opening. In addition, $[Ca^{2+}]_i$ did not alter during stomatal opening stimulated by fusicoccin.

(5) $[Ca^{2+}]_i$ did not increase in a uniform manner throughout the cytoplasm. Transient changes occurred in localised regions around the nucleus and vacuole. Free Ca^{2+} in the nucleus itself remained at a constant low level in all treatments, except after prolonged exposure to Br-A23187.

(6) Increases in $[Ca^{2+}]_i$ still occurred in response to decreasing $[K^+]_e$ and $InsP_3$ in the presence of Ca^{2+} channel blockers or after chelation of $[Ca^{2+}]_e$ with EGTA.

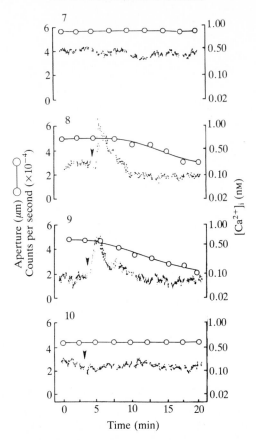

Figs 7–10. Photometric traces showing the effects of caged $InsP_3$ on $[Ca^{2+}]_i$ and stomatal aperture (circles) from guard cells previously microinjected with Fluo-3. Fig. 7. No photolysis (control) of caged $InsP_3$ in the pressure of La^{3+}. Fig. 10. UV photolysis (arrow) of caged ATP (control). Note that the release of ATP from its caged form does not influence stomatal aperture (data from Gilroy *et al.* 1990b, with permission).

The time course of increases in $[Ca^{2+}]_i$ was slower and lower in magnitude, but suggests that release of Ca^{2+} from intracellular stores had occurred.

(7) ABA caused stomatal closure, but only rarely in less than 25 % of guard cells analysed ($n=54$) was there an associated change in $[Ca^{2+}]_i$ (Figs 11, 12). This is in contrast to a recent report in which ABA resulted in an increase in guard cell $[Ca^{2+}]_i$ preceding closure in 8 out of 10 cells examined (McAinsh *et al.* 1990).

These results indicate a far greater complexity in the signal transduction pathways of guard cells than previously envisaged. Stomatal closure may be specifically initiated by increases in $[Ca^{2+}]_i$ above a threshold point. However, there would appear to exist both Ca^{2+}-dependent and -independent pathways resulting in the closure response. $[Ca^{2+}]_i$ may be modulated by $InsP_3$-stimulated release from internal stores and possibly by net influx across the plasma

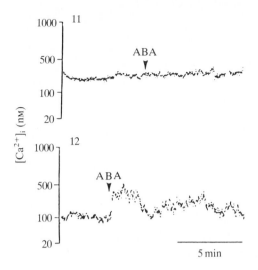

Figs 11, 12. Ratio photometric traces of [Ca²⁺]ᵢ in guard cells microinjected with Indo-1 after which ABA was added (arrow). Note that in Fig. 11 there is no change after ABA addition but the experiments shown in Fig. 12, irregular oscillations in [Ca²⁺]ᵢ occur. Stomatal closure occurred with similar kinetics in both treatments (data from Gilroy *et al.* 1990*a*, with permission).

membrane. There is no evidence, however, for $[Ca^{2+}]_i$ operating as a signal in stomatal opening.

Levels of $[Ca^{2+}]_i$ are maintained in dynamic equilibrium through the action of channels, pumps and buffering by Ca^{2+}-binding proteins and phospholipids (Carafoli, 1987). Binding of Ca^{2+} also reduces its mobility in the cytoplasm and permits formation of short range gradients in Ca^{2+} (Hodgkin and Keynes, 1957). The observed Ca^{2+} 'hot spots' (Fig. 3) may arise from altered fluxes across either the plasma membrane or from intracellular stores, or reflect differential Ca^{2+} buffering capacity in the cytoplasm with a uniform influx. In the following paragraphs we will discuss the current evidence for the first two of these possible mechanisms. The alternative possibility that there is a differential distribution of Ca^{2+}-binding proteins and hence buffering capacity in the cytoplasm, possibly reflecting a spatial organisation of one or more transduction pathways, is purely hypothetical at present.

Evidence for Ca²⁺ fluxes across the plasma membrane

It is clear from the data of MacRobbie (1989) that a significant influx of Ca^{2+} $(0.3–2.3\,\mathrm{pmol\,cm^{-2}\,s^{-1}})$ occurs across the guard cell plasma membrane, even before stimulation. As $[Ca^{2+}]_i$ remains constant under these conditions (Gilroy *et al.* 1990*a*,*b*, 1991), an equivalent efflux is also presumed to occur to maintain a steady state equilibrium. Rapid elevation of $[Ca^{2+}]_e$ is known to cause stomatal closure (De Silva *et al.* 1985*a*,*b*; Schwartz, 1985; Inoue and Katoh, 1987; Schwartz *et al.* 1988; Smith and Willmer, 1988) although $[Ca^{2+}]_i$ increased less than five fold, despite a fifty-fold increase in $[Ca^{2+}]_e$ Gilroy *et al.* 1991), indicating a new steady-state level was reached. It is not clear to what extent modulation of $[Ca^{2+}]_e$ or Ca^{2+} influx is a critical step in signal transduction *in vivo*. Indeed, MacRobbie (1989) found effects of ABA on $^{45}Ca^{2+}$-influx ranged from stimulation, through no effect to inhibition, and concluded that ABA did not cause any long term (>2 min) elevation in Ca^{2+}-influx, although more rapid fluctuations might still occur.

Other evidence for plasma membrane fluxes of Ca^{2+} in stomatal closure involve restriction of Ca^{2+}-uptake during stimulation by either lowering $[Ca^{2+}]_e$ or application of impermeant Ca^{2+}-channel blockers. It is difficult to interpret these results unequivocally, however, as the inhibitions observed are rarely 100%, or occur at levels when non-specific effects might be anticipated. Inhibition of stomatal closure induced by ABA ranged from 80% (La^{3+}) to 40% (verapamil and nifedipine) (De Silva *et al.* 1985*a*), whilst Brindley (1990) found virtually no effect of methoxyverapamil, nifedipine or bepridil on dark-induced closure in *Vicia faba*. Our data indicates that La^{3+} slows the elevation of $[Ca^{2+}]_i$ induced by raising $[Ca^{2+}]_e$, but does not completely abolish it or the ensuing stomatal closure (Gilroy *et al.* 1991). Similarly, MacRobbie (1989) found La^{3+} reduced Ca^{2+} influx about eight-fold, but again did not abolish it completely.

Results from experiments where $[Ca^{2+}]_e$ was reduced, usually by chelation with EGTA, are equally variable. An absolute requirement for extracellular Ca^{2+} has been reported for closure induced by darkness, lowering $[K^+]_e$, by CO_2 (Schwartz *et al.* 1988) or ABA (De Silva *et al.* 1985*b*). In contrast, Smith and Willmer (1988) found ABA caused shrinkage of guard cell protoplasts (analagous to stomatal closure) in the absence of Ca^{2+}, whilst Curvetto and Delmastro (1990) report an interference between Ca^{2+} and ABA rather than the synergism reported by De Silva *et al.* (1985*b*).

The available evidence indicates that $[Ca^{2+}]_i$ is linked to $[Ca^{2+}]_e$, with levels maintained in part by cycling of Ca^{2+} across the plasma membrane. An explanation for the heterogeneity of $[Ca^{2+}]_i$ after stimulation could be that there is a physical clustering of Ca^{2+} pumps, channels or their control systems (e.g. receptors or membrane depolarisation). Voltage-gating for plasma membrane Ca^{2+} channels, which open on membrane depolarisation in guard cells, has been reported (MacRobbie, 1989) and would be expected to result in feed-forward stimulation of further Ca^{2+} influx, once triggered.

Ca^{2+} release from internal stores

There are two lines of evidence for release of Ca^{2+} from internal stores:

(1) Elevation of $[Ca^{2+}]_i$ still occurs in the absence of external Ca^{2+} or in the presence of Ca^{2+}-channel blockers when plasma membrane influx of Ca^{2+} should be severely reduced (e.g. Smith and Willmer, 1988; Gilroy *et al.* 1991).

(2) Elevation of Ca^{2+} can be induced by release of $InsP_3$ from its caged form directly into the cytoplasm in the presence of Ca^{2+}-channel blockers externally

(Fig. 9; Gilroy *et al.* 1990*b*). InsP_3 normally stimulates Ca^{2+} release from internal Ca^{2+} stores (Berridge and Irvine, 1989; Einspahr and Thompson, 1990).

Identity of the intracellular Ca^{2+} store(s)

The vacuole, endoplasmic reticulum and specialised vesicles (calciosomes) have all been suggested as potential stores of Ca^{2+} in plant cells. InsP_3 has been shown to release $^{45}Ca^{2+}$ from both vacuoles (Ranjeva *et al.* 1988) and tonoplast vesicles (Schumaker and Sze, 1987), and the electrical properties of this release have been characterised using patch-clamp techniques on intact vacuoles (Alexandre *et al.* 1990; Alexandre and Lasselles, 1990). No similar information is presently available for Ca^{2+} release from the endoplasmic reticulum or calciosomes in plants. Elevation of $[Ca^{2+}]_i$ induced by intracellular release of InsP_3 from its caged form (Gilroy *et al.* 1990*b*) was the first evidence for a link *in vivo* between InsP_3 and intracellular Ca^{2+} release in plant cells, although the site of action of InsP_3 was not examined in detail.

We have started to use CSLM approaches to analyse the contributions, if any, of Ca^{2+} release from the endoplasmic reticulum or vacuole. The first approach involves correlation of the site of Ca^{2+} release with specific structures in the guard cell. This exploits the optical sectioning and 3-D reconstruction capabilities of CSLM to determine the 3-D spatial organisation of the endomembrane systems present in the guard cell. Typical results for the vacuole and endoplasmic reticulum are shown in Figs 13 and 14. The second approach involves direct measurement of Ca^{2+} dynamics with Fluo-3 at higher spatial resolution than is possible with our current, non-confocal fluorescence ratio imaging system.

The acridine orange-stained vacuole is a single compartment running throughout the guard cell in >90 % of the cells so far examined (Fig. 13). The nucleus is usually positioned near the midpoint of the thickened wall adjacent to the stoma and the chloroplasts are distributed in the peripheral cytoplasm. The DiOC$_6$-stained endoplasmic reticulum is concentrated around the nucleus, around the periphery of the cell, and in cytoplasmic strands passing through the vacuole (Fig. 14). When stains such as DiOC$_5$ or DiOC$_7$ are used the mitochondria are more clearly seen as discrete fluorescent spots which are significantly smaller than the chloroplasts (data not shown). Unfortunately we have been unable to unequivocally correlate localised Ca^{2+} increases using fluorescence ratio imaging (Figs 2 and 3) with either the vacuole or endoplasmic reticulum imaged by CLSM (Figs 11 and 12). However, there is a tendency for 'hot spots' of Ca^{2+} to be localised in cytoplasm around the nucleus. This may reflect localised Ca^{2+} release from either part of the endoplasmic reticulum network or vacuole in the nuclear region.

It is clear that the spatial resolution of our non-confocal fluorescence ratio imaging system is insufficient to directly visualise the origin of the Ca^{2+} leading to these changes. In addition, the fluorescence measured using this technique is derived both from excitation of dye in the focal plane midway through the cell and,

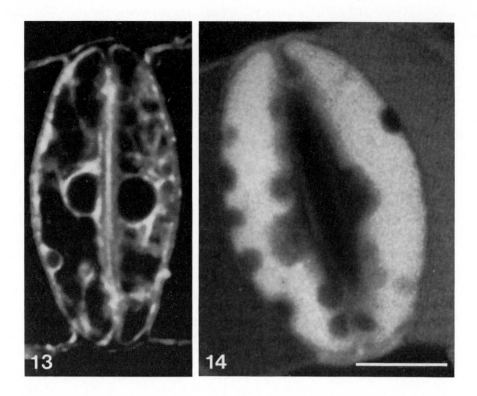

Figs 13, 14. Confocal fluorescence micrographs of guard cells. Fig. 13. Stained with DiOC$_6$ to visualize internal membranes. Projection of ten 1 μm thick optical sections through median plane of stoma. Fig. 14. Stained with acridine orange to visualise vacuole. Single 1 μm optical section from median plane through stoma. Bar=10 μm.

to a lesser extent, in the cones of illumination extending above and below this plane. Care must be taken, therefore, when interpreting spatial positions in the final 2-D image, given the poorly defined 3-D nature of the sampling achieved.

It should be possible to achieve greater precision in the spatial localization of the origin of Ca^{2+} using CLSM and the single wavelength Ca^{2+} indicator Fluo-3. We have recently initiated a study of this but it is too early to report our results. In the longer-term it will be possible to combine fluorescence ratio imaging with CLSM of Ca^{2+} but this will require either operating with UV excitation of Indo-1 (the technology for doing this is still in its infancy) or using Ca^{2+} ratio dyes at longer wavelengths; the latter dyes are not yet available.

The link between the kinetics of Ca^{2+} transients and the closure response

Two areas need defining as far as the kinetics of the Ca^{2+} transients are concerned: (1) The mechanism controlling the magnitude, duration and frequency (if appropriate) of the Ca^{2+} transients induced in response to each stimulus, and

(2) which features of the Ca^{2+} transient are important in transduction of the signal and the final response of the guard cell.

In animal cells, information is encoded in graded, threshold and frequency form. Different effectors stimulate elevation of Ca^{2+}, but with different kinetics and by different mechanisms (Berridge and Irvine, 1989). Initial reports in guard cells using photometry measurements have suggested that 'oscillations' in $[Ca^{2+}]_i$ can occur (McAinsh *et al.* 1990; Gilroy *et al.* 1991). Although these results may indicate an overall trend in $[Ca^{2+}]_i$, detailed interpretation of photometry measurements in guard cells is not possible because intensities at each wavelength are spatially averaged over the whole cell prior to calculation of the ratio, and the signal at each wavelength is dominated by fluorescence from the nucleus (Gilroy *et al.* 1990*a*, 1991). Results, therefore, tend to underestimate dynamic changes in the cytoplasm, particularly transients in restricted areas. Imaging approaches have revealed the heterogeneous nature of the Ca^{2+} response and suggest that short term transients and oscillations are prevalent in more restricted areas of the cell (Gilroy *et al.* 1991).

Release of Ca^{2+} directly into the cytoplasm from its caged form suggests that a threshold in Ca^{2+} triggers closure (Figs 4–6: Gilroy *et al.* 1990*b*). The pulse of Ca^{2+} delivered by photolysis should occur only during exposure to the UV light source. In many cells examined $[Ca^{2+}]_i$ declined more or less exponentially to resting levels (Fig. 5). It seems likely that this reflects the removal of Ca^{2+} from the cytoplasm to intracellular stores or across the plasma membrane and indicates the normal operation of a Ca^{2+}-homeostasis mechanism. Similarily release of caged $InsP_3$ stimulated a rise in $[Ca^{2+}]_i$, which commonly also decayed exponentially (Fig. 8). However, in a number of cases $[Ca^{2+}]_i$ was maintained or even increased further following its initial elevation after termination of the UV pulse. $[Ca^{2+}]_i$ decreased exponentially thereafter (Fig. 6). These results may indicate Ca^{2+}-induced Ca^{2+} release from other Ca^{2+} stores although it apparently does not always occur.

Loss of turgor (equivalent to stomatal closure) only occurred in the guard cells containing the photoactivated caged compound and continued for a considerable period after the disappearance of the elevated Ca^{2+} (Figs 6, 8 and 9). The loss of turgor was reversible upon application of fusicoccin.

With the exogenous stimuli we have so far used the time course, averaged over the whole cell for the increase in $[Ca^{2+}]_i$, is much longer than that observed with the release of caged Ca^{2+} or $InsP_3$: the elevated levels were sometimes maintained for in excess of 20 min. However, the imaging data indicates that in any given region of the cell transient increases are of much shorter duration, although they may continue well after the closing response is initiated (Gilroy *et al.* 1991). Imaging clearly demonstrates the spatial and temporal heterogeneity of $[Ca^{2+}]_i$, which is not evident from the photometry data.

In a series of experiments involving several different closing stimuli, the rate of stomatal closure was relatively constant despite markedly different profiles for changes in $[Ca^{2+}]_i$ (Gilroy *et al.* 1991). We interpret this data as indicating: (1)

elevation of $[Ca^{2+}]_i$ above a threshold for a short period (possibly 3–5 min) is sufficient to initiate closure, (2) interaction with other parts of the signal transduction pathway maintains the response in the absense of further detectable increases in $[Ca^{2+}]_i$, and (3) the rate of response is finally governed by other factors (e.g. the maximum rate of ion efflux).

Signal transduction in guard cells *in vivo*

An important distinction has to be made between stimuli used to dissect the components of the putative signal transduction pathway and the behaviour of that pathway to stimuli *in vivo*. In virtually all experiments on guard cells in epidermal strips, stimuli are presented as stepwise increments in concentration, often of several orders of magnitude. Such abrupt perturbations may allow identification of possible signal transduction elements, but do not necessarily permit us to conclude how these stimuli operate *in vivo*. In relation to this we will consider firstly, ABA as a stimulus, and secondly, whether changes in $[Ca^{2+}]_e$ constitute a physiologically relevant signal.

ABA levels increase *in vivo* over a period of hours to days in response to water stress. Experimentally, levels are increased over a period of a few seconds. Apart from any role in signal transduction, ABA is a weak acid and partitions into the cell where it is trapped by the change in pH (Baier and Hartung, 1988). An immediate consequence would be an expected acidification of the cytoplasm that in turn may trigger a whole variety of cellular responses including some apparently contradictory to the role of ABA as a closing stimulus (e.g. stimulation of the plasma membrane H^+-ATPase).

Levels of $[Ca^{2+}]_e$ have been suggested to influence the maximum opening of stomata *in vivo* (Atkinson *et al.* 1989) and account for the differential behaviour of adaxial and abaxial stomata in *C. communis* (De Silva *et al.* 1986). Ca^{2+} is bound to negatively charged groups or coordinated by hydroxyls present in wall polymers. The total cation-exchange capacity of the wall has been estimated as 300–500 mol m^{-3} (Raschke, 1979), but the apoplastic Ca^{2+} in the vicinity of the guard cells is not known. It can be predicated, however, that a decrease in the apoplast pH during opening (Edwards *et al.* 1988) should displace Ca^{2+} from bound sites in the wall. The increased free $[Ca^{2+}]$ may act directly as a negative-feedback signal limiting the rate or extent of opening. Alternatively, high rates of transpiration may result in an accumulation of Ca^{2+} in the region of the guard cells, increasing stomatal sensitivity to other closing stimuli. In this respect, it is noteworthy that *C. communis* is remarkably sensitive to externally applied Ca^{2+}, whilst at least ten-fold higher levels are required for *Vicia faba*. It is possible that part of this discrepancy can be attributed to adaption to the prevailing $[Ca^{2+}]_e$. In *Commelina* substantial amounts of calcium oxalate are deposited in the vacuoles of mesophyll and epidermal cells, in contrast to *Vicia*, possibly maintaining lower $[Ca^{2+}]_e$. It is necessary to measure $[Ca^{2+}]_e$ *in vivo*, to determine the potential role of the apoplast in the regulation of signal transduction in guard cells.

Conclusions

The power of new analytical imaging techniques has revealed a vast complexity of intracellular signal transduction in terms of spatial organisation and kinetics. This complexity is not unexpected and we believe that a simple linear signal transduction pathway is unlikely to be able to integrate the wide diversity of stimuli to give the coherent physiological responses that we observe. Our data is so far consistent with a major but not exclusive role for Ca^{2+} in closing responses. Calcium employed for closure comes predominantly from intracellular sources but there may be a component from the apoplast. A variety of mechanisms probably operate in mobilising this Ca^{2+} (e.g. plasma membrane Ca^{2+}-channels/pumps and $InsP_3$- or Ca^{2+}-release from internal stores). Unravelling the complexity of interactions during signal transduction in guard cells will clearly provide an exciting challenge for the future.

This work was funded by grants from the Gatsby Foundation, Nuffield Foundation, Royal Society, AFRC and SERC.

References

ALEXANDRE, J. AND LASSALLES, J. P. (1990). Effect of D-myo-inositol 1,4,5-triphosphate on the electrical properties of the red beet vacuole membrane. *Pl. Physiol.* **93**, 837–840.

ALEXANDRE, J., LASSALLES, J. P. AND KADO, R. T. (1990). Opening of Ca^{2+} channels in isolated red beet vacuole membrane by inositol 1,4,5-triphosphate. *Nature* **343**, 567–570.

ATKINSON, C. J., MANSFIELD, T. A., KEAN, A. M. AND DAVIES, W. J. (1989). Control of stomatal aperture by calcium in isolated epidermal tissue and whole leaves of *Commelina communis* L. *New Phytol.* **111**, 9–17.

BAIER, M. AND HARTUNG, W. (1988). Movement of abscisic acid across the plasmalemma and the tonoplast of guard cells of *Valerianella locusta*. *Bot. Acta* **101**, 332–337.

BERRIDGE, M. J. AND IRVINE, R. F. (1989). Inositol phosphates and cell signalling. *Nature* **341**, 197–205.

BRINDLEY, H. M. (1990). Effects of light/dark and calcium-channel drugs on fluxes of $^{86}Rb^{+}$ in 'isolated' guard cells of *Vicia faba* L. *Planta* **181**, 440–447.

CARAFOLI, E. (1987). Intracellular calcium homoeostasis. *A. Rev. Biochem.* **56**, 395–433.

COBBOLD, P. H. AND RINK, T. J. (1987). Fluorescence and bioluminescence measurement of cytoplasmic free calcium. *Biochem. J.* **248**, 313–328.

CURVETTO, N. AND DELMASTRO, S. (1990). A biochemical and physiological proposal for stomatal movement: possible involvement of adenosine 3′,5′-cyclic monophosphate. *Pl. Physiol. Biochem.* **28**, 367–378.

DE SILVA, D. L. R., COX, R. C., HETHERINGTON, A. M. AND MANSFIELD, T. A. (1985*a*). Suggested involvement of calcium and calmodulin in the response of stomata to abscisic acid. *New Phytol.* **101**, 555–563.

DE SILVA, D. L. R., COX, R. C., HETHERINGTON, A. M. AND MANSFIELD, T. A. (1986). The role of abscisic acid and calcium in determining the behaviour of adaxial and abaxial stomata. *New Phytol.* **104**, 41–51.

DE SILVA, D. L. R., HETHERINGTON, A. M. AND MANSFIELD, T. A. (1985*b*). Synergism between calcium ions and abscisic acid in preventing stomatal opening. *New Phytol.* **100**, 473–482.

EDWARDS, M. C., SMITH, G. N. AND BOWLING, D. J. F. (1988). Guard cells extrude protons prior to stomatal opening – A study using fluorescence microscopy and pH micro-electrodes. *J. Exp. Bot.* **39**, 1541–1547.

EINSPAHR, K. J. AND THOMPSON, G. A., JR (1990). Transmembrane signalling *via* phosphatidylinositol 4,5-bisphosphate hydrolysis in plants. *Pl. Physiol.* **93**, 361–366.

FRICKER, M. D., GILROY, S. AND TREWAVAS, A. J. (1990). Signal transduction in plant cells and the calcium message. In *Signal perception and transduction in plant cells*. (ed. R. Ranjeva and A. Boudet). Springer-Verlag, Berlin, Heidelberg. Pp. 89–102.

FRICKER, M. D. AND WHITE, N. (1990). Volume measurement of guard cell vacuoles during stomatal movements using confocal microscopy. *Trans. Roy. Microsc. Soc.* **1**, 345–348.

GILROY, S., FRICKER, M. D., READ, N. AND TREWAVAS, A. J. (1990a). Fluorescent ratio imaging and photometry of calcium in living plant cells. *Trans Roy. Microsc. Soc.* **1**, 475–478.

GILROY, S., READ, N. AND TREWAVAS, A. J. (1990b). Elevation of cytoplasmic calcium or caged inositol triphosphate initiates stomatal closure. *Nature* **346**, 769–771.

GILROY, S., FRICKER, M. D., READ, N. AND TREWAVAS, A. J. (1991). The role of cytosolic Ca^{2+} in signal transduction in stomatal guard cells of *Commelina communis*. *Pl. Cell* **3** (in press).

GRYNKIEWICZ, G., POENIE, M. AND TSIEN, R. Y. (1985). A new generation of Ca^{2+} indicators with greatly improved fluorescence properties. *J. biol. Chem.* **260**, 3440–3450.

HODGKIN, A. L. AND KEYNES, R. D. (1957). Movements of labelled calcium in squid giant axons. *J. Physiol.* **138**, 253–281.

INOUE, H. AND KATOH, Y. (1987). Calcium inhibits ion stimulated opening in epidermal strips of *Commelina communis* L. *J. exp. Bot.* **38**, 142–149.

MACROBBIE, E. A. C. (1989). Calcium influx at the plasmalemma of isolated guard cells of *Commelina communis*. *Planta* **178**, 231–241.

MCAINSH, M. R., BROWNLEE, C. AND HETHERINGTON, A. M. (1990). Abscisic acid-induced elevation of guard cell cytosolic Ca^{2+} precedes stomatal closure. *Nature* **343**, 186–188.

RANJEVA, R., CARRASCO, A. AND BOUDET, A. (1988). Inositol trisphosphate stimulates the release of calcium from intact vacuoles isolated from *Acer* cell. *FEBS Lett.* **230**, 137–141.

RASCHKE, K. (1979). Movements of stomata. In *Encyclopedia of Plant Physiology. New Series* **7**, 381–441 (ed. W. Haupt and M. E. Feinleib). Springer-Verlag, Berlin.

SCHUMAKER, K. S. AND SZE, H. (1987). Inositol 1,4,5-trisphosphate releases Ca^{2+} from vacuolar membrane vesicles of oat roots. *J. biol. Chem.* **262**, 3944–3946.

SCHWARTZ, A. (1985). Role of Ca^{2+} and EGTA on stomatal movements in *Commelina communis* L. *Pl. Physiol.* **79**, 1003–1005.

SCHWARTZ, A., ILAN, N. AND GRANTZ, D. A. (1988). Calcium effects on stomatal movement in *Commelina communis* L. *Pl. Physiol.* **87**, 583–587.

SHOTTON, D. M. (1989). Confocal scanning optical microscopy and its application for biological specimens. *J. Cell Sci.* **94**, 175–206.

SMITH, G. N. AND WILLMER, C. M. (1988). Effect of calcium and abscisic acid on volume changes of guard cell protoplasts of *Commelina*. *J. exp. Bot.* **39**, 1529–1539.

Printed in Great Britain © *Society for Experimental Biology 1991*   191

REGULATORY ELEMENTS REQUIRED FOR LIGHT-MEDIATED EXPRESSION OF THE *PETROSELINUM CRISPUM* CHALCONE SYNTHASE GENE

BERND WEIßHAAR, ANNETTE BLOCK, GREGORY A. ARMSTRONG, ANETTE HERRMANN, PAUL SCHULZE-LEFERT and KLAUS HAHLBROCK

MPI für Züchtungsforschung, Abteilung Biochemie, D-5000 Köln 30, Germany

Summary

Chalcone synthase (CHS) catalyzes the committed enzymatic step in flavonoid biosynthesis. In parsley (*Petroselinum crispum*), CHS is encoded by a single gene locus. Transcriptional activation of the gene in response to UV-containing white light has been demonstrated. Analysis of the CHS gene promoter by *in vivo* footprinting revealed four short sequences, designated Boxes I, II, III, and IV, which contain guanosine residues with altered reactivity to the methylating agent dimethylsulfate in UV-treated *versus* untreated parsley cells.

Studies were performed to characterize the functional components of the CHS gene promoter using a parsley protoplast transient expression system. By deletion and block-mutation analyses it was shown that Boxes I and II act together as a *cis*-acting unit and are necessary components of the minimal, light-responsive CHS gene promoter. The Box II sequence, which is similar to the conserved G Box sequence defined in promoters of ribulose 1,5-bisphosphate carboxylase small subunit (RBCS) genes, has been subjected to detailed analysis by site-directed mutagenesis. The heptameric sequence 5'-ACGTGGC-3' has been defined as the critical core of Box II required for light induction in the context of the CHS gene minimal promoter. Box II is functionally equivalent to a second, sequence-related element (Box III) that can replace Box II in an orientation-dependent manner.

Chimaeric promoter-fusion constructs to the GUS reporter gene demonstrated that Boxes I and II, together constituting a *cis*-acting unit, are necessary and sufficient for light-mediated activation of the CHS gene promoter.

Introduction

Throughout their development and life cycle, organisms are exposed to various biotic and abiotic stress factors. Plants have evolved an array of mechanisms to protect themselves against these stresses. As a response to the potentially damaging effects of the abiotic stress factor UV light, plants accumulate UV-absorbing substances in the exposed tissue. The main compounds deposited,

Key words: promoter structure, chalcone synthase, *cis*-regulatory elements, light induction, ACGT factors.

which absorb light of 230 to 380 nm, are flavonoids (Hahlbrock *et al.* 1982). Flavonoids are a diverse class of substances that occur in all higher plants in the structurally related forms of anthocyanins, flavones, flavanones, flavonols and isoflavones, among others. The basic biochemical structure in all cases is a chalcone scaffold (Fig. 1). In addition to their role as UV protectants, flavonoids were also found to be involved in flower and fruit pigmentation (Harborne and Turner, 1984) and as phytoalexins in plant-pathogen interactions (Dixon, 1986). The increased synthesis of flavonoids in response to UV light is preceded by the transient, coordinated expression of the enzymes catalyzing the formation of these compounds (Chappell and Hahlbrock, 1984; Scheel *et al.* 1987). Chalcone synthase catalyzes the committed enzymatic step in the flavonoid-specific branch of phenylpropanoid metabolism (Hahlbrock and Grisebach, 1979). CHS enzymatic activity leads to the stepwise condensation of three acetate residues from malonyl-CoA with 4-coumaroyl-CoA to give 4,2',4'6'-tetrahydroxy-chalcone (Heller and Hahlbrock, 1980; Ebel and Hahlbrock, 1982). The main substrate of CHS, 4-coumaroyl-CoA, is provided by the general phenylpropanoid metabolism (Fig. 1).

Extensive studies using cell suspension cultures and intact leaves of parsley have revealed that flavonoids are only present in minor amounts in dark-grown cells or etiolated leaves. Upon irradiation with UV light or with UV-containing white light, flavonoids accumulated in the vacuole of cultured parsley cells (Matern *et al.* 1983). In leaves exposed to UV light the vacuolar accumulation is restricted in a tissue-specific manner to epidermal cells (Schmelzer *et al.* 1988).

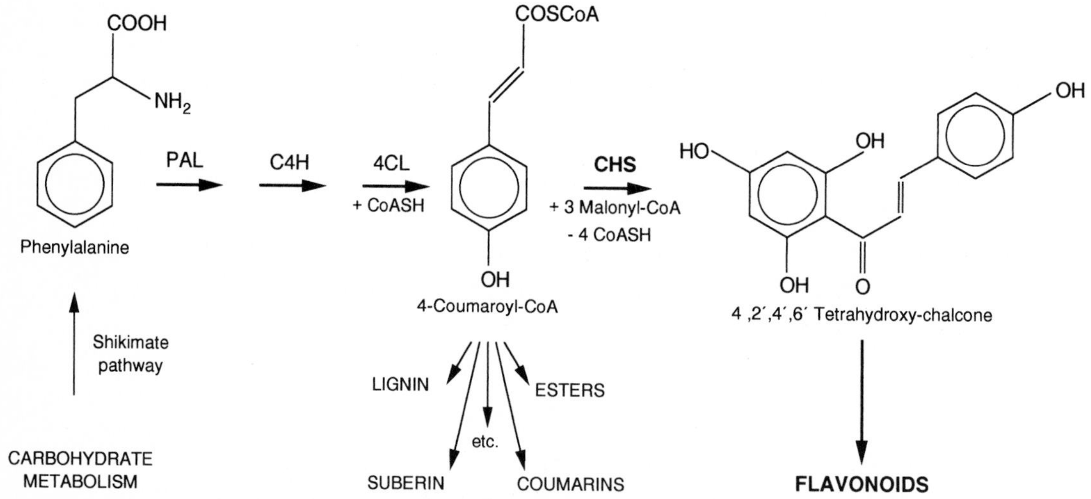

Fig. 1. Schematic representation of a selected portion of the flavonoid biosynthetic pathway. The enzymes of general phenylpropanoid metabolism and the key step towards flavonoid biosynthesis are shown: phenylalanine ammonia-lyase (PAL), cinnamic acid 4-hydroxylase (C4H), 4-coumarate:CoA ligase (4CL), chalcone synthase (CHS). The branch-point reaction for the formation of flavonoids is catalyzed by CHS and is specifically induced by UV light in parsley.

The need for various phenylpropanoid products varies during plant development and in response to changing environmental conditions, including stresses like pathogen attack or short-wavelength light. As a consequence, direction and intensity of metabolic fluxes through general phenylpropanoid and subsequent biochemical pathways are strictly regulated (see Hahlbrock and Scheel (1989) for review of phenylpropanoid metabolism). We focus our interest on the regulatory events necessary for light-dependent gene expression. Because of its clear response to light and its central role in flavonoid biosynthesis, CHS is a key candidate for molecular dissection of regulatory elements mediating light-dependent gene regulation in the biochemically well defined parsley system.

Induction of CHS gene expression by light

In many plant species CHS gene expression is strongly induced by more than one signal, e.g. floral development or light in *Petunia*, elicitor or light in french bean and soybean (see Dangl *et al.* 1989, for review). In cultured parsley cells, maximal CHS expression is UV-dependent, although blue light, red light, diurnal rhythm and developmental state of the tissue have additional modulating effects (Kreuzaler *et al.* 1983; Ohl *et al.* 1989). Cloning of cDNAs encoding CHS (Reimold *et al.* 1983) permitted experiments to determine the kinetics of CHS mRNA accumulation *in vivo* (Kreuzaler *et al.* 1983). Maximal transcriptional activity of the respective gene, as measured by run-on transcription in isolated nuclei, occurs a few hours after onset of irradiation with UV-containing white light (Chappell and Hahlbrock, 1984). There is a lag period of approximately 2 h before the first increase in transcriptional activity of the CHS gene is detectable. This lag period is modulated by light quality and can be circumvented by a pretreatment with blue light. This effect lasts for at least 20 h after blue light treatment (Ohl *et al.* 1989).

CHS transcripts are located in the epidermis of irradiated leaves, as shown by *in situ* hybridization experiments using the CHS cDNA as a probe (Schmelzer *et al.* 1988). The same tissue-specific localization was found for CHS protein and for the end products of flavonoid biosynthesis. Importantly, the kinetics of CHS induction in these *in planta* studies paralleled the results obtained with parsley cell suspension cultures.

CHS is encoded by one gene in parsley

Genes encoding CHS have been structurally characterized in various plant species. The deduced amino acid sequences have been well conserved during evolution (Niesbach-Klösgen *et al.* 1987). In parsley, CHS is encoded by a single genetic locus (Herrmann *et al.* 1988). Two allelic forms have been cloned and sequenced. Allele *PcCHS^a*, in contrast to the allele *PcCHS^b*, contains a 927-bp long transposon-like insertion at position −586 relative to the transcription start site, which is defined as position +1. No other differences were detected between

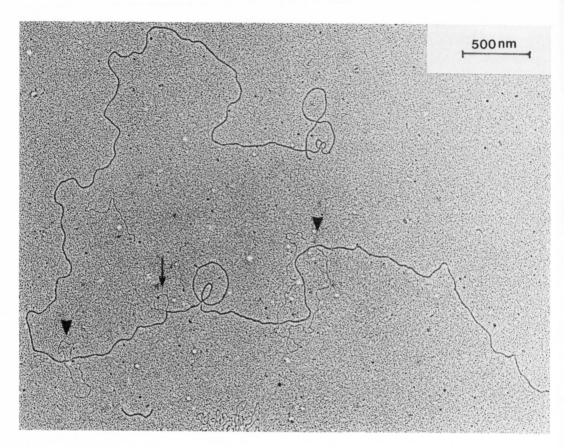

Fig. 2. Heteroduplex analysis of genomic λ-clones containing fragments from *PcCHS^a* and *PcCHS^b*. The arrow marks the position of the DNA loop corresponding to the unique segment in *PcCHS^a*, the arrowheads indicate loops caused by the different size of fragments adjacent to the λ-vector arms. The heteroduplex analysis was performed according to Davis *et al.* (1971).

the two alleles, either by heteroduplex analysis (Fig. 2) or by comparison of the available sequence information. Since both alleles showed light-inducible CHS expression in cell cultures derived from homozygous plants, the insertion in the promoter of allele *PcCHS^a* does not abolish correct light inducibility (Herrmann *et al.* 1988). The gene structure (Fig. 3) resembles that of most known genes encoding CHS in other species. One intron is present at a position conserved during evolution. Upon splicing of the primary transcript, a cysteine codon (position 65) is formed at the splice site (Niesbach-Klösgen *et al.* 1987). This cysteine codon was found in all CHS genes so far examined (Schröder and Schröder, 1990; Epping *et al.* 1990).

Parsley protoplasts: a transient assay system for inducible gene expression

Parsley protoplasts retain the responsiveness of dark-grown suspension culture

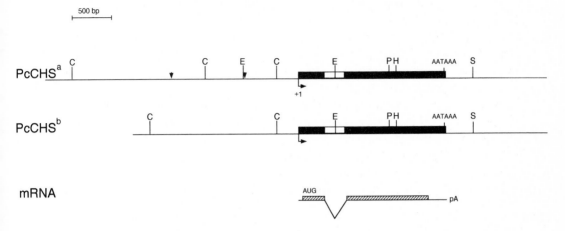

Fig. 3. Diagram of the parsley CHS gene. Some structural features of the two alleles of the parsley chalcone synthase gene, *PcCHS^a* and *PcCHS^b*, and of the common mRNA are shown. Filled boxes indicate the two exons, the open box marks the 263-bp intron. The region between the two arrowheads in *PcCHS^a* represents a transposon-like insertion. From the two mapped start sites (Herrmann *et al.* 1988) only the major one is shown and designated +1 as reference for nucleotide positions in the CHS gene promoter. The cutting sites of the following restriction enzymes are indicated: E, *Eco*RI; C, *Cla*I; H, *Hin*dIII; P, *Pst*I; S, *Sma*I. Positions of the putative poly(A) addition signal, of the poly(A) tail, and of the AUG are given. The hatched boxes on the bar representing the mRNA indicate the coding region.

cells to light or fungal elicitor (Dangl *et al.* 1987). The transcriptionally regulated genes involved in the formation of flavonoid glycosides and coumarin phytoalexins, respectively, are not activated during protoplasting or by treatment of the protoplasts with polyethylene glycol (PEG) to introduce foreign DNA. Thus, chimaeric gene constructs can be introduced into parsley protoplasts to analyze the effects of promoter manipulations on the expression of a reporter gene (Lipphardt *et al.* 1988). The application of the *Escherichia coli* β-D-glucuronidase (GUS) reporter gene (Jefferson *et al.* 1987) proved very useful in the parsley protoplast system (Schulze-Lefert *et al.* 1989a). This reporter gene offered the advantage of low background and rapid and sensitive fluorometric detection of the product formed by the GUS enzyme.

Chimaeric CHS gene promoter constructions (see below) were translationally fused to the GUS reporter gene in the background of a pRT99-based vector, pRT99 GUS JD (Schulze-Lefert *et al.* 1989a), which also contains a cauliflower mosaic virus (CaMV) 35S promoter-driven neomycin phosphotransferase (NPTII) gene. The construction of pUC-based GUS expression vectors also allowed the testing of elicitor-activated promoters (van de Löcht *et al.* 1990) under further optimized experimental conditions, e.g. linearization of the plasmid DNA (Ballas *et al.* 1988). The basic CHS promoter–GUS fusions tested in the pUC-based vectors respond to light qualitatively in the same fashion as in the more complex pRT99-based vectors (B. Weißhaar and K. Hahlbrock, unpublished results).

To broaden the usage of this system, the vector pUC GUS BT-2 (Fig. 4) was constructed that allows the easy cloning of *cis*-acting regulatory elements 5′ to a heterologous core promoter (see below). An important part of such a vector is the target promoter chosen. A 35S promoter truncated to −90 was used in experiments showing evidence that the binding site of the nuclear factor GT-1 confers light responsiveness in transgenic plants (Lam and Chua, 1990). The region between −46 and −90 of the 35S promoter, which contains an activation sequence factor 1 (ASF 1) binding site (positions −83 to −63, Katagiri *et al.* 1989), has been shown to have modulating effects on other promoter elements (Fang *et al.* 1989). For this reason, in pUC GUS BT-2 the CaMV 35S promoter from positions −46 to +8 is used as the target for *cis*-acting element-dependent transcription initiation. Because in this vector a transcriptional fusion is used, the sequence surrounding

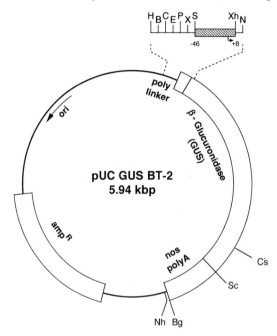

Fig. 4. Schematic representation of pUC GUS BT-2. The nucleotide sequences of pUC GUS BT-2 have the following origin: The modified ATG and the GUS coding sequence between the *Xho*I site and the *Csp*45I site originate from the plasmid pRT103 (Töpfer *et al.* 1988). The rest of the GUS coding sequence, the nos poly(A) addition sequence, and the pUC9 vector backbone are derived from pUC GUS (van de Löcht *et al.* 1990). Sequences including the polylinker in front of the promoter, the promoter itself and the polylinker after the nos poly(A) site were introduced as synthetic oligonucleotides. The length of the CaMV 35S promoter fragment was chosen according to Fang *et al.* (1989). The *Nhe*I and *Bgl*II sites, which are *Xba*I and *Bam*HI compatible, respectively, allow the easy subcloning of the hybrid promoter-GUS construction into other vectors (e.g. plant transformation vectors). All cloning steps were controlled by sequencing. The actual sequence of the whole cassette is available upon request. The hatched region represents the 35S promoter part, the numbers indicate the positions relative to the start site. Abbreviations for restriction enzyme sites are: H, *Hind*III; B, *Bam*HI; C, *Cla*I; E, *Eco*RI; P, *Pst*I; X, *Xba*I; S, *Sal*I; Xh, *Xho*I; N, *Nco*I; Cs, *Csp*45I; Sc, *Sac*I; Bg, *Bgl*II; Nh, *Nhe*I.

the ATG start codon of the GUS open reading frame was introduced in a modified version following the Kozak rules for optimized eukaryotic translational initiation (Kozak, 1983). Details of the construction are given in the legend to Fig. 4.

Genomic footprinting to the parsley CHS gene promoter

To gain insight into the mechanisms governing light-regulated gene expression, experiments were initiated to identify *cis*-acting elements within the parsley CHS gene promoter. Of special interest were elements that are necessary and sufficient for the light-dependent activation of this promoter. Characterization of light-induced protein-DNA interactions in the parsley CHS gene promoter provided a necessary basis for later experiments to define the DNA sequences and protein factors involved in these interactions.

To find possible *cis*-acting sequences, the method of *in vivo* footprinting (Church and Gilbert, 1984; Nick and Gilbert, 1985) was applied. Intact cells were exposed to the strong methylating agent dimethylsulfate (DMS). The concentration of DMS and the length of the treatment were optimized to allow partial methylation of guanosine residues at the N^7 position of the guanine ring. This chemical reaction is believed to be influenced by protein molecules that are in close contact with the DNA, thereby leading to hyper- or hypomethylation of G residues at the protein binding site or in the vicinity thereof. The product of this chemical reaction is a DNA molecule that is cleavable at the position of the modified residue by piperidine. Differences in the cleavage at certain G residues are then compared between *in vivo* -treated and control DNA.

To increase the sensitivity of detection, a restriction fragment containing the promoter region of interest is enriched by centrifugation of appropriately cut genomic DNA through a sucrose density gradient (Schulze-Lefert *et al.* 1989a). After creating a defined end point in the DNA fragment by a second restriction enzyme cut, the Maxam-Gilbert piperidine reaction is carried out to yield a series of genomic DNA fragments, starting at the reference cut and ending at the various methylated G residues. The position of the reference cut is crucial for the display of the resulting sequence ladder. The DNA fragments are then separated on a polyacrylamide gel and detected after electroblotting by indirect end labeling. A strand-specific probe of high specific activity is prepared by primer elongation on a single-stranded template (Church and Gilbert, 1984).

Suspension-cultured parsley cells homozygous for the *CHS^a* allele were used in the genomic footprinting experiments, excluding the problem of detection of multiple sequences by the probe. This problem may arise not only when heterozygous eukaryotic cells are used, but also when gene families are analyzed. The entire CHS gene promoter region from +40 to −615 (see Fig. 3) was analyzed for the appearance of light-induced *in vivo* footprints. Four short sequences showing differential reactivity to DMS in dark-grown *versus* UV-irradiated cells were detected (Schulze-Lefert *et al.* 1989a; Schulze-Lefert *et al.* 1989b). These differences were taken as indications of light-inducible DNA-protein interactions.

The four regions were named Boxes I through IV. They were defined by the outermost residues showing altered reactivity to DMS *in vivo* (see Fig. 5 for a summary of these results).

A time course of light-induced changes in the *in vivo* reactivity of the G residues to DMS more clearly showed the light-dependence of the four footprints in the CHS gene promoter. The footprints appeared approximately one hour after the onset of irradiation and were maintained during the period investigated (Schulze-Lefert *et al.* 1989*a*, 1989*b*). The notion that the four boxes are of functional relevance is strengthened by the fact that the timing of appearance of the footprints is in good agreement with the previously established kinetics of transcriptional activation of the CHS gene (Chappell and Hahlbrock, 1984; Ohl *et al.* 1989).

The four boxes are functionally necessary for light-dependent CHS gene promoter activity

The development of a transient expression system in parsley protoplasts allowed the rapid analysis of the parsley CHS gene promoter in the homologous system. A *PcCHS*^{a} promoter fragment, containing sequences up to -615, was fused translationally to the GUS reporter gene. Serial deletions of 5' sequences from the basic construct 041 (Fig. 6) showed that sequences starting at position -226 relative to the transcriptional start site (construct 061) retain all the necessary information to direct light-regulated GUS expression. Further deletion to -100 resulted in a complete loss of light responsiveness. Therefore, construct 061 operationally defined the promoter sequences from -226 to $+147$ as the minimal light-responsive promoter (Schulze-Lefert *et al.* 1989*a*; see below and Fig. 9). By comparing constructs 061 and 071 it became clear that sequences between positions -100 and -226, which contain Box I and Box II, might include an element(s) needed for light-dependent expression (Fig. 6). To address this question, clustered point mutations were introduced into Boxes I and II. Ten basepairs were chosen from each sequence element and mutated to unrelated sequences creating diagnostic restriction sites. Each block mutation, when tested in the context of the minimal promoter, abolished light-regulated GUS expression in the transient assay system. It is important to note that mutation of either sequence element results in loss of light responsiveness (Fig. 6). Therefore, Box I and Box II are necessary *cis*-acting elements for the light response in the context of the minimal CHS promoter (Schulze-Lefert *et al.* 1989*a*). Since both sequences had to be intact for inducibility of expression, they were defined as a light-responsive *cis*-acting unit (Unit 1).

This interpretation was supported by the results of additional experiments carried out by Block *et al.* (1990). In the region between Box I and Box II a restriction site was introduced by site-directed mutagenesis. The changes in the promoter sequence at these positions were shown not to affect light inducibility. This restriction site was then used for the deletion or introduction of four

in vivo footprinting of the parsley CHS promoter

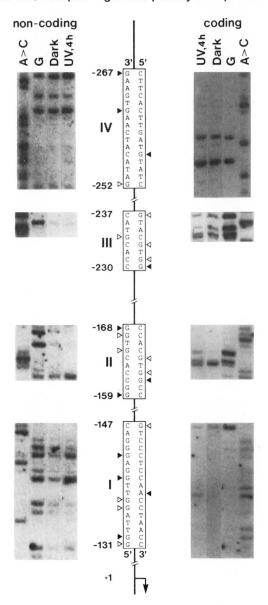

Fig. 5. Genomic footprints defining four protein contact regions in the parsley CHS gene promoter. The results from Schulze-Lefert *et al.* (1989*a,b*) are summarized. The filled triangles point to G residues showing hypermethylation, the open triangles indicate hypomethylation. The nucleotide sequences of the boxes and their positions are given. Results of an *in vivo* footprinting experiment comparing light-treated (UV, 4 h) and control (dark) cells are shown for the coding and the non-coding strands. The lanes labeled A>C and G show the Maxam-Gilbert reactions performed with cloned DNA.

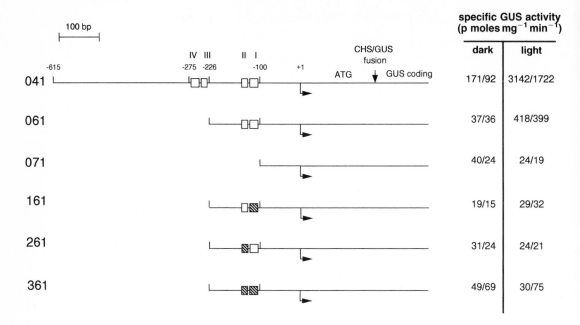

Fig. 6. GUS expression data from various CHS gene promoter-GUS deletion and block-mutation constructs. Results in this Figure are taken from Schulze-Lefert *et al.* (1989*a*). Names of CHS gene promoter-GUS fusions are listed on the left and the results from GUS assays on the right. The amount of fluorescent 4-methylumbelliferone formed is given as specific activity (pmoles product formed min^{-1}) of GUS enzyme mg^{-1} protein and is taken as a measure for promoter activity. The results from two typical experiments, each represented by the average of two transformations, are shown. End points of the CHS gene promoter sequences and their positions relative to the transcriptional start site are displayed. Boxed regions and roman numbers mark the positions of the *in vivo* footprints; hatched boxes indicate block mutations.The same translational fusion is shared by all constructs; 20 amino acids are encoded by *CHS* sequences and 10 amino acids are encoded by polylinker nucleotides.

nucleotides. Both mutations essentially destroyed light-regulated GUS expression, suggesting a strict spacing requirement within the functional *cis*-acting unit.

Functional redundancy in the parsley CHS gene promoter

The light-responsive unit contained in construct 061 generated only partial GUS activity when compared with the longer construct 041. The presence of Box III and Box IV further upstream in the promoter motivated the construction of CHS gene promoter-GUS fusions carrying these upstream elements in the context of various other parts of the CHS gene promoter (Schulze-Lefert *et al.* 1989*b*). Fig. 7 shows schematically the various promoter constructions and the results from GUS assays with these constructs. The data demonstrate that a second light-responsive element is contained in the CHS gene promoter in addition to Unit 1. In analogy to

Unit 1, Boxes III and IV were named Unit 2. Interestingly, a striking sequence similarity exists between Box II and Box III (see below). The presence of Unit 2 compensates partly for the loss of Unit 1 (see construct 351). Unit 2 can enlarge the light responsiveness of Unit 1 (compare constructs 051 and 061) and the 49-basepair fragment encompassing Boxes III and IV is required for the light response mediated through CHS gene promoter sequences from −226 to −615 (constructs 091 and 0101). Unit 2 in combination with sequences farther upstream is able to partially replace Unit 1 (compare constructs 341, 0101 and 061 in Fig. 7). It was concluded that Unit 2 is a weak light-responsive *cis*-acting unit. It is clear from these data that the elevated response to light of Unit 1 (construct 061) in conjunction with further upstream regions (construct 341) does not regenerate the complete activity of the full-length promoter construct 041. Therefore, the synergistic action of the two *cis*-active units controls the expression of CHS (Schulze-Lefert *et al.* 1989*b*). Interestingly, the elements are separated by approximately full numbers of turns of the B-DNA helix (see below; Fig. 9), although recent experiments have shown that 49 basepairs between Unit 1 and Unit 2 could be deleted without effect on light-regulated expression

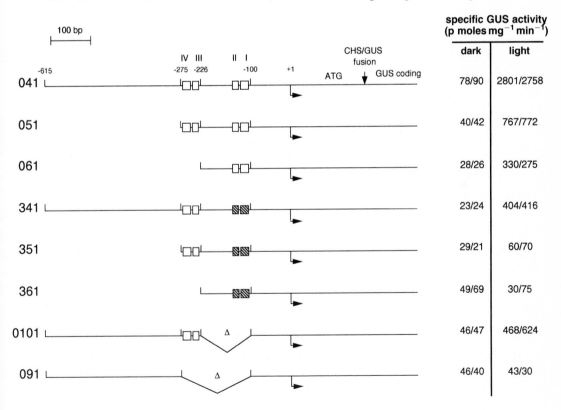

Fig. 7. Summary of GUS expression data demonstrating the existence of a second light responsive unit. Results are taken from Schulze-Lefert *et al.* (1989*b*). See Fig. 6 for description of symbols and data calculation. Deletions within the promoter region in constucts 0101 and 091 are indicated.

(Block *et al.* 1990). Taken together, these and other data show that parsley CHS expression in response to light is regulated by at least two separable light-responsive *cis*-acting units and one upstream region that is able to enhance the response of each of the units (Schulze-Lefert *et al.* 1989*b*).

Unit 1 is sufficient for light-dependent expression

The data so far discussed demonstrate that Unit 1 is necessary for light responsiveness in the context of the minimal CHS promoter. But is this Unit able to confer light inducibility on a heterologous promoter? To answer this question, the vector pUC GUS BT-2 was used (Fig. 4). A synthetic oligonucleotide comprising the sequence of Unit 1 was cloned in both orientations into the polylinker of pUC GUS BT-2 as single copy, as a dimer, and as a tetramer. The basic construct containing only the 35S promoter start site was shown to be minimally active in the transient expression system. GUS expression directed by this construct is not increased after irradiation of the protoplasts. Possibly, the expression is slightly repressed by light. As a control, a derivative of this vector was made, lacking the 35S promoter sequences (construct pUC GUS BT-Δ; Fig. 8). This construct served as a negative control in transient expression experiments and is transcriptionally inactive; it shows only the endogeneous background in the fluorometric determination of the product formed by the GUS enzyme.

The results obtained with the oligonucleotide-containing constructs showed that either orientation of Unit 1 is able to confer light responsiveness on a heterologous transcription start site (Weißhaar *et al.* 1991). From this it can be concluded that Unit 1 is not only necessary, but also sufficient for directing light-dependent expression in the homologous system (see Fig. 9). As mentioned above, co-action of both Box I and Box II at a defined distance is needed for light-mediated expression (Block *et al.* 1990). This finding is reinforced by additional results

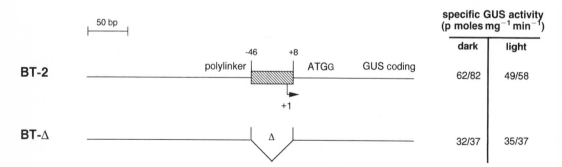

Fig. 8. Schematic delineation and results from GUS assays of pUC GUS BT-2 and its deletion derivative pUC GUS BT-Δ. The CaMV 35S promoter fragment contained in pUC GUS BT-2 (BT-2) is represented by the hatched region. These sequences were deleted from plasmid pUC GUS BT-2 to give pUC GUS BT-Δ (BT-Δ). See Fig. 6 for description of other symbols and data calculation.

demonstrating that tetramers of either of these boxes cloned 5′ to the truncated 35S promoter in pUC GUS BT-2 do not confer light inducibility on the constructs (Weißhaar, B. and Hahlbrock, K., unpublished results).

Box II of the parsley CHS gene promoter belongs to the G Box family of *cis-acting* elements

Many features of the transcriptional machinery of higher eukaryotes are functionally related and evolutionarily conserved. In the case of Box II (5′-CCACGTGGCC-3′) from the parsley CHS gene promoter, similar sequences have been found in many other promoters in a variety of plant species, including light-regulated promoters such as those of several CHS and RBCS genes. Also, in promoters known not to be light-regulated, sequences similar to Box II were found (Schulze-Lefert *et al.* 1989*a*). In the parsley CHS gene promoter, this sequence element is present in two related copies. Both Box II and Box III were detected by *in vivo* footprinting and found to be components of Units 1 and 2, respectively. Boxes II and III are functionally related as well in parsley (see below).

A sequence similar to Box II was originally identified as a conserved region among promoters of RBCS genes (Giuliano *et al.* 1988). These authors identified a factor in nuclear extracts from *Arabidopsis* and tomato, which binds to the conserved sequence, the so called G Box. More recently, a G Box binding factor (GBF) has been characterized in fractionated *Arabidopsis* extracts by gel retardation and *in vitro* DNAse I footprinting (DeLisle and Ferl, 1990). The G Box sequence is of functional relevance in the expression of RBCS genes, as demonstrated by mutation of the G Box present in the *Arabidopsis rbcS-1A* promoter. Also in this promoter, a second sequence (I Box) required for full RBCS gene promoter activity in transgenic plants has been identified (Donald and Cashmore, 1990).

Single base substitutions within Box II define a functional core of 7 bases

The Box II sequence has been recently subjected to detailed analysis by site-directed mutagenesis (Block *et al.* 1990). The goal of these experiments was to define the functionally relevant bases within the Box II/G Box *cis*-acting element. Each base within the 10-basepair region defined as Box II (Schulze-Lefert *et al.* 1989*b*) was substituted separately. All mutations were tested for their ability to direct light-dependent GUS expression in the context of the minimal promoter construct 061 (Fig. 9). The specific GUS activity yielded by the reference construct 061 was set to 100 % when evaluating the results (Fig. 10a). Single base mutations at each position between −160 and −166 virtually eliminated light-controlled expression (constructs 061/4 to 061/10). Upstream and downstream of this core, base substitutions had either no effect or retained at least half of the GUS expression of the controls (constructs 061/2, /3, and /11). The data show the heptameric sequence 5′-ACGTGGC-3′ to be the critical core of Box II required

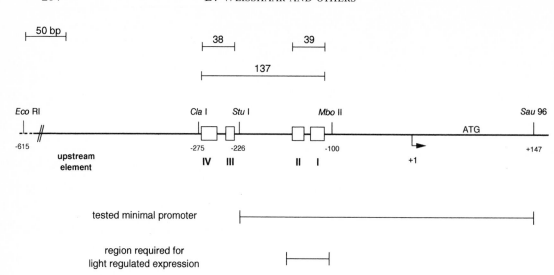

Fig. 9. Functional representation of the parsley chalcone synthase promoter. The positions and spacing of the promoter elements defined thus far in the parsley *CHS* *(PcCHS^a)* promoter are indicated. Positions of restriction sites used in the construction of the CHS gene promoter-GUS fusions are marked. The broken line at the left end of the drawing shows the position of the transposon-like element in the *PcCHS^a* allele. Above, the sizes and the distances of the *cis*-acting units are given in basepairs. As borders, the outermost nucleotides with changed reactivity to DMS in the *in vivo* footprinting experiments are given. At the bottom, the extension of the minimal promoter used in transient expression experiments and the region encompassed by the synthetic oligonucleotide cloned into pUC GUS BT-2 are indicated.

for light-induction in the context of the CHS minimal promoter. Together with data defining the borders of Box II by the introduction of clustered point mutations, it was concluded that flanking nucleotides had little or no relevance, provided that Box I or other *cis*-acting elements were not affected (Block *et al.* 1990). Obviously, it cannot be ruled out that point mutations other than those tested (11 out of 30 possibilities were examined) may lead to the definition of additional functionally important bases in Box II. In general, the data yielded by these analyses fully agree with the results of the *in vivo* footprinting experiments (see Block *et al.* 1990 for discussion).

Functional replacement of Box II by Box III

The nucleotide sequence of Box II is related to that of Box III, raising the question as to whether these sequences are functionally equivalent. The construction of the GUS-fusion plasmids 061/H and 061/I addressed this question (Block *et al.* 1990). As shown in Fig. 10b, replacement of Box II by Box III in the wild-type 5′ to 3′ orientation resulted in the restoration of light inducibility, whereas the reverse orientation did not. Comparison of the nucleotides changed by replacing Box II with Box III in this experiment with the results of the point-

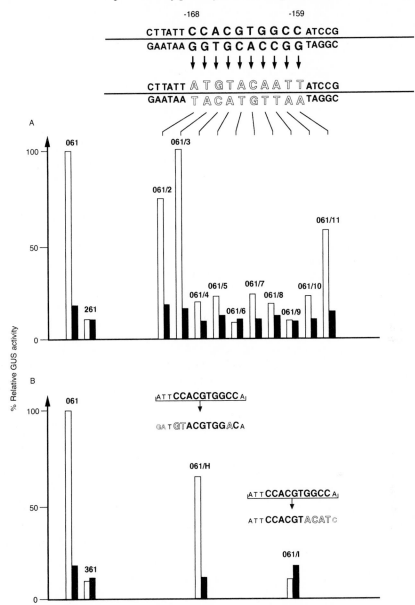

Fig. 10. Results of point-mutation analyses of Box II and of replacement experiments of Box II by Box III. Data presented are taken from Block *et al.* (1990). The nucleotides constituting the wild-type sequence of Box II are represented in bold and surrounding sequences are given in smaller characters. Mutated positions are indicated by outline letters. All constructs were made in the 061 background and the names of the respective 061 derivatives are listed. The open bars show the specific GUS activity for the given construct relative to the reference construct 061 after light treatment of the transformed protoplasts. Filled bars show the corresponding data for the dark control of each transformation assay. The data were calculated from at least six independent transformation experiments.

mutation series revealed that functionally important bases were substituted in both cases. It was concluded that Box III can functionally substitute for Box II in an orientation-dependent manner (Block *et al.* 1990). It is interesting to note that in Unit 1, Box II is located 5′ to Box I although Box III is located 3′ to Box IV. Since the individual *cis*-acting elements were unable to mediate light responsiveness, the interaction between factors binding to Boxes I and II (and to Box III in the construct 061/H) is necessary. This result suggests that the asymmetry of nucleotides surrounding the Box II/Box III palindrome influences the binding of protein factor(s). The asymmetry may be mandatory for the proper interaction with Box I, the other *cis*-acting element of the unit conferring light responsiveness.

How are differential effects of related *cis*-acting elements controlled by *trans*-acting factors?

Box II contains a palindromic sequence, as is often observed to be the case in *cis*-acting elements. It has been interpreted as a reflection of the fact that DNA-binding proteins are in many cases active as homodimers or tetramers, binding symmetrically to DNA matching the symmetry of the binding site (Takeda *et al.* 1983).

On the other hand, the functional core of Box II is asymmetrical. In animal systems it was hypothesized that the formation of heterodimers between different polypeptides constituting *trans*-acting factors may increase dramatically the combinatorial possibilities for regulation (Busch and Sassone-Corsi, 1990), thereby offering one possible explanation of asymmetric binding sites. If this concept holds true for plant *cis*- and *trans*-acting elements, it might explain the finding of the asymmetrical core region of Box II and the results of the replacement experiments with Box II and Box III.

An interesting observation was made when the heptameric sequence defined as the critical core was compared with some other published sequences of *cis*-acting elements (Fig. 11). This heptameric sequence is present not only in the G Box (Giuliano *et al.* 1988), but also in the *cis*-acting element called *hex* in histone gene promoters (Tabata *et al.* 1989) and in the abscisic acid responsive element (ABRE) defined in *rab* gene promoters (Marcotte *et al.* 1989; Mundy *et al.* 1990). This family of related elements may be named common plant regulator (CPR).

The data collected so far indicate a diverse class of similar *cis*-acting elements and cognate binding factors that are involved in the regulation of various promoters. In fact, the occurrence of the CPR sequence in genes controlled by anaerobic conditions, developmental state (e.g. *adh* or *patatin*, respectively; see Schulze-Lefert *et al.* 1989*a*) or in the promoter of a parsley polyubiquitin gene (P. Kawalleck, I. Somssich, K. Hahlbrock and B. Weißhaar, unpublished data) strongly supports the argument in favour of an involvement of this family of *cis*-acting elements and cognate binding factors in promoter activation by several different kinds of stimulus.

It remains to be shown whether the results of the point-mutation analysis of Box

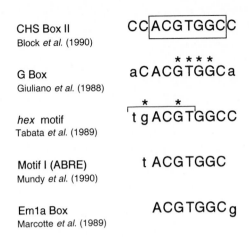

Fig. 11. Sequence elements identical to the functional core of Box II. The boxed region in the Box II sequence indicates the critical core. In the sequence surrounding the *hex* motif the nucleotides defined as the hexamer are marked. The asterisk indicates nucleotide positions that were mutated simultaneously and used as controls in the references given. In addition to the heptameric sequences identical to the Box II core there are also others matching 6 out of 7 positions (see for example Box III, Fig. 10)

II are transferable to the CPR family in a more general fashion. Interestingly, the mutations introduced into the *hex* motif or the *Arabidopsis adh*-promoter G Box, which abolish factor binding and/or *cis*-activation, all change the heptameric sequence at one position at least (Tabata *et al.* 1989; Donald *et al.* 1990). In the parsley minimal CHS promoter the specificity is brought about by a *cis*-acting Unit (Box I plus Box II), which is sufficient for light regulation. A sequence called I Box and the G Box are both important for expression from the *Arabidopsis rbcS-1A* promoter (Donald and Cashmore, 1990). In the case of the ABA-responsive promoter of the wheat *Em* gene a second element (Em2) was identified in addition to Em1a in the 50-basepair region shown to confer ABA responsiveness on a truncated 35S promoter (Marcotte *et al.* 1989). Also, the artificial combination of regulatory *cis*-acting elements from the *RBCS-3A* gene and from a heat shock gene created a novel specificity, namely a light-dependent heat shock response (Strittmatter and Chua, 1987).

The co-action of two *cis*-acting elements as one unit (and the synergistic action of two *cis*-active units), together with the hypothesis of heterodimer formation of *trans*-acting factors binding to the class of elements similar to Box II, may explain the experimental results obtained for the light response of the parsley CHS gene promoter and serve as a working hypothesis.

We and others have so far not detected differences in the binding of nuclear factors to the Box II/G Box sequence between extracts prepared from light-treated *versus* untreated or dark-adapted plant sources (Armstrong, G. A., Weißhaar, B. and Hahlbrock, K., unpublished, Giuliano *et al.* 1988). One possible explanation of these findings is that the gel retardation assay used in these studies is not sensitive to important differences between the bound proteins. Either the differences were lost during extract preparation or modified versions of the same binding activity differ functionally. The question of how Box II is involved in light-dependent gene regulation remains to be answered. The cloning of factor(s) binding to this or related sequences would help to address this question.

We thank R. Töpfer for the vector pRT103, and I. Somssich, P. Schweizer, and O. da Costa e Silva for critically reading the manuscript. This material is based in part upon work supported by the North Atlantic Treaty Organisation under a grant awarded in 1989 to G. A. and by a fellowship of the Fritz Thyssen Stiftung to P. S.-L.

References

BALLAS, N., ZAKAI, N., FRIEDBERG, D. AND LOYTER, A. (1988). Linear forms of plasmid DNA are superior to supercoiled structures as active templates for gene expression in plant protoplasts. *Pl. molec. Biol.* **11**, 517–527.

BLOCK, A., DANGL, J. L., HAHLBROCK, K. AND SCHULZE-LEFERT, P. (1990). Functional borders, genetic fine structure, and distance requirements of cis elements mediating light responsiveness of the parsley chalcone synthase promoter. *Proc. natn. Acad. Sci. U.S.A.* **87**, 5387–5391.

BUSCH, S. J. AND SASSONE-CORSI, P. (1990). Dimers, leucine zippers and DNA-binding domains. *Trends Genet.* **6**, 36–40.

CHAPPELL, J. AND HAHLBROCK, K. (1984). Transcription of plant defense genes in response to UV light or fungal elicitor. *Nature* **311**, 76–78.

CHURCH, G. M. AND GILBERT, W. (1984). Genomic sequencing. *Proc. natn. Acad. Sci. U.S.A.* **81**, 1991–1995.

DANGL, J. D., HAHLBROCK, K. AND SCHELL, J. (1989). Regulation and structure of chalcone synthase genes. In *Cell structure and somatic cell genetics of plants*, vol. 6 (ed. J. U. Vasil and J. Schell), pp. 155–173. Academic Press, New York.

DANGL, J. D., HAUFFE, K. D., LIPPHARDT, S., HAHLBROCK, K. AND SCHEEL, D. (1987). Parsley protoplasts retain differential responsiveness to UV light and fungal elicitor. *EMBO J.* **6**, 2551–2556.

DAVIS, R. W., SIMON, M. AND DAVIDSON, N. (1971). Electron microscope heteroduplex methods for mapping regions of base sequence homology in nucleic acids. *Methods Enzymol.* **21**, 413–428.

DELISLE, A. J. AND FERL, R. J. (1990). Characterisation of the *Arabidopsis Adh* G-Box binding factor. *Pl. cell* **2**, 547–557.

DIXON, R. A. (1986). The phytoalexin response: Elicitation, signaling and control of host gene expression. *Biol. Rev.* **61**, 239–291.

DONALD, R. G. K. AND CASHMORE, A. R. (1990). Mutation of either G Box or I Box sequences profoundly affects expression from the *Arabidopsis rbcS-1A* promoter. *EMBO J.* **9**, 1717–1726.

DONALD, R. G. K., SCHINDLER, U., BATSCHAUER, A. AND CASHMORE, A. R. (1990). The plant G box promoter sequence activates transcription in Saccharomyces cerevisiae and is bound *in vitro* by a yeast activity similar to GBF, the plant G box binding factor. *EMBO J.* **9**, 1727–1735.

EBEL, J. AND HAHLBROCK, K. (1982). Biosynthesis. In *The Flavonoids* (eds. Harborne, J. B. and Mabry, T. J.), pp. 641–675. Chapmann and Hall, London.

EPPING, B., KITTEL, M., RUHNAU, B. AND HEMLEBEN, V. (1990). Isolation and sequence analysis of a chalcone synthase cDNA of *Matthiola incana* R. Br. *(Brassicaceae). Pl. molec. Biol.* **14**, 1061–1063.

FANG, R.-X., NAGY, F., SIVASUBRAMANIAM, S. AND CHUA, N.-H. (1989). Multiple cis regulatory elements for maximal expression of the cauliflower mosaic virus 35S promoter in transgenic plants. *Pl. Cell* **1**, 141–150.

GIULIANO, G., PICHERSKY, E., MALIK, V. S., TIMKO, M. P., SCOLNIC, P. A. AND CASHMORE, A. R. (1988). An evolutionary conserved protein binding sequence upstream of a plant light-regulated gene. *Proc. natn. Acad. Sci. U.S.A.* **85**, 7089–7093.

HAHLBROCK, K. AND GRISEBACH, H. (1979). Enzymatic controls in the biosynthesis of lignin and flavonoids. *Rev. Pl. Physiol.* **30**, 105–130.

HAHLBROCK, K., KREUZALER, F., RAGG, H., FAUTZ, E. AND KUHN, D. N. (1982). Regulation of flavonoid and phytoalexin accumulation through mRNA and enzyme induction in cultured plant cells. In *Biochemistry of Differentiation and Morphogenesis* (ed. L. Jaenicke) pp. 34–43. Springer, Berlin.

HAHLBROCK, K. AND SCHEEL, D. (1989). Physiology and molecular biology of phenylpropanoid metabolism. *Annu. Rev. Plant Physiol. Plant Mol. Biol.* **40**, 347–369.

HARBORNE, J. B. AND TURNER, B. L. (1984). *Plant Chemosystematics*. Academic Press, London.

HELLER, W. AND HAHLBROCK, K. (1980). Highly purified 'flavanone synthase' from parsley catalyzes the formation of naringenin chalcone. *Archs Biochem. Biophys.* **200**, 617–619.

HERRMANN, A., SCHULZ, W. AND HAHLBROCK, K. (1988). Two alleles of the single-copy chalcone synthase gene in parsley differ by a transposon-like element. *Molec. gen. Genet.* **212**, 93–98.

JEFFERSON, R. J. (1987). Assaying chimaeric genes in plants: The GUS gene fusion system. *Pl. molec. Biol. Rep.* **5**, 387–405.

KATAGIRI, F., LAM, E. AND CHUA, N.-H. (1989). Two tobacco DNA-binding proteins with homology to the nuclear factor CREB. *Nature* **340**, 727–729.

KOZAK, M. (1983). Comparison of initiation of protein synthesis in prokaryotes, eucaryotes, and organelles. *Microbiol. Rev.* **47**, 1–45.

KREUZALER, F., RAGG, H., FAUTZ, E., KUHN, D. N. AND HAHLBROCK, K. (1983). UV-induction of chalcone synthase mRNA in cell suspension cultures of *Petroselinum hortense*. *Proc. natn. Acad. Sci. U.S.A.* **80**, 2591–2593.

LAM, E. AND CHUA, N.-H. (1990). GT-1 binding site confers light responsive expression in transgenic tobacco. *Science* **248**, 471–474.

LIPPHARDT, S., BRETTSCHNEIDER, R., KREUZALER, F., SCHELL, J. AND DANGL, J. D. (1988). UV-inducible transient expression in parsley protoplasts identifies regulatory *cis*-elements of a chimeric *Antirrhinum majus* chalcone synthase gene. *EMBO J.* **7**, 4027–4033.

MARCOTTE, W. R., JR, RUSSELL, S. H. AND QUATRANO, R. S. (1989). Abscisic acid-responsive sequences from the *Em* gene of wheat. *Pl. Cell* **1**, 969–976.

MATERN, U., HELLER, W. AND HIMMELSPACH, K. (1983). Conformational changes of apigenin 7-*O*-(6-*O*-malonylglucoside), a vacuolar pigment from parsley, with solvent composition and proton concentration. *Eur. J. Biochem.* **133**, 439–448.

MUNDY, J., YAMAGUCHI-SHINOZAKI, K. AND CHUA, N.-H. (1990). Nuclear proteins bind conserved elements in the abscisic acid-responsive promoter of a rice *rab* gene. *Proc. natn. Acad. Sci. U.S.A.* **87**, 1406–1410.

NICK, H. AND GILBERT, W. (1985). Detection *in vivo* of protein-DNA interactions within the *lac* operon of *Escherichia coli*. *Nature* **313**, 795–798.

NIESBACH-KLÖSGEN, U., BARTIN, E., BERNHARDT, J., ROHDE, W., SCHWARZ-SOMMER, Z., REIF, H. J. AND SAEDLER, H. (1987). Chalcone synthase genes in plants: A tool to study evolutionary relationships. *J. molec. Evol.* **26**, 213–225.

OHL, S., HAHLBROCK, K. AND SCHÄFER, E. (1989). A stable blue-light-derived signal modulates ultraviolet-light-induced activation of the chalcone-synthase gene in cultured parsley cells. *Planta* **177**, 228–236.

REIMOLD, U., KRÖGER, M., KREUZALER, F. AND HAHLBROCK, K. (1983). Coding and 3′ non-coding nucleotide sequence of chalcone synthase mRNA and assignment of amino acid sequence of the enzyme. *EMBO J.* **2**, 1801–1805.

SCHEEL, D., DANGL, J. L., DOUGLAS, C., HAUFFE, K. D., HERRMANN, A., HOFFMANN, H., LOZOYA, E., SCHULZ, W. AND HAHLBROCK, K. (1987). Stimulation of phenylpropanoid pathways by environmental factors. In *Plant Molecular Biology* (ed. von Wettstein, D. and Chua, N.-H.) pp. 315–316, Plenum Press.

SCHMELZER, E., JAHNEN, W. AND HAHLBROCK, K. (1988). *In situ* localization of light-induced chalcone synthase mRNA, chalcone synthase, and flavonoid end products in epidermal cells of parsley leaves. *Proc. natn. Acad. Sci. U.S.A.* **85**, 2989–2993.

SCHRÖDER, J. AND SCHRÖDER, G. (1990). Stilbene and chalcone synthases: Related enzymes with key functions in plant-specific pathways. *Z. Naturforsch.* **45c**, 1–8.

SCHULZE-LEFERT, P., BECKER-ANDRE, M., SCHULZ, W., HAHLBROCK, K. AND DANGL, J. L. (1989*b*). Functional architecture of the light-responsive chalcone synthase promoter from parsley. *Pl. Cell* **1**, 707–714.

SCHULZE-LEFERT, P., DANGL, J. L., BECKER-ANDRE, M., HAHLBROCK, K. AND SCHULZ, W. (1989*a*). Inducible *in vivo* DNA footprints define sequences necessary for UV light activation of the parsley chalcone synthase gene. *EMBO J.* **8**, 651–656.

STRITTMATTER, G. AND CHUA, N.-H. (1987). Artificial combination of two cis-regulatory elements generates a unique pattern of expression in transgenic plants. *Proc. natn. Acad. Sci. U.S.A.* **84**, 8986–8990.

TABATA, T., TAKASE, H., TAKAYAMA, S., MIKAMI, K., NAKATSUKA, A., KAWATA, T., NAKAYAMA, T. AND IWABUCHI, M. (1989). A protein that binds to a *cis*-acting element of wheat histone genes has a leucine zipper motif. *Science* **245**, 965–967.

TAKEDA, Y., OLLENDORF, D. H., ANDERSON, W. F. AND MATTHEWS, B. W. (1983). DNA-binding proteins. *Science* **221**, 1020–1026.

TÖPFER, R., PRÖLS, M., SCHELL, J. AND STEINBIß, H. H. (1988). Transient gene expression in tobacco protoplasts: II. Comparision of the reporter gene systems for CAT, NPT II, and GUS. *Pl. Cell Rep.* **7**, 225–228.

VAN DE LÖCHT, U., MEIER, I., HAHLBROCK, K. AND SOMSSICH, I. (1990). A 125 bp promoter fragment is sufficient for strong elicitor-mediated gene activation in parsley. *EMBO J.* **9**, 2945–2950.

WEIßHAAR, B., ARMSTRONG, G. A., BLOCK, A., DA COSTA E SILVA, O. AND HALLBROCK, K. (1991). Light-inducible and constitutively expressed DNA-binding proteins recognizing a plant promoter element with functional relevance in light responsiveness. *EMBO J.*, in press.

AN *ARABIDOPSIS THALIANA* G-BOX-BINDING PROTEIN SIMILAR TO THE WHEAT LEUCINE ZIPPER PROTEIN IDENTIFIED AS HBP-1

ULRIKE SCHINDLER, JOSEPH R. ECKER and ANTHONY R. CASHMORE

Plant Science Institute, Department of Biology, University of Pennsylvania, Philadelphia, PA 19104, USA

Summary

G-box (CCACGTGG) like sequences are present in a variety of plant promoters and in many cases they have been demonstrated to be required for maximal expression of the corresponding gene. A nuclear protein, GBF, interacts specifically with the G-box motif of several RBCS and CAB promoters. Here we describe the isolation of a cDNA from *Arabidopsis thaliana* that encodes a protein, designated GBF-1, with DNA binding properties similar to GBF. GBF-1 is characterized by a basic/leucine zipper motif which is strikingly similar to the wheat protein identified as HBP-1. GBF-1 also interacts with an oligonucleotide derived from the wheat histone 3 promoter containing the binding site (hexamer, TGACGT) for HBP-1. This DNA element also contains a G-box-like motif, modification of which results in loss in binding of GBF-1.

Introduction

We have shown previously that promoters for *RbcS* (ribulose-bisphosphate carboxylase) genes are often characterized by conserved sequences that we have referred to as L-, I- and G-boxes (Giuliano *et al.* 1988). These sequences commonly reside 200 to 300 bp upstream from the start site of transcription. In keeping with this strong conservation of sequence we have demonstrated that there is a clear requirement for both the G- and I-boxes for expression from the *Arabidopsis thaliana rbcS-1A* promoter (Donald and Cashmore, 1990). These studies were performed by examining GUS expression driven by a 'full-length' 1.7 kb *rbcS-1A* promoter in transgenic tobacco plants.

The G-box sequence is not restricted to *RbcS* promoters. A G-box-like sequence is found in the *cab-E* gene from *Nicotiana plumbaginifolia* (Castresana *et al.* 1988) and mutation of this sequence also results in a substantial loss in expression (Bringmann and Cashmore, unpublished results). Similarly, chalcone synthase promoters are often characterized by G-box sequences (Schulze-Lefert *et al.* 1989; Staiger *et al.* 1989) and, at least in the case of the parsley promoter, these sequences are required for UV light-induced expression. The *Arabidopsis thaliana*

Key words: *Arabidopsis*, G-box binding protein, HBP-1.

adh promoter, which mediates anaerobically-enhanced root expression, also contains a G-box (Ferl and Laughner, 1989). ABA-induced genes have also been shown to contain functional G-box sequences within their promoters (Guiltinan *et al.* 1990).

We have characterized a factor, GBF, from nuclear extracts of tomato and *Arabidopsis* that binds specifically to G-box promoter sequences (Giuliano *et al.* 1988; Schindler and Cashmore, 1990). Also, GBF-like activity has been characterized in extracts from *Antirrhinum*, *Petunia*, and *Nicotiana tabacum* (Staiger *et al.* 1989). *In vivo* footprinting studies demonstrate that a GBF-like factor binds to the chalcone synthase G-box sequence and this binding is altered by UV-induction (Schulze-Lefert *et al.* 1989). Similarly, *in vivo* footprinting was used to characterize a factor bound to the *Arabidopsis ADH* promoter (Ferl and Laughner, 1989). We have demonstrated that G-box sequences are active as upstream activating sequences in yeast and we have characterized a yeast factor with *in vitro* binding properties similar to GBF (Donald *et al.* 1990).

In view of the essential role that GBF plays in the photoregulated expression of certain plant genes and the widespread occurrence of GBF-like factors in plant tissues and other species, it is obviously of interest to further characterize this factor. Questions of interest include: (a) what is the sequence of this factor; (b) are there multiple GBF-like factors present in different tissues and mediating quite distinct expression characteristics and (c) what is the nature of associated factors that we presume are involved, along with GBF, in mediating specific expression.

A recombinant phage expresses a protein that binds specifically to G-box promoter sequences

We screened an *Arabidopsis thaliana* λZAP cDNA expression library for recombinant phage expressing proteins that bound oligomerized G-box sequences. Two positive phages were plaque-purified and the specificity of their DNA binding properties was examined. Protein produced by these phages was shown to bind strongly to the synthetic oligonucleotide containing the tomato *rbcS-3A* G-box sequence ACACGTGG but not to the mutant G-box oligonucleotide containing the sequence ACACtgtt (Fig. 1).

GBF-1 is a leucine zipper protein

In vivo excision yielded two recombinant Bluescript plasmids containing the cDNAs *gbf-1a* and *gbf-1b*. Southwestern studies and electrophoretic mobility shift assays with *E. coli* extracts confirmed the G-box-binding specificities of the proteins encoded by these cDNAs (Fig. 2). Restriction analysis of the plasmids showed that the two cDNA sequences were very similar but differed in size; *gbf-1a* was 1.42 kb and *gbf-1b* was 0.82 kb. Sequence studies showed that the two clones contained identical sequences differing only in length. The sequences were indicative of a leucine zipper protein – which we called GBF-1–containing a basic domain adjacent to a heptameric repeat of leucines (Fig. 3).

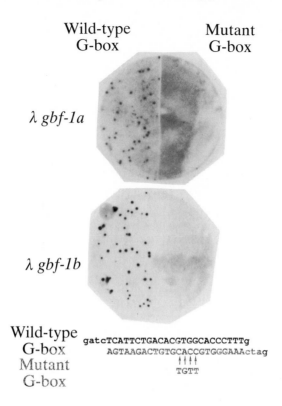

Fig. 1. Oligonucleotide binding properties of proteins encoded by λgbf recombinant phage. An *Arabidopsis thaliana* cDNA expression library in λZAP was screened with a ^{32}P-labeled catenated DNA fragment containing eight copies of a synthetic oligonucleotide representing the tomato *rbcS-3A* G-box motif. Two positive phage (λgbf-1a and λgbf-1b) were isolated from a primary screen of 4×10^5 phage plaques. The proteins encoded by the two purified recombinant phage were shown to bind specifically to the G-box sequence (left) but not the mutant derivative (right).

The basic DNA binding domain of GBF-1 shows striking sequence homology to a protein previously identified as HBF-1

We compared the basic putative DNA-binding domain of GBF-1 with other plant leucine zipper proteins (Fig. 3). In a region of 30 amino acids GBF-1 showed some similarity to TGA1a and TGA1b (43 % and 40 % respectively; Katagiri *et al.* 1989) and was even more similar to O2 (63 %; Schmidt *et al.* 1990) and OCSBF-1 (Singh *et al.* 1990). Both TGA1a and TGA1b bind to cauliflower mosaic virus 35S promoter elements containing the sequence TGACG and OCSBF-1 binds to octopine synthase promoter elements containing similar sequences; this sequence has some similarity to the core ACGT sequence of the G-box. A striking similarity was observed on comparing GBF-1 with two wheat proteins: EmBP-1 (Guiltinan *et al.* 1990) and a protein encoded by a cDNA sequence identified as HBP-1. This latter protein binds to the hexameric ACGTCA sequence found in many histone gene promoters (Tabata *et al.* 1989). The complement of this hexamer sequence

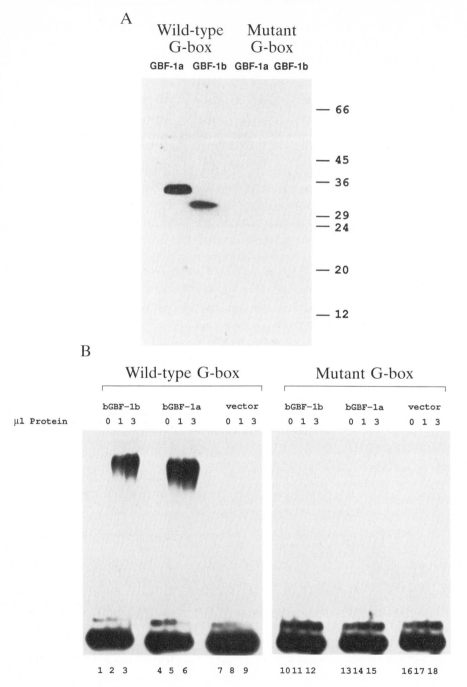

(TGACGT) is also similar to the core G-box sequence. GBF-1 and the sequence identified as HBP-1 exhibited 93 % homology in the 30 amino acid basic putative DNA-binding domain. This striking sequence similarity raises the possibility that GBF-1 and the protein identified as HBP-1 may be evolutionary homologues and this prospect in turn raises questions concerning the true function of these related

Fig. 2. (A) Southwestern analyses. *E. coli* cells (DH5α) transformed with recombinant plasmids containing the *gbf-1a* and *gbf-1b* cDNA fragments were grown to an OD_{600} of 0.5. Protein synthesis was induced by adding 5 mM IPTG and growth was continued for an additional hour. The cells (1 ml aliquots) were pelleted, washed and after another centrifugation, resuspended in SDS loading buffer. After sonication the proteins were electrophoretically separated on a 15% SDS–PAGE and electroblotted onto nitrocellulose filters. The filters were treated as described by Vinson *et al.* (1988). DNA binding was performed with a DNA fragment containing eight copies of the synthetic oligonucleotide representing the wild-type or mutant *rbcS-3A* G-box motif. (B) Gel mobility shift assay. *E. coli* cells (DH5α) were grown and protein synthesis was induced as described above. The cells were collected by centrifugation and lysed in buffer containing 6 M guanidine–HCl. The proteins were renatured in solution by dialysis overnight. DNA binding assays were performed with radiolabeled wild-type and mutant G-box oligonucleotides as described by Schindler and Cashmore (1990). Proteins isolated from *E. coli* cells transformed with the vector plasmid (vector) only were used as a negative control.

---

proteins. It is quite clear that the *in vivo* role of any transcription factor is difficult to assign in the absence of genetic studies. However, in evaluating the relationship between these closely-related DNA-binding proteins, we note that the oligonucleotide used to screen for the wheat clone identified as HBP-1 somewhat fortuitously contained an overlapping partial G-box sequence (in the form TGACGTGG). In the mutant oligonucleotide (TtACtTGG) used for these studies (Tabata *et al.* 1989) the G-box sequence, as well as the hexamer sequence, was

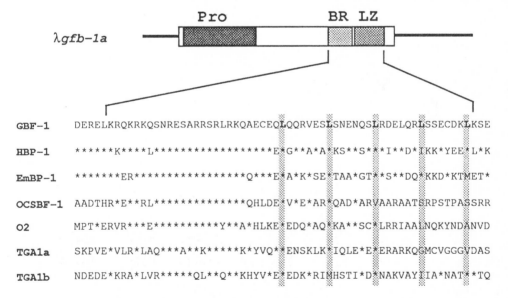

Fig. 3. Comparison of putative DNA binding domains of GBF-1 and other plant leucine zipper proteins. A schematic representation of the cDNA encoding GBF-1 is shown at the top of the figure. The proline rich region, the basic putative DNA binding domain (basic region, BR), and leucine zipper (LZ) are indicated. The amino acid sequence within the basic region is aligned with the corresponding regions of other plant DNA binding proteins (Katagiri *et al.* 1989; Tabata *et al.* 1989; Schmidt *et al.* 1990; Singh *et al.* 1990; Guiltinan *et al.* 1990).

altered and thus these experiments did not distinguish binding of the two sequences. We have repeated these studies with GBF-1 and obtained similar results. Furthermore, we have observed that binding does not occur to a mutant oligonucleotide containing the sequence TGACGTtt in which the G-box sequence, but not the hexamer sequence, has been destroyed (data not shown). In view of these observations we conclude that GBF-1 binds a G-box sequence and

probe: *gbf-1a*

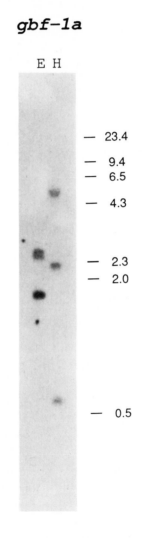

Fig. 4. Southern blot analyses with *gbf-1*. *Arabidopsis thaliana* genomic DNA (6 μg) was digested with *Hind*III (H) and *Eco*RI (E). The DNA fragments were electrophoretically separated and blotted onto nitrocellulose filters. Hybridization was performed with the ^{32}P-labeled *Eco*RI cDNA fragment of λgbf-1a under the following conditions: 5×SSPE, 0.2% PVP, 0.2% BSA, 0.2% Ficoll, 0.5% SDS, 10 μg ml^{-1} sonicated salmon sperm DNA, for 12 h at 42°C. The filters were washed in 0.5×SSC at 65°C and autoradiographed. Lambda DNA digested with *Hind*III was used as a relative molecular mass standard.

not the hexamer sequence. Furthermore, as the putative DNA-binding domain of the protein encoded by the wheat cDNA is so similar to GBF-1, we consider it likely that it is also a G-box-binding protein and that it is not HBP-1. Here it should be noted that we are using the term HBP-1 as it was originally used to describe the protein present in wheat nuclear extracts – this DNA binding protein has been clearly demonstrated to bind to a hexamer sequence distinct from the G-box (Mikami *et al.* 1989). We predict that the sequence of the true HBP-1 is likely to be less similar to GBF-1 than it is to the hexamer binding proteins TGA1a, TGA1b and OCSBF-1. Somewhat in keeping with this suggestion is the observation that the putative DNA-binding domain of EmBP-1, a protein that binds to a G-box-containing abscisic acid response element from wheat (Guiltinan *et al.* 1990), is more similar to GBF-1 than it is to the hexamer binding proteins (Fig. 3).

Genomic southern blots show that *gbf-1* is present as a single or low-copy sequence

In order to examine the complexity of the DNA sequences encoding GBF-1, we performed Southern blot hybridization studies with genomic *Arabidopsis* DNA (Fig. 4). A simple pattern of hybridization was observed indicating that *gbf-1* was present as a single or low-copy sequence. The isolation of genomic clones will be required to determine the exact copy number of this sequence.

Concluding remarks

We have isolated a cDNA encoding a 'leucine zipper' DNA-binding protein that we have designated GBF-1. The binding properties of GBF-1 are similar to those of GBF, which we have previously characterized in nuclear extracts of tomato and *Arabidopsis thaliana* (Giuliano *et al.* 1988; Schindler and Cashmore, 1990). A determination of the precise relationship between these two activities will require further studies. The sequence of the putative DNA-binding domain of GBF-1 is strikingly similar to the protein encoded by a wheat cDNA sequence and identified as the histone promoter binding factor HBP-1 (Tabata *et al.* 1989). Both the sequence and the published binding properties of the factor encoded by this wheat cDNA prompt us to suggest that it may have been misidentified as HBP-1 and that it may be a homologue of GBF-1.

This work was supported by grants to A.R.C. from NIH (GM-38409) and DOE (DE-FG02-87ER13680) and grants to J.R.E. from NIH (GM-38894 and HG-00322).

References

CASTRESANA, C., GARCIA-LUQUE, I., ALONSO, E., MALIK, V. S. AND CASHMORE, A. R. (1988).

Both positive and negative regulatory elements mediate expression of a photoregulated CAB gene from *Nicotiana plumbaginifolia*. *EMBO J.* **7**, 1929–1936.

DONALD, R. G. K. AND CASHMORE, A. R. (1990). Mutation of either G-box or I-box sequences profoundly affects expression from the *Arabidopsis rbcS-1A* promoter. *EMBO J.* **9**, 1717–1726.

DONALD, R. G. K., SCHINDLER, U., BATSCHAUER, A. AND CASHMORE, A. R. (1990). The plant G-box promoter sequence activates transcription in *Saccharomyces cerevisiae* and is bound *in vitro* by a yeast activity similar to GBF, the plant G-box binding factor. *EMBO J.* **9**, 1727–1735.

FERL, R. J. AND LAUGHNER, B. H. (1989). *In vivo* detection of regulatory factor binding sites of *Arabidopsis thaliana Adh*. *Plant molec. Biol.* **12**, 357–366.

GIULIANO, G., PICHERSKY, E., MALIK, V. S., TIMKO, M. P., SCOLNIK, P. A. AND CASHMORE, A. R. (1988). An evolutionarily conserved protein binding sequence upstream of a plant light-regulated gene. *Proc. natl. Acad. Sci. U.S.A.* **85**, 7089–7093.

GUILTINAN, M. J., MARCOTTE, W. R. AND QUATRANO, R. S. (1990). A plant leucine zipper protein that recognizes an abscisic acid response element. *Science* **250**, 267–271.

KATAGIRI, F., LAM, E. AND CHUA, N.-H. (1989). Two tobacco DNA-binding proteins with homology to the nuclear factor CREB. *Nature* **340**, 727–730.

MIKAMI, K., NAKAYAMA, T., KAWATA, T., TABATA, T. AND IWABUCHI, M. (1989). Specific interaction of nuclear protein HBP-1 with the conserved hexameric sequence ACGTCA in the regulatory region of wheat histone genes. *Pl. Cell Physiol.* **30**, 107–119.

SCHINDLER, U. AND CASHMORE, A. R. (1990). Photoregulated gene expression may involve ubiquitous DNA binding proteins. *EMBO J.* **9**, 3415–3429.

SCHMIDT, R. J., BURR, F. A., AUKERMAN, M. J. AND BURR, B. (1990). Maize regulatory gene opaque-2 encodes a protein with a 'leucine-zipper' motif that binds to zein DNA. *Proc. natl. Acad. Sci. U.S.A.* **87**, 46–50.

SCHULZE-LEFERT, P., DANGL, J. L., BECKER-ANDRE, M., HAHLBROCK, K. AND SCHULZ, W. (1989). Inducible *in vivo* DNA footprints define sequences necessary for UV light activation of the parsley chalcone synthase gene. *EMBO J.* **8**, 651–656.

SINGH, K., DENNIS, E. S., ELLIS, J. G., LLEWELLYN, D. J., TOKUHISA, J. G., WAHLEITHNER, J. A. AND PEACOCK, W. J. (1990). OCSBF-1, a maize ocs enhancer binding factor: isolation and expression during development. *Pl. Cell.* **2**, 891–903.

STAIGER, D., KAULEN, H. AND SCHELL, J. (1989). A CACGTG motif of the *Antirrhinum majus* chalcone synthase promoter is recognized by an evolutionarily conserved nuclear protein. *Proc. natn. Acad. Sci. U.S.A.* **86**, 6930–6934.

TABATA, T., TAKASE, H., TAKAYAMA, S., MIKAMI, K., NAKATSUKA, A., KAWATA, T., NAKAYAMA, T. AND IWABUCHI, M. (1989). A protein that binds to a cis-acting element of wheat histone genes has a leucine zipper motif. *Science* **245**, 965–967.

VINSON, CH. R., LAMARCA, K. L., JOHNSON, P. F., LANDSCHULZ, W. H. AND MCKNIGHT, S. L. (1988). In situ detection of sequence-specific DNA binding activity specified by a recombinant bacteriophage. *Genes Dev.* **2**, 801–806.

Printed in Great Britain © Society for Experimental Biology 1991

RFLP ANALYSIS OF COMPLEX TRAITS IN CROP PLANTS

S. R. BARNES

ICI Seeds, Jealott's Hill Research Station, Bracknell, Berkshire, RG12 6EY, UK

Summary

Detailed genetic maps, based upon molecular markers (in particular, on restriction fragment length polymorphisms – RFLPs) have now been constructed for a number of crop plant species, and permit a range of genetic analyses hitherto considered impossible. The availability of such maps has made it possible to approach the dissection and manipulation of both simply-inherited and complex characteristics. Even those characters that show apparently 'quantitative' inheritance (displaying essentially continuous variation within a segregating family) can frequently be resolved into a handful of major gene effects. Once tagged with molecular markers, the genes can be assembled in any desired combination, permitting the testing of hypotheses on gene action/interaction, or the construction of varieties of plant species with improved agronomic performance.

The next technological challenge is to 'walk' from RFLP markers to isolate the actual genes responsible for the complex trait, by a combination of genetical and physical mapping techniques. Such analyses will begin to clarify our picture of the relationship between genetic and physical maps, of recombination, and of the arrangement of diverse families of DNA sequences in plant genomes.

Introduction

The development of a genetic map for virtually any plant species that undergoes sexual reproduction is now a relatively simple matter, thanks to the availability of techniques for the detection of polymorphism at the DNA level. The most frequently used technique is the examination of restriction fragment length polymorphism (RFLP) at specific loci. The polymorphism in restriction sites in the region of a specific DNA probe is visualised by observing the molecular size of plant genomic fragments that hybridise to the probe, after separation by electrophoresis, and Southern blotting. The advantages of this and other systems for detection of polymorphism will be discussed in the following sections, with reference to their usefulness in a range of applications. However, it is clear that the availability of detailed markers for any desired species opens up new possibilities for the genetic analysis of characters formerly considered inaccessible. Furthermore, the use of *physical* markers (RFLPs look at structure, rather than function) in genetic analysis will permit relationships between physical and genetical maps to be explored and exploited.

Key words: RFLP, crop plants, genetic mapping.

Systems for the analysis of DNA polymorphism

Benefits from the use of DNA marker systems

The possibility of working at the genome level offers a number of significant advantages for genetic analysis.

(1) The frequency of available polymorphism that can be examined in each segregating population is increased considerably; the proportion of loci displaying polymorphism between a given pair of individuals is clearly not fixed, and is a function of the loci themselves and the genetic distance between the individuals. This latter factor varies considerably between species of plant, with maize (an outbreeder), for example, showing considerably higher levels of polymorphism than wheat (an inbreeder).

(2) The density of genetic markers that can be placed on the map may be increased by orders of magnitude over what would be achievable using morphological markers. Essentially, this is because there is no practical limit to the number of different sites that can be examined in a genome of around a billion base pairs. The limit is effectively determined by the effort taken in screening and mapping new probes, and by need – for most practical applications saturation is unnecessary. However, it is unusual to find a genetic map where a completely even spacing of markers has been achieved. This may in part be due to random sampling effects, but could equally be the result of uneven distribution of recombination with respect to the physical map.

(3) The markers (with some qualification – see later) behave in a codominant fashion; thus complete classification of the progeny from crosses is possible, and the information retrieved from each individual is maximised.

(4) The markers can be scored unequivocally at any stage in plant development – no problems can occur with penetrance or expressivity. Barring scoring errors, heritability of an RFLP allele should equal unity; this is in direct contrast to the experience of many generations of quantitative geneticists working with complex agronomic traits, where measurement error/environmental influences can have dramatic effects on the reliability of many phenotypic measurements.

Technology for RFLP analysis

A number of approaches have been developed for the analysis of polymorphism at the DNA level, and are described briefly below; although other methods (including unstacking gels and genomic sequencing) are feasible, in this review consideration is restricted to those systems that are practically usable in programmes of genetic analysis of crop plants.

(i) Single-locus hybridisation probes – Cloned genomic or cDNA sequences are used as hybridisation probes against restriction digests of genomic samples. Allelic variation is seen as differences in the size of homologous fragments, which can be due to point mutation (in the restriction site), insertions or deletions. When identifying new probes, a favoured approach has been to enrich for single-copy sequences by cloning hypomethylated regions that can be digested by methylation-sensitive enzymes such as *Pst*I. Specific PCR (polymerase chain reaction) tests can be developed to test for defined alleles.

(ii) Multi-locus hybridisation probes – These fall into a number of classes, including functional regions (e.g. storage protein genes) and sequences of unknown function. The class demonstrated to be the most useful in animal genomes are the so-called minisatellite sequences (Jeffreys *et al.* 1985). These comprise families of repeats, found in tandem clusters at a number of genomic sites. Polymorphism is generated by variation in repeat number at any particular locus, leading to differences in the size of restriction fragments containing the arrays, and visualised on Southern blots by repeats used as hybridisation probes. Thus a single probe can provide a 'snapshot' of the whole genome, and the high frequency of polymorphism for copy number has rendered this type of probe of great utility for forensic and paternal identity applications. Individual loci can also be amplified using PCR.

Homologous minisatellites have been demonstrated (Dallas, 1989) and sequenced (Daly *et al.* 1991) in plant species, where they have been shown to be polymorphic. To date, however, no genetic analysis has succeeded in demonstrating a genuinely widely distributed family of these elements in a plant genome.

(iii) Simple-sequence variation and dinucleotide repeats – The use of PCR to amplify fragments containing arrays of dinucleotide repeats has revealed that in both mammalian and plant genomes, length variation is due to differences in numbers of repeats. Polymorphism is revealed by electrophoresis following PCR; blotting and hybridisation are not required.

(iv) Random oligonucleotides/PCR – The usefulness of 9- and 10-base primers in PCR tests with plant genomic DNAs has recently been demonstrated by a number of groups. Under optimal conditions, and with selected primers, up to 10 bands may be produced and visualised by electrophoresis. Polymorphisms in band number and size are observed at useful frequencies. This approach has the advantage that each primer will give information on a number of loci simultaneously, and that the same set of primers can be used in a number of different species. Linkage relationships between markers must be determined in each cross that is analysed; for trait analysis in species that have no existing sets of molecular markers, this probably represents the most rapid approach currently available.

Molecular markers and trait analysis

The genetic analysis of traits using molecular markers is essentially no different from conventional linkage studies. Each plant from a segregating population (typically F_2 or backcross generations) is measured for the trait concerned; this may entail determining the phenotype of progeny families from the individual plants, in cases where single-plant measurements are not sufficiently precise (but see below). The genotypes of the segregating plants are then determined at a set of

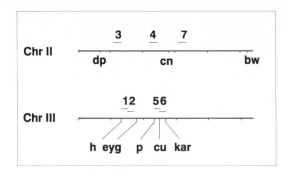

Fig. 1. Analysis of sternopleural chaeta number in *Drosophila melanogaster*. The
positions of loci (numbered 1–7) segregating for differences in sternopleural chaeta
number were identified by comparing the phenotype of individuals recombinant with
respect to defined morphological markers. (*Redrawn from Thoday, 1961.*)

polymorphic loci that provide an evenly-spread coverage of the genome. 10 to
20 cM spacing of probes can normally be considered sufficient for a preliminary
screen (50–100 probes for a species such as maize) (Lander and Botstein, 1989).

The identification of genomic regions influencing the character under analysis
may be achieved in a number of ways. When the character is simple (i.e. it is
discontinuous – a simple colour marker, for example) it can be handled as a simple
mendelian factor, and linked molecular markers may be identified as those that do
not show independent segregation from the trait. Linkage distances are calculated
in the normal way.

When complex characters are being analysed, however, it might be expected
that several genes, each of different effect and magnitude, may be involved. In
practice, this generally appears to be true, and traits displaying continuous
variation in a segregating population are frequently referred to as 'polygenic' or
'quantitative' traits (with the loci responsible as QTL – quantitative trait loci).

It is important, however, to appreciate that a complex, or continuously-varying
phenotype does not necessarily indicate that *many* genes are involved. The classic
example is that of sternopleural chaeta number in *Drosophila melanogaster*, which
was long considered to be a clear case of 'polygenic' or quantitative inheritance.
Thoday (1961) showed that the variation could be explained by a small number of
genes (or, rather, loci) located on two chromosomes – 3 genes on chromosome II,
and 4 on chromosome III (Fig. 1). Thus a 'polygenic' character was shown to be
'oligogenic'; the question remains, however, as to the proportion of quantitative
characters that can be explained by a similarly small number of loci.

The approach taken by Thoday was to compare the phenotype of different
genotypic classes (genotype, in this case, determined by morphological markers),
and thereby to dissect the genome into regions of major effect on the trait being
observed. Essentially the same approaches are available for the analysis of
complex traits in plant systems, although until recently the lack of well-saturated
maps has been a major obstacle. Genetic maps based on RFLPs and other DNA-

based markers should supply ample variation and definition to dissect a wide range of complex plant characters.

Marker-based analysis of complex traits in plants

The analysis of quantitative traits in segregating plant populations has essentially followed that advocated by Thoday, although a number of statistical approaches have been adopted. The simplest form of analysis is a locus-by-locus analysis of variance; the phenotypic distributions of the different genotypic classes (AA, AB, BB) at each marker locus are compared using analysis of variance. The relative phenotypes of these three classes permit conclusions to be drawn concerning the type of gene action that is operating at the trait locus linked to the marker; Fig. 2 shows the type of results that could be obtained by this method.

It is clear that while such simple analyses can yield much useful information, the conclusions are subject to a number of complicating factors, namely that the size of the effect seen at each marker locus will be determined by: (a) The absolute effect of the allelic differences at the trait locus. (b) The proportion of the total variation contributed by this locus. (c) The distance between the marker and trait loci. (d) The precision of trait measurement. (e) The possibility that two loci of opposite effect may occur in close proximity.

More sophisticated statistical methods may be applied to reduce the effects of some of these. Lander and Botstein (1989) and Paterson *et al.* (1988) have shown that resolution can be improved by combining the use of maximum likelihood estimation and interval mapping. Interval mapping makes use of information on the position of the markers, by comparing the phenotypic effects of chromosomal segments (the intervals between markers). The incorporation of positional information in this way permits the construction of 'QTL likelihood maps' (QTL=quantitative trait locus), which display the probability that one or more

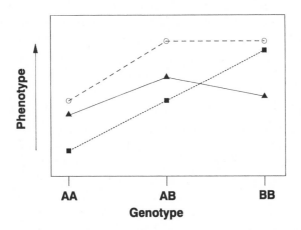

Fig. 2. Analysis of dominance effects using molecular markers. ■---■, No dominance; ○— —○, complete dominance; ▲——▲, overdominance.

QTLs lie at any defined point in the genome. Knapp *et al.* (1990) have recently described models to generate estimates of both magnitude of effect and position of QTLs, using a range of statistical approaches, applied to a wide range of types of segregating population.

How many loci are involved?

None of the above analytical methods permit the definition of the number of *genes* involved in the regulation of a complex trait. So far in this review, the term trait 'locus' has been used to describe a region of the genome that has an effect on the phenotype under consideration. The number or nature of structural or regulatory genes that make up that locus cannot be determined by the methods described above. Those methods simply identify *regions* of the genome that contain one *or more* genes affecting the phenotype. As pointed out above, there could easily be several genes within such a region – either with complementary or antagonistic effects; tightly-linked loci with opposite and equal effects would go undetected in this type of analysis.

A more refined approach to defining the number and effects of genes involved in the regulation of complex differences between plant lines involves the use of 'substitution mapping' (Paterson *et al.* 1990). Chromosomal segments from one genotype are introgressed into a common background; the breakpoints of the substitutions are defined precisely by RFLP analysis, and the phenotypes of plants carrying overlapping and non-overlapping substitutions can be compared at all points of the genome. The resolution of the technique – potentially very powerful – is limited by the ability to find the desired recombinants from meiotic segregations, by the density of markers in the region(s) of interest, and by the ability to measure phenotype accurately on the available material. The latter aspect is the key to all trait-marker associations; unless the individuals whose genotypes are determined are inbred, or the trait extremely easy to measure precisely on single plants, it will normally be necessary to obtain replication, either by backcrossing the individuals to a common genetic tester, or to determine phenotype from the progeny produced by selfing the individual. The potential to produce large numbers of dihaploid plants in some species offers a clear route to accelerated progress with this approach.

Numbers of plants and markers required to detect linkage

A number of workers have estimated the numbers of plants needed to detect linkage of markers to QTLs; the most useful contribution to date is that of Lander and Botstein (1989) who calculated sample numbers for F_2 and backcross families, using conventional analysis (random individuals from the population) and 'selective genotyping' (RFLP analysis only on individuals from the extremes of the phenotypic distribution). By sampling extreme phenotypes, it is possible to increase greatly the efficiency of detection of loci with small phenotypic effect. For example, by sampling 5% tails of a backcross population, where the original parents differed by 3 s.d.s (standard deviations), only 100 progeny would need to be fingerprinted to detect 4 loci affecting the trait; if the parents differed by only half this amount, 400 progeny should be fingerprinted. With F_2 populations, these figures are approximately halved. The calculations assume marker loci evenly-spaced at 20 cM intervals. Using conventional methods, these numbers would need to be increased many-fold.

Successful dissection of complex traits in plants

Despite the relatively recent development of the required technology there is already a growing number of examples of complex characters that have been dissected using molecular markers. One of the earliest RFLP maps available is for tomato, *Lycopersicon esculentum* (Helentjaris *et al.* 1986; Bernatzky and Tanksley, 1986); as a result, some of the most complete stories can be told for that species. Osborn *et al.* (1987), and later Tanksley and Hewitt (1988) and Paterson *et al.* (1988) analysed the genetics of high soluble solids and other 'quantitative' characters in tomato, the traits being introgressed from the wild species, *Lycopersicon chmielewskii*. Using interval mapping, Paterson *et al.* (1988) mapped six loci controlling fruit mass, four affecting soluble solids content, and five with effects on fruit pH. Further dissection of some of the regions involved with these traits has been performed by Paterson *et al.* (1990), by comparing selected overlapping segments of 'donor' chromosome in an otherwise equivalent genetic background. Using such approaches, it is possible to resolve a number of tightly-linked loci into their separate (and often opposing) effects. A key advantage for plant breeders that is offered by marker-assisted breeding is the ability to determine the quantity of donor germplasm that has been introgressed along with the trait gene of interest (Young and Tanksley, 1989); provided the region of interest is sufficiently saturated with polymorphic markers, it is possible to minimise the amount of unwanted material to a few centimorgans within 2 generations of backcrossing. Within tomato, it has also been possible to map loci associated with other complex traits, notably water use efficiency (Martin *et al.* 1989).

A considerable amount of trait dissection is currently underway in the major crop species, in particular maize, where a number of early publications give clear indications that very complex traits such as yield are approachable (e.g. Edwards *et al.* 1987, who identified associations between marker locus alleles and genes affecting 82 different traits in maize).

The use of marker-based selection in practical plant breeding programmes is still in its infancy, but will certainly make a major contribution to increasing the efficiency of such programmes (Lande and Thompson, 1990), and the agronomic performance of varieties issuing from them. Until now, effort has been concentrated on the use of single-locus probes and isozymes. The use of multilocus probing systems is clearly a possibility (see Hillel *et al.* 1990, for a mammalian parallel) provided that suitable probes are available. The development of PCR-based systems – possibly making use of random oligonucleotides as primers, as discussed above, could also open up the genetics of a number of complex traits in a wide range of hitherto uncharacterised species.

Use of marker-trait linkages to construct defined genotypes

In addition to the genetic dissection of complex traits using molecular markers, it is clearly possible to use the marker-trait associations to construct specific genotypes. The codominant nature of RFLP markers permits the differentiation of homo- and heterozygotes. Thus by constructing an appropriate cross, and thence a segregating population, RFLPs can be used to identify individuals with the required genotype. This need not only include the *trait* loci of interest (which may be identified in homozygous or heterozygous condition) but also markers defining the required genetic background. When assembling genes of interest it is clearly desirable to use markers that flank the trait locus; in this way the probability of the trait and marker loci being separated by an unobserved recombination event becomes very small. If markers are positioned at r (recombination fraction) to each side of the trait gene, the probability of 'losing' the trait gene by a double recombination event would be r^2 – or 0.01, where markers were 0.10 (or 10 cM) to each side. In fact, the overall probability would be much less than 0.01, because double recombination events in such close proximity would be greatly inhibited by crossover interference. Furthermore, in most trait-tagging programmes linkage tighter than 0.1 would be sought.

Thus it is possible to construct, from complex genetic crosses and importing genes from a number of diverse sources, genotypes with any combination of genes affecting a complex character, in what can be essentially an isogenic background. It is therefore possible to test interactions between genes, and more precisely to elucidate the regulatory mechanisms involved.

From marker to gene

It is clear from all that has been said above that RFLP and other DNA marker techniques should permit us to dissect the genetics of a wide range of complex traits, and to construct novel genotypes, either with the purpose of testing genetic hypothesis, or of producing a new and useful/desirable phenotype. The *technological* challenge for the next decade, however, is to find efficient means of moving from linkage information to find the actual genes that have been tagged.

The past five years have seen considerable developments in the technology for cloning large pieces of genomic DNA. The progression from cloning in bacteriophages, cloning around 20 kb per clone, to YAC (yeast artificial chromosome), cloning around 300 kb per clone, has been highly significant. It is important to appreciate, however, that enormous distances may have to be 'walked' in plant genomes. A walk of 10 cM, in a plant genome containing 5×10^9 bp and 1500 cM, will on average correspond to 30 000 kb, or 100 YAC clones, each of 300 kb. (These calculations assume that recombination is random with respect to DNA sequence; in fact, this is probably not true, and the inequality will work to the investigator's advantage and disadvantage in ways that cannot yet be predicted. We do know that large parts of the genome are heterochromatic or inactive; it is not yet clear how these regions are arranged with respect to the molecular markers currently in use, and the genes involved in interesting complex traits.) The task of chromosome 'walking' may be greatly simplified by the use of chromosome 'jumping' and 'linking' libraries, which enable large distances to be spanned without the need to clone all of the intervening DNA. The difficulty or ease of such a task in a complex plant genome is hard to estimate at this stage; most plant genomes have a large complement of repeated sequences that could easily lead to the walker losing his or her way along the chromosome. The first results of walks in complex plant genomes will yield a wealth of invaluable information, and indicate the areas of technology where additional development is needed.

In addition to the problems of moving from a genetic marker towards a gene of interest, an even greater challenge can be seen when the walk nears its end: when walking to a gene whose only known effect is on the plant's phenotype, how will we know when we have arrived? In the few mammalian examples published so far, the identification of the final gene has been complex, and has involved the comparison of candidate sequences with respect to their representation in different cDNA libraries, their evolutionary conservation, and finally to differences between genotypes. It is difficult to envisage a parallel with genes encoding components of highly complex plant traits (for example, yield). To clone out all candidate regions and to test them individually by transformation (complementation) would mean that such technology could only be applied to the most important of traits. A second major challenge for the near future, then, is to develop efficient ways of identifying active genomic regions, and of testing their function.

Conclusions

RFLP and other DNA markers offer the opportunity to dissect both simply-inherited and complex traits at the genetic level, and to construct desired genotypes, both for the testing of the genetic control of the character, and for the generation of plant lines or varieties with improved agronomic characterisitics. For a molecular dissection of the genes themselves, it will be necessary to develop techniques for identifying plant genes by their physical position and biological function.

References

BERNATZKY, R. AND TANKSLEY, S. D. (1986). Towards a saturated linkage map in tomato based on isozymes and random cDNA sequences. *Genetics* **112**, 887–898.

DALY, A., KELLAM, P., BERRY, S. T., CHOJECKI, A. J. S. AND BARNES, S. R. (1991). The isolation and characterisation of plant sequences homologous to human hypervariable minisatellites. In *DNA Fingerprinting: Approaches and Applications* (ed. T. Burke, G. Dolf, A. J. Jeffreys and R. Wolff) (in press).

EDWARDS, M. D., STUBER, C. W. AND WENDEL, J. F. (1987). Molecular-marker-facilitated investigations of quantitative-trait loci in maize. I. Numbers, genomic distribution and types of gene action. *Genetics* **116**, 113–125.

Helentjaris, T., Slocum, M., Wright, S., Schaefter, A. and Nienhuis, J. (1986). Construction of genetic linkage maps in maize and tomato using restriction fragment length polymorphisms. *Theor. appl. Genet.* **72**, 761–769.

Hillel, J., Schaap, T., Haberfeld, A., Jeffreys, A. J., Plotzky, Y., Cahaner, A. and Lavi, U. (1990). DNA fingerprints applied to gene introgression in breeding programs. *Genetics* **124**, 783–789.

Jeffreys, A. J., Wilson, V. and Thein, S. L. (1985). Hypervariable 'minisatellite' regions in human DNA. *Nature* **314**, 67–73.

Knapp, S. J., Bridges, W. C. Jr and Birkes, D. (1990). Mapping quantitative trait loci using molecular marker linkage maps. *Theor. appl. Genet.* **79**, 583–592.

Lande, R. and Thompson, R. (1990). Efficiency of marker-assisted selection in the improvement of quantitative traits. *Genetics* **124**, 743–756.

Lander, E. S. and Botstein, D. (1989). Mapping mendelian factors underlying quantitative traits using RFLP linkage maps. *Genetics* **121**, 185–199.

Martin, B., Nienhuis, J., King, G. and Schaeffer, A. (1989). Restriction fragment length polymorphisms associated with water use efficiency in tomato. *Science* **243**, 1725–1728.

Osborn, T. C., Alexander, D. C. and Forbes, J. F. (1987). Identification of restriction fragment length polymorphisms linked to genes controlling soluble solids content in tomato fruit. *Theor. appl. Genet.* **73**, 350–356.

Paterson, A. H., Lander, E. S., Hewitt, J. D., Peterson, S., Lincoln, S. E. and Tanksley, S. D. (1988). Resolution of quantitative traits into mendelian factors using a complete linkage map of restriction fragment length polymorphisms. *Nature* **335**, 721–726.

Paterson, A. H., DeVerna, J. W., Lanini, B. and Tanksley, S. D. (1990). Fine mapping of quantitative trait loci using selected overlapping recombinant chromosomes, in an interspecies cross of tomato. *Genetics* **124**, 735–742.

Tanksley, S. D. and Hewitt, J. (1988). Use of molecular markers in breeding for soluble solids content in tomato – a re-examination. *Theor. appl. Genet.* **75**, 811–823.

Thoday, J. M. (1961). Location of polygenes. *Nature* **191**, 368–370.

Young, N. D. and Tanksley, S. D. (1989). RFLP analysis of the size of chromosomal segments retained around the *Tm-2* locus of tomato during backcross breeding. *Theor. appl. Genet.* **77**, 353–359.

Printed in Great Britain © *Society for Experimental Biology 1991*   229

MOLECULAR ANALYSIS OF GENE REGULATION AND FUNCTION DURING MALE GAMETOPHYTE DEVELOPMENT

SHEILA McCORMICK, DAVID TWELL, GUY VANCANNEYT and JUDY YAMAGUCHI

Plant Gene Expression Centre, USDA/ARS-UC-Berkeley, 800 Buchanan Street, Albany, CA 94710, USA

Summary

We have characterised three pollen-expressed genes (*LAT52, LAT56* and *LAT59*) from tomato in order to determine their role in pollen development, and to determine the DNA sequences responsible for gene expression in pollen. *LAT52* encodes a protein that shows amino acid sequence similarity to a protein encoded by a pollen-specific cDNA clone (*pZmc13*) isolated from maize, and both proteins have amino acid sequence similarity to Kunitz trypsin inhibitors of soybean and winged bean. The proteins encoded by *LAT56* and *LAT59* genes are 54% identical at the amino acid level, and show significant sequence similarity to bacterial pectate lyases and to a fungal pectin lyase. Additionally, regions of LAT56 and LAT59 show significant sequence similarity to tryptic peptides of ragweed and Japanese cedar pollen allergens.

Preliminary results suggest that plants harboring antisense constructs of the *LAT52* coding region show defects in pollen germination and fertilisation; no obvious phenotype was seen in plants harboring antisense constructs of the *LAT59* coding region.

Promoter fragments of these three *LAT* genes were fused to the reporter gene *GUS* and assayed using both a transient system and stably transformed plants. We have identified relatively short regions of the LAT promoters that are important for pollen expression, and are attempting to isolate *trans*-acting factors that interact with these *cis*-acting sequences, using both molecular and classical genetic approaches.

Introduction

Microsporogenesis offers an ideal model for studying gene expression, cell division, and cell–cell communication during an important, but non-essential developmental sequence. The formation and function of pollen requires gene expression in both the sporophytic and gametophytic cells of the anther (reviewed by Mascarenhas, 1975). For example, a transitory layer, the tapetum (part of the

Key words: allergen, antibody, antisense, *Arabidopsis*, microprojectile bombardment, pectate lyase, proteinase inhibitor.

sporophytic tissue of the anther) is thought to provide metabolites required for the development of the gametophyte. Mutations that affect tapetal metabolism can affect the viability of the gametophytic generation (pollen), as exemplified by certain recessive male sterile mutations that have been found in higher plants (Kaul, 1988). Gametophytic gene expression is known to play a role in self-incompatibility in many plant families (Ebert *et al.* 1989), and molecular and biochemical evidence suggests that gametophytic gene expression is required for some aspects of pollen growth through the sporophytic female tissue (reviewed by Mascarenhas, 1989).

The first mitotic division after meiosis is a unique asymmetric division that gives rise to the vegetative and generative nuclei. The generative nucleus later divides to give rise to the two sperm nuclei, one of which will fertilize the central cell in the megagametophyte, eventually forming the endosperm, while the other will fertilize the egg, forming the zygote. Several lines of evidence (reviewed by Mascarenhas, 1975) suggest that the generative nucleus is essentially quiescent, while the vegetative nucleus is transcriptionally active, although the expression of specific genes in these nuclei has not been examined. The isolation of genes that are expressed gametophytically around the time of this mitotic division will provide tools with which to examine the regulation of this asymmetrical mitosis.

One of our objectives is to understand the regulatory circuits that control differential gene expression in pollen. Our approaches towards this objective include: (1) the isolation and characterisation of pollen-expressed promoter sequences, in order to determine the *cis*- and *trans*-acting factors that regulate pollen gene expression, and (2) the characterisation of the gene products from several pollen-expressed cDNA clones, in an attempt to assign functions to these genes. Transgenic plants expressing chimeric genes composed of these pollen-expressed promoters, or coding regions, should allow the manipulation of pollen development and function for basic and applied studies.

Several cDNA clones were isolated from a library prepared from mature anthers of tomato (McCormick *et al.* 1987). These clones were further characterised by DNA and RNA hybridisations, sequence analyses, and by *in situ* localization analysis (Ursin *et al.* 1989; Twell *et al.* 1989a; Wing *et al.* 1989; and our unpublished data). By northern analysis, it was found that *LAT52*, *LAT56* and *LAT59* (Late Anther Tomato) are expressed most strongly in pollen, although *in situ* analyses indicate that the RNA is also present in the sporophytic cells of the anther. The genes corresponding to these cDNAs are single copy in the tomato genome, and have been mapped to their chromosomal location using RFLP (restriction fragment length polymorphism) analysis.

Several other groups have isolated anther- or pollen-expressed cDNAs from maize (Stinson *et al.* 1987; Hanson *et al.* 1989), from sunflower (Herdenberger *et al.* 1990) from *Oenothera* (Brown and Crouch, 1990) and from tobacco (Goldberg, 1988). With molecular tools now available from these diverse species, it is our hope that common themes will emerge that will allow an understanding of the mechanisms regulating gene expression during pollen development. These molecular tools may also allow us to address interesting but less well characterised phenomena occurring during pollen tube growth and gamete fusion.

Gene function

When cDNAs are isolated solely on the basis of tissue-specific expression and without regard to function, the question remains as to whether the corresponding gene products, to these tissue-specific cDNAs, are important to the development or function of the organ from which they were isolated. We have used database searches to determine if the *LAT* genes are similar to known sequences. We have also raised antibodies against the different LAT proteins, in order to determine the cellular localisation and biological properties of the *LAT* gene products. Lastly, we have made antisense constructs with the *LAT* coding regions to attempt to correlate a phenotype with the loss of *LAT* gene function.

Sequence similarities of plant and bacterial pectin degrading enzymes

LAT56 and *LAT59* encode proteins that show sequence similarities to the pectate lyases of the bacterium *Erwinia* (Wing *et al.* 1989). Gysler *et al.* (1990) have recently cloned a pectin lyase gene from *Aspergillus niger*. Fig. 1 is an expanded version of the previously published sequence alignments (Wing *et al.* 1989) that includes the *Aspergillus* protein and additional similarities. The overall similarity between the *LAT56* or *LAT59* proteins and any of the other sequences is 75 % over a region of 142 amino acids. The conservation of these regions in bacterial, fungal and plant proteins might suggest that the *LAT56* and *LAT59* proteins encode a pectin-degrading enzyme. Moreover, these regions of similarity might delineate important regions for the functioning of these enzymes. Gysler *et al.* (1990) speculate that the regions of homology between the *Erwinia* pectate lyases and the *Aspergillus* pectin lyase might be involved in substrate binding. A pollen-expressed cDNA isolated from *Oenothera* shows 35 % amino acid identity (54 % similarity) to the polygalacturonase from tomato fruit (Brown and Crouch, 1990). It is not yet known whether the *Oenothera* cDNA clone encodes a protein with polygalacturonase enzyme activity, although pectin-degrading enzyme activity has been detected in pollen of some plant species (Pressey and Reger, 1989). Hinton *et al.* (1990) and Saarilahti *et al.* (1990) have sequenced polygalacturonase-encoding genes from *Erwinia* and have pointed out that the protein sequences have a 26 % identity to the tomato polygalacturonase amino acid sequence. Huang and Schell (1990) cloned a polygalacturonase gene from *Pseudomonas solanacearum* that shows 41 % similarity to the tomato polgalacturonase in a portion of the sequence (amino acids 176–324 in the tomato sequence). Similarly, blocks of conserved amino acids can be found when pectin methyl esterase amino acid sequences from tomato (Ray *et al.* 1988), *Erwinia* (Plastow, 1988) and *Aspergillus* (Khanh *et al.* 1990) are compared. It seems possible to us that the *LAT56* and *LAT59* genes do in fact encode proteins that play some role in pectin metabolism, considering the extent of sequence similarities between known pectin-degrading enzymes. We

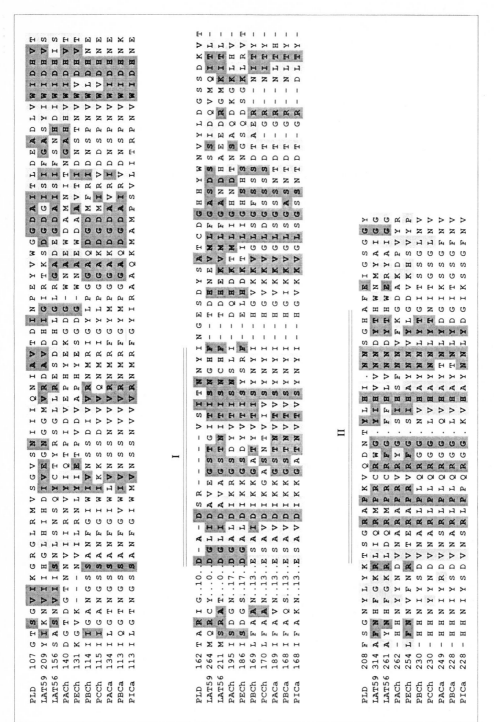

Fig. 1. Amino acid sequence similarities between *LAT56*, *LAT59* and pectate and pectin lyase gene products. Boxes are drawn around amino acids that are identical (dark shading) or similar (light shading) in either *LAT56* or *LAT59* and at least one of the pectate or pectin lyases. PLD is *Aspergillus niger* pectin lyase, PACh, PBCh, PCCh and PECh are pectate lyases A,B,C, and E from *Erwinia chrysanthemi*, PACa, PBCa and PICa are pectate lyases A, B and I from *Erwinia carotovora*. Conserved amino acids substitutions are grouped as follows: (K,R), (E,D), (Q,N), (S,T), (F,Y,W,H), (L,I,V,A), (P,G) and (M,C).

have not been able, as yet, to demonstrate that the *LAT56* or *LAT59* gene products have pectate lyase activity, perhaps because the exact substrate or co-factor requirements are lacking.

LAT56 and LAT59 show sequence similarities to pollen allergens

We noted previously (Wing *et al.* 1989) that a region of the *LAT56* and *LAT59* proteins showed sequence similarity to the partial N-terminal peptide sequence of an allergen isolated from Japenese cedar pollen (Taniai *et al.* 1988). From this observation, we suggested that the *LAT56* and *LAT59* proteins might be secreted into the pollen wall, the expected location of pollen allergens. The *LAT56* and *LAT59* proteins also show significant sequence similarity (65 % and 72 %) to the amino acid sequences of two tryptic peptides of the allergen Amb a I isolated from ragweed pollen (Smith, J. J. *et al.* 1988). The sequence similarities to the Japanese cedar and ragweed allergen tryptic peptide sequences are shown in Fig. 2.

Tomato, cedar and ragweed are quite unrelated taxonomically. Because of the significant amino acid similarities between regions of LAT56 and LAT59 and the partial amino acid sequences that are thus far available for the ragweed and Japanese cedar allergens, we speculate that these allergens may be the ragweed and Japanese cedar LAT56 or LAT59 counterparts. Further information obtained *via* antibody cross-reactivity, or *via* cloning of the cDNAs from ragweed that correspond to the Amb a I allergen may clarify this issue.

Pollen proteins that are allergens presumably play some role in pollen development or function. Breiteneder *et al.* (1989) recently noted that the birch pollen allergen Betv1 showed significant sequence similarity to a disease resistance response gene from pea, and they suggested that the birch allergen might play a role in pathogen resistance in birch pollen. Alternatively, such proteins on the surface of pollen grains may play roles in pollen–pistil interactions (Clarke and Gleeson, 1981).

```
Cedar   Cry j I   D  N  P  I  D  S  ?  W  R  G  D  S  N  W  A  Q  N  R  M  K
LAT56   (46)      V  N  S  I  K  K  C  W  R  C  D  P  F  W  A  E  R  C  G  Q
LAT59   (99)      T  N  P  I  D  K  C  W  R  C  D  P  N  W  A  D  N  R  K  K


Ragweed Amb a I   G  M  L  A  T  V  A  F  N  T  F  T  D  N  V  D  Q  R
LAT56   (255)     G  M  K  I  T  L  A  Y  N  H  F  G  K  R  L  D  Q  R
LAT59   (308)     V  M  Q  I  T  L  A  F  N  H  F  G  K  R  L  I  Q  R


Ragweed Amb a I   R  W  G  S  Y  A  I  G  G  S  A  S  P  W  I  L  T  Q  T  N
LAT56   (290)     H  W  E  R  Y  A  I  G  G  S  S  G  A  T  I  I  S  Q  G  N
LAT59   (343)     H  W  N  M  Y  A  I  G  G  S  M  H  P  T  I  I  H  Q  G  N
```

Fig. 2. Amino acid similarities between either LAT56 or LAT59 and the partial amino acid sequences of Japanese cedar allergen *Cry j I* and ragweed allergen Amb a I. Boxes are drawn around amino acids that are identical (dark shading) or similar (light shading) in either LAT56 or LAT59 and the allergen peptide sequences.

Sequence similarity of LAT52 and Zm13 to proteinase inhibitors

The coding regions of *LAT52* (Twell *et al.* 1989*a*) and pZmc13, a corn pollen-expressed cDNA clone (Hanson *et al.* 1989), are 33 % identical at the amino acid level, with blocks of conserved and non-conserved regions. Interestingly, most of the cysteine residues are conserved, as shown in Fig. 3. Fig. 3 also shows that the LAT52 and pZmc13 proteins show sequence similarity to portions of the amino acid sequences of Kunitz trypsin inhibitors from soybean (Kim *et al.* 1985; Jofuku and Goldberg, 1989) and to the Ti trypsin inhibitor from winged bean (Yamamoto *et al.* 1983). Within this region of overlap (aa37–97 in LAT52), the Zmc13 or *LAT52* proteins show 58 % amino acid similarity to the trypsin inhibitors.

Kunitz trypsin inhibitors are abundantly expressed in soybean embryos. Jofuku and Goldberg (1989) transferred the *KTi1* and *KTi2* genes into tobacco to assess the tissue specificity of their gene expression in a heterologous system. These genes were expressed in tobacco seeds, and in the stem and leaf, but not in the root. Jofuku and Goldberg (1989) did not report whether they detected any expression of the *KTi1/2* genes in flower parts of transgenic tobacco. It is not known whether the LAT52 and Zmc13 proteins exhibit proteinase inhibitor activity. Proteinase inhibitor family members that do not exhibit proteinase inhibitor activity may serve, alternatively, as storage proteins (reviewed by Richardson, 1977).

Antibodies to LAT proteins

We are interested in characterizing the *LAT52*, *LAT56* and *LAT59* derived proteins at the biochemical and cytological level, and in determining if their transcripts are translated prior to pollen germination. To this end, we expressed portions of the coding regions as truncated translational fusions to *trp E*

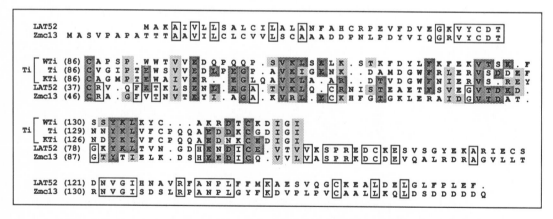

Fig. 3. Sequence similarities between LAT52, Zmc13, and Kunitz trypsin inhibitors. WTi1 is the winged bean trypsin inhibitor, Ti is the major soybean trypsin inhibitor Ti[a], and KTi1 is a minor soybean trypsin inhibitor. Boxes are drawn around amino acids that are identical in the LAT52 and Zmc13 sequences. Dark shading denotes amino acid identity between either LAT52 or Zmc13 and one of the trypsin inhibitors, while light shading denotes conserved amino acid changes.

(Dieckmann and Tzagoloff, 1985). These proteins were purified from *Escherichia coli* and used as antigens for the production of polyclonal antisera in mice.

Antiserum against LAT52 cross-reacts with pollen extracts, but not with leaf extracts. The LAT52 antibody reacts with a protein of approximately $16 \times 10^3 M_r$, which corresponds to the relative molecular mass predicted from the cDNA sequence, assuming the predicted N-terminal signal sequence is cleaved. Preliminary experiments indicate that the *LAT52* protein is heat stable. Since most proteinase inhibitors are relatively insensitive to heat (reviewed by Richardson, 1977) and are in the relative molecular mass range of 10 to $20 \times 10^3 M_r$, these data support the idea of LAT52 being related to trypsin inhibitors.

Antiserum against LAT56 cross-reacts with pollen extracts, but not with leaf extracts. The antiserum reacts with a protein of the expected relative molecular mass, assuming that the predicted N-terminal signal sequence is cleaved. The *LAT56* protein appears to be quite unstable, because proteinase inhibitors are needed in order to detect the protein on western blots. Antisera against the *LAT59* protein have not yet been tested.

Mature pollen contains 'stored' mRNA that is translated upon pollen germination, and whose protein products may be required during pollen germination and pollen tube growth (Mascarenhas, 1989). Using antibodies, Brown and Crouch (1989) demonstrated that the polygalacturonase-like proteins encoded by the P2 family of pollen-expressed cDNAs of *Oenothera* were already present in mature pollen.

By RNA blot analysis it was found that *LAT52* is expressed in petals and anther walls (Twell *et al.* 1989*a*; Twell *et al.* 1990), and by using transgenic plant analysis of *LAT52–GUS* (*β*-glucuronidase) fusions (Twell *et al.* 1991), was also found in the endosperm. Using RNA blot analysis, *LAT59* was found to be expressed in anther wall (Twell *et al.* 1990), and by using transgenic plant analysis of *LAT59–GUS* fusions (Twell *et al.* 1991) in the root cap, seed coat and endosperm. The antibodies raised against the *LAT* proteins will allow us to determine whether these sporophytic patterns of expression reflect the endogenous distribution of the *LAT52* and *LAT59* gene products.

Staff *et al.* (1990) were able to localise pollen allergens of rye grass to the exine, and show their absence in the intine, using antibodies raised against the allergens; antibodies to the *LAT* proteins will similarly be used for immunolocalisations in tomato pollen. We also expect that these antibodies will prove useful in the analysis of the plants harboring antisense constructs of the *LAT* cDNAs (see below).

Lastly, it may also be possible to use these antibodies in *in vitro* pollen germination tests, to test whether the *LAT* proteins play important roles in pollen tube growth, analogous to the use of purified *S*-glycoproteins as inhibitors of self-incompatible pollen tube growth (Jahnen *et al.* 1989).

Antisense constructs with the LAT52 *and* LAT59 *coding regions*

One approach to determining the function of a gene product is to identify a phenotype correlated with the antisense inhibition of gene expression. Successful uses of antisense inhibition of genes with known functions include polygalacturonase (Smith, C. J. S. *et al.* 1988) and chalcone synthase (van der Krol *et al.* 1988). Recently, Hamilton *et al.* (1990) used antisense inhibition of the fruit expressed cDNA pTOM13 to correlate the expression of the endogenous gene with the ethylene biosynthetic pathway. In *Drosophila* antisense inhibition of a ribosomal protein gene that is specifically translated during early embryogenesis led to a defect in oogenesis (Qian *et al.* 1988).

If gene products of the *LAT52*, *LAT56* or *LAT59* genes are essential for pollen formation or function, then, in plants with reduced expression of these genes, we might expect a reduction of the respective transcripts and proteins, and hence an altered phenotype during pollen development or pollen germination. Our initial experiments tested the *in vitro* pollen germination of the pollen of R_0 antisense plants, and scored the transmission of kanamycin resistance in self-pollinated R_0 plants, and in outcrosses with untransformed tomatoes.

We initially constructed transgenic plants harbouring the cDNA antisense constructs of *LAT56* and *LAT52*, directed by the CaMV35S promoter. Examination of the pollen of these plants showed no obvious phenotypic effects, and transmission of the linked kanamycin resistance gene was normal. We later discovered that CaMV35S is weakly expressed in pollen (Twell *et al.* 1989*b*), relative to the endogenous promoters of *LAT52* and *LAT59*. For this reason additional antisense experiments were performed using the *LAT52* and *LAT59* promoters (Twell *et al.* 1990). We made antisense constructs of the *LAT52* and *LAT59* coding regions, resulting in three constructs, termed 52/52AS, 52/59AS and 59/59AS. Transgenic tomato plants were generated with these three constructs and their phenotypes examined. Fifteen independent plants of 52/52AS, eight independent plants of 52/59AS and fourteen independent plants of 59/59AS were generated.

52/59AS and 59/59AS plants

At the whole plant level, there were no obvious phenotypic differences between the transgenic plants with the antisense genes and the wild-type controls. The 59/59AS plants developed normal flowers and R_1 seeds were obtained from seven of the eight 52/59AS plants. Four of these plants showed an expected 3:1 segregation ratio for kanamycin resistance, and homozygotes were obtained in the R_2 generation. The other three plants showed a 2:1 or 1:1 segregation upon self-pollination; further analysis of these plants is required to determine if the altered segregation is related to the expression of the antisense construct. R_1 seeds were obtained from ten of the fourteen 59/59AS plants. Eight of these plants showed the expected 3:1 segregation for kanamycin resistance, and homozygotes were obtained in the R_2 generation. Two of the 59/59AS plants showed 1:1 segregation for kanamycin resistance and require further analysis to determine if there is a correlation with the antisense construct.

To summarise, although the 52/59AS and 59/59AS plants have not been fully analysed, the results to date indicate that the majority of these plants show no obvious defects in pollen germination *in vitro* or *in vivo*, and give normal fruit and seed set, with the expected transmission of the linked kanamycin resistance marker. From some of these R_1 plants we have identified homozygous plants, perhaps the best indication that there was no detrimental effect of the antisense expression of *LAT59* in the pollen of these plants. It is possible that screening additional plants, or more detailed analysis of the progeny of the plants that showed aberrant transmission of the kanamycin resistance, may allow us to identify 59AS plants with discernible pollen phenotypes. Because *LAT56* and *LAT59* have sequence similarity it is also possible that we will not be able to detect a phenotypic effect unless the expression of both *LAT56* and *LAT59* is suppressed using methods such as antisense RNA. A further possibility is that there is no phenotypic consequence of down regulation of *LAT59* expression in pollen.

52/52AS plants

Our preliminary data suggest that there is some phenotypic effect on pollen function owing to the expression of the antisense DNA to the *LAT52* gene. Of the 15 independent transformants harbouring the 52/52AS construct, none set fruit without hand pollination, whereas the cultivar VF36, used for transformations, normally does not require hand pollination for fruit and seed set. Eight of the plants eventually died before any seed could be obtained. Flowers on the remaining seven plants were hand-pollinated to obtain fruit and seed set. Additionally, pollen from these seven plants was used to pollinate non-transformed tomato plants, and for *in vitro* pollen germination tests. Two of these plants had approximately 50% aborted pollen, but the apparently normal pollen in these plants germinated *in vitro*. The pollen from the five other plants germinated *in vitro*, although approximately 50% of the grains had corkscrew pollen tubes. The growth pattern of these pollen tubes somewhat resembled the corkscrew pattern seen when tobacco pollen is germinated in the presence of Triton X-100 (Rao and Kirsten, 1990) and may reflect an osmotic or membrane defect. All seven of these 52/52AS plants will set seed if hand self-pollinated, but the transmission of the linked kanamycin resistance marker is aberrant, and suggests that the kanamycin marker is only being transmitted through the female. Additionally, all the R_1 plants we have obtained from the self-pollinations of the R_0 plants are heterozygous for the kanamycin resistance marker (that is, no homozygous R_2 plants have been obtained). The simplest explanation of these results is that 50% of the pollen (those grains that are 52/52AS, Kan^R) does not function.

Over 100 pollinations of non-transformed tomato (cv. VF36) were attempted with pollen from the seven 52/52AS plants, but to date we have obtained only seven fruit, which have yielded no kanamycin-resistant progeny.

Although these results suggest that a defect in the 52/52AS, Kan^R pollen exists, several unanswered questions remain. The pollen from the primary transformants of 52/52AS plants does not function normally in fruit and seed set on outcrosses.

This result is not predicted with pollen of primary transformants; because *LAT52* gene expression is correlated with first microspore mitosis (Twell *et al.* 1990), 50% of the R_0 pollen should not carry, and therefore should not express the 52/52AS gene. These pollen grains should function and result in seed set, albeit kanamycin sensitive seeds, in the R_1 fruit. The data available so far might suggest that the pollen carrying the 52/52AS construct can somehow influence the germination or fertilisation efficacy of the pollen that does not carry the 52/52AS gene; but apparently most severely in outcrosses, rather than in self pollinations. Transmission of the Kan^R marker in the selfed seed of the primary 52/52AS transformants does occur, presumably *via* female transmission of the Kan^R marker.

To summarise, we have not yet obtained molecular proof that the apparent defects in the pollen of the 52/52AS plants correlate with the expression of the antisense RNA. However, the fact that 15 independent transformants harboring this construct show a similar phenotype suggests that antisense expression of the *LAT52* coding region can alter pollen function. Experiments are in progress to assess whether pollen from the 52/52AS plants can germinate *in vivo*, and if so, when and where the 52/52AS pollen stops during pollen germination in the style.

Gene regulation

Sequence motifs important for pollen expression

We showed (Twell *et al.* 1990) that the 5′ flanking regions of the *LAT52* and *LAT59* genes, when fused to the *GUS* reporter gene, are sufficient to direct expression in an essentially pollen-specific manner in transgenic tomato, tobacco and *Arabidopsis* plants. We have used stably transformed tomato plants, and microprojectile bombardment of tobacco pollen (Twell *et al.* 1989*b*) to assay 5′ deletion derivatives of these promoters (McCormick *et al.* 1991; Twell *et al.* 1991). Additionally, we have assayed 5′ deletion derivatives of the LAT56 promoter with the transient assay.

5′ Deletion analysis of the *LAT52* and *LAT59* promoters in transgenic tomato plants demonstrated that only minimal (less than 200 base pairs) regions are required for developmentally regulated expression in pollen (McCormick *et al.* 1991; Twell *et al.* 1991). In addition, we found that these minimal promoter regions direct distinct, but overlapping patterns of expression in specific cell-types of the sporophyte.

Comparisons of the *LAT52* and *LAT59* promoter sequences revealed no obvious regions of sequence similarity. However, when the *LAT56* promoter sequence was also analysed, we were able to identify a 12 base pair sequence that is identical to the *LAT52* and *LAT56* promoters, and a different 10 base pair region of homology between the *LAT56* and *LAT59* promoters (Twell *et al.* 1991). The 52/56 'box' shows homology to the GT-1 box, found in many light-regulated genes (reviewed by Gilmartin *et al.* 1990) and our functional analyses of this region in the *LAT52* and *LAT56* promoters show that this region is required for high level expression in pollen. Similarly, functional analyses of the 56/59 'box' indicate that this region is important for pollen gene expression (Twell *et al.* 1991). We have noted sequence similarity between both the 52/56 box and the 56/59 box and regions of the pollen and anther expressed chalcone synthase *chiA* P_{A2} promoter of *Petunia* (Twell *et al.* 1991; van Tunen *et al.* 1990).

Approaches to isolate trans-*acting factors that interact with the* cis-*elements of the* LAT *promoters*

DNA fragments corresponding to the functionally important regions of the *LAT* promoters are currently being used in gel retardation experiments, and as probes for screening a phage expression library that was prepared from tobacco pollen mRNA. With these approaches, we hope to isolate cDNA clones corresponding to pollen-expressed DNA binding proteins that interact with selected regions of the *LAT* promoters.

We have shown (Twell *et al.* 1990) that the *LAT52* and *LAT59* promoters exhibit the same tissue-specific and developmental patterns of gene expression in tomato, tobacco and *Arabidopsis*. Because of the attractive features of *Arabidopsis* for mutagenesis and gene isolation, we are initiating experiments to isolate genes that act in the signal transduction pathway of pollen gene expression. To this end, we have generated large quantities of homozygous *LAT52–GUS* and *LAT59–GUS* transgenic *Arabidopsis* seed and have begun EMS (ethyl methanesulfonate) mutagenesis and screening of the M_1 plants for potential mutations that affect expression from the *LAT* promoters. We have made the assumption that *trans*-acting factors are required to interact with *cis*-elements in the *LAT52* and *LAT59* promoters; absence of these factors should result in the apparent reduction of *LAT52* or *LAT59* promoter activity, that is, a lack or reduction of GUS expression in pollen of the M_1 plants.

As mentioned above, our evidence from transgenic plant analyses indicate that the *LAT52* and *LAT59* promoters are expressed post meiotically (Twell *et al.* 1990). Therefore, if a mutation occurs in a putative *trans*-acting factor that interacts with the *LAT52* or *LAT59* promoters (*Tf*→*tf*), then such a mutation might be scorable in 50% of the pollen of a mutant sector in a M_1 plant. It is possible that the gene product of *LAT52* or *LAT59*, or the putative *Tf* gene product is required for pollen viability; if this proves to be the case, the mutant pollen of the sector in the M_1 plant will not function and it will be impossible to obtain homozygous *GUS/GUS*; *tf/tf* plants. However, the M_2 from the M_1 sector with 50% blue pollen should again generate 1/2 *GUS/GUS*; *Tf/tf* plants.

In summary, we hope to identify genetic loci whose products interact directly or indirectly with the promoter regions of the *LAT* genes. With this screening scheme, we should be able to distinguish between T-DNA-linked coding regions or promoter mutations and unlinked *trans*-acting factor mutations.

Pollen promoters direct tissue-specific and developmental regulated expression

Because the *LAT* genes are highly expressed in a relatively narrow spatial and temporal window, their promoters are useful in directing the expression of reporter and other interesting genes for studying pollen development.

Overproduction of Agrobacterium plant hormone genes

Klee *et al.* (1987) introduced the use of regulated promoters to direct plant hormone biosynthesis to specific tissues at specific times in development. Using this approach to investigate the effects of cytokinin overproduction on pollen development, we fused the *LAT59* promoter to the *ipt* gene. In attempts to create a conditional lethal phenotype (male sterile) through IAA (indole acetic acid) overproduction in the presence of IAM (indoleacetamide), we fused the *LAT59* promoter to the *iaaH* gene. We used these constructs to transform tomato and tobacco separately. Pollen-specific expression of cytokinins might be expected to have some influence on seed or fruit set, especially with reference to parthenocarpy (Mapelli *et al.* 1978; Mapelli, 1981), while auxin production in pollen might be expected to have detrimental effects on pollen function.

The *iaaH* gene converts the non-toxic precursor indolacetamide (IAM) into indole acetic acid (IAA). This activity can be used in a seedling selection scheme to identify seedlings that express *iaaH*. We previously had shown that the *LAT59–GUS* fusion shows low activity in roots (Twell *et al.* 1990); and our results with a seedling screen of transgenic *LAT59–iaaH* tobacco and tomato, using IAM-containing medium, suggest that the root expression of the *LAT59* promoter might be responsible for our ability to identify *LAT59–iaaH* expressing seedlings. *In vitro* tomato pollen germination is severely inhibited by 10^{-4} M IAA, and no germination is observed at 10^{-3} M IAA. IAM shows some inhibition of pollen tube growth at 10^{-3} M, but no effects at 10^{-4} M; similar results were found with tobacco pollen germination. However, in the transformants thus far examined, no differences in *in vitro* pollen germination in the presence of IAM were noted, even when comparing pollen from homozygous *LAT59–iaaH* plants with pollen from non-transformed plants. Thus the *LAT59–iaaH* chimeric gene does not appear to be useful for the construction of a conditional male sterile plant.

The *LAT59–ipt* construct shows extreme phenotypic effects in transgenic tobacco, but the transgenic tomatoes expressing this construct show no whole plant phenotypic effects when compared to wild-type control plants (Twell *et al.* in preparation). Using RNA blot analysis on two independent tobacco transformants, we estimated that the *LAT59–ipt* transcript was at least 100-fold more abundant in pollen than in leaves. We also showed that the pollen of these plants contained elevated levels of cytokinins (Twell *et al.* in preparation). Seed germination of non-transformed tobacco and tomato seeds in the presence of exogenous cytokinin suggests that the lack of an altered phenotype in the *LAT–ipt* tomato plants may in part be due to a difference in sensitivity to cytokinin. Tobacco seed germination is severely inhibited (seedlings form calli and show typical cytokinin-excess phenotypes) in the presence of 0.1 mg l^{-1} zeatin, but tomato seedlings show little or no growth inhibition at this concentration, but only show significant inhibition of shoot and root growth at 1 mg l^{-1} zeatin.

Transcription in pollen

The literature suggests that the generative nucleus is quiescent, while the vegetative nucleus is transcriptionally active. For example, the low number of nuclear pores in the generative nucleus is thought to be correlated with lower transcriptional activity relative to the generative nucleus (Wagner *et al.* 1990). Gene expression directed by the *LAT52* and *LAT59* promoters is first detected about the time of first microspore mitosis (Twell *et al.* 1990); thus this expression could be transcription from the vegetative nucleus, from the generative nucleus, or from both nuclei. If the LAT genes can be shown to be expressed in the generative nucleus, then experiments designed towards the selection of stable transformation of pollen *via* microprojectile bombardment (Twell *et al.* 1989*b*) will be worth attempting. We are also interested in using the LAT promoters to drive expression of transposase, towards increasing the frequency of new germinally transmitted integration sites (Hehl and Baker, 1990). Methods for distinguishing the cellular location of nascent transcripts have recently been published (Lawrence *et al.* 1989; Lawrence *et al.* 1990; O'Farrell *et al.* 1989). We plan to adapt these techniques to determine whether the generative nucleus can transcribe the LAT genes.

Reporter gene fusions to mark cellular events

Because the *LAT52* and *LAT59–GUS* fusions can be stored for gene expression from first pollen grain mitosis through pollen germination, as well as in the seeds and/or roots of the resulting seed generation, these reporter gene fusions can serve diverse functions. The inclusion of *LAT–GUS* chimeric genes in transformation vectors will allow the identification of plants that are homozygous for the linked gene of interest in the pollen of the R_1 plants, rather than in the R_2 seedlings, which will therefore save time and greenhouse space. The *LAT–GUS* constructs may also be useful as markers for studying pollen competition or pollen mentor effects in mixed pollinations, and would allow an examination of these phenomena during the early stages of pollen germination and tube growth. These promoter–*GUS* constructs could also be used to follow gene flow and/or pollen dispersal. We do not know whether the *LAT* promoters will function in a wide range of plant species, but promoters of homologous genes could be used in other species if necessary. The *LAT–GUS* constructs might also be developed as markers to distinguish differences in gene expression due to paternal *versus* maternal transmission (Allen *et al.* 1990).

We thank Keith Hamby and Brian Osborne for comments on the manuscript. This work was supported by USDA-ARS CRIS 5335–22230–002–00D and by the NSF Center for Plant Developmental Biology, UC-Berkeley, DIR-8719933.

References

ALLEN, N. D., NORRIS, M. L. AND SURANI, M. A. (1990). Epigenetic control of transgenic expression and imprinting by genotype-specific modifiers. *Cell* **61**, 853–861.

Breiteneder, H., Pettenburger, K., Bito, A., Valenta, R., Kraft, D., Rumpold, H., Scheiner, O. and Breitenbach, M. (1989). The gene coding for the major birch pollen allergen *Bet v I* is highly homologous to a pea disease resistance response gene. *EMBO J.* **8**, 1935–1938.

Brown, S. M. and Crouch, M. L. (1990). Characterisation of a gene family abundantly expressed in *Oenothera organensis* pollen that shows sequence similarity to polygalacturonase. *Plant Cell* **2**, 263–274.

Clarke, A. E. and Gleeson, P. A. (1981). Molecular aspects of recognition and response in the pollen-stigma interaction. *Rec. Adv. Phytochem.* **15**, 161–211.

Dieckmann, C. L. and Tzagoloff, A. (1985). Assembly of the mitochondrial membrane system. *CBP6*, a yeast nuclear gene necessary for synthesis of cytochrome b. *J. biol. Chem.* **260**, 1513–1520.

Ebert, P. R., Anderson, M. A., Bernatzky, R., Altschuler, M. and Clarke, A. E. (1989). Genetic polymorphism of self-incompactibility in flowering plants. *Cell* **56**, 255–262.

Gilmartin, P., Sarokin, L., Memlink, J. and Chua, N.-H. (1990). Molecular light switches for plant genes. *Plant Cell* **2**, 369–378.

Goldberg, R. B. (1988). Plants: novel developmental processes. *Science* **240**, 1460–1467.

Gysler, C., Harmsen, J. A. M., Kester, H. C. M., Visser, J. and Heim, J. (1990). Isolation and structure of the pectin lyase D-encoding gene from *Aspergillus niger*. *Gene* **89**, 101–108.

Hamilton, A. J., Lycett, G. W. and Grierson, D. (1990). Antisense gene that inhibits synthesis of the hormone ethylene in transgenic plants. *Nature* **346**, 284–287.

Hanson, D. D., Hamilton, D. A., Travis, J. L., Bashe, D. M. and Mascarenhas, J. P. (1989). Characterisation of a pollen-specific cDNA clone from *Zea mays* and its expression. *Plant Cell* **1**, 173–179.

Hehl, R. and Baker, B. (1990). Properties of the maize transposable element *Activator* in transgenic tobacco plants: a versatile inter-species genetic tool. *Plant Cell* **2**, 709–721.

Herdenberger, F., Evrard, J-L., Kuntz, M., Tessier, L-H., Klein, A., Steinmetz, A. and Pillay, D. T. N. (1990). Isolation of flower specific cDNA clones from sunflower. *Pl. Sci.* **69**, 111–122.

Hinton, J. C. D., Gill, D. R., Lalo, D., Plastow, G. S. and Salmond, G. P. C. (1990). Sequence of the *peh* gene of *Erwinia carotovora*: homology between *Erwinia* and plant enzymes. *Molec. Microbiol.* **4**, 1029–1036.

Huang, J. and Schell, M. A. (1990). DNA sequence analysis of *pglA* and mechanism of export of its polygalacturonase product from *Pseudomonas solanacearum*. *J. Bact.* **172**, 3879–3887.

Jahnen, W., Lush, W. M. and Clarke, A. E. (1989). Inhibition of *in vitro* pollen tube growth by isolated S-glycoproteins of *Nicotiana alata*. *Plant Cell* **1**, 501–510.

Jofuku, K. D. and Goldberg, R. B. (1989). Kunitz trypsin inhibitor genes are differentially expressed during the soybean life cycle and in transformed tobacco plants. *Plant Cell* **1**, 1079–1093.

Kaul, M. L. H. (1988). *Male Sterility in Higher Plants. Monographs on Theoretical and Applied Genetics*, vol 10. Berlin: Spring-Verlag.

Khanh, N. Q., Albrecht, H., Ruttowski, E., Loffler, F., Gottschalk, M. and Jany, K-D. (1990). Nucleotide and derived amino acid sequence of a pectinesterase cDNA isolated from *Aspergillus niger* strain RH 5344. *Nucl. Acids Res.* **18**, 4262.

Kim, S.-H., Hara, S., Ikenaka, T., Toda, H., Kitamura, K. and Kaizuma, N. (1985). comparative study on amino acid sequences of Kunitz-type soybean trypsin inhibitors, Ti^a, Ti^b and Ti^c. *J. Biochem.* **98**, 435–448.

Klee, H. J., Horsch, R. B., Hinchee, M. A., Hein, M. B. and Hoffmann, N. L. (1987). The effects of overproduction of two *Agrobacterium tumefaciens* T-DNA auxin biosynthetic gene products in transgenic petunia plants. *Genes and Dev.* **1**, 86–96.

Lawrence, J. B., Marselle, L. M., Byron, K. S., Johnson, C. H., Sullivan, J. L. and Singer, R. H. (1990). Subcellular localisation of low-abundance human immunodeficiency virus nucleic acid sequences visualised by fluorescence *in situ* hybridisation. *Proc. natn. Acad. Sci. U.S.A.* **87**, 5420–5454.

Lawrence, J. B., Singer, R. H. and Marselle, L. M. (1989). Highly localised tracks of specific transcripts within interphase nuclei visualized by *in situ* hybridisation. *Cell* **57**, 493–502.

MAPELLI, S. (1978). Changes in cytokinin in fruits of parthenocarpic and normal tomatoes. *Plant Sci. Lett.* **22**, 227–233.

MAPELLI, S., FROVA, C., TORTI, G. AND SORESSI, G. P. (1981). Relationship between set, development and activities of growth regulators in tomato fruits. *Pl. Cell Physiol.* **19**, 1281–1288.

MASCARENHAS, J. P. (1975). The biochemistry of angiosperm pollen development. *Bot. Rev.* **41**, 259–314.

MASCARENHAS, J. P. (1989). The male gametophyte of flowering plants. *Plant Cell* **1**, 657–664.

MCCORMICK, S., SMITH, A., GASSER, C., SACHS, K., HINCHEE, M., HORSCH, R. AND FRALEY, R. (1987). Identification of genes specifically expressed in reproductive organs of tomato. In *Tomato Biotechnology* (ed. D. Nevins and R. Jones), pp. 255–265. Alan R. Liss: New York.

MCCORMICK, S., YAMAGUCHI, J. AND TWELL, D. (1991). Deletion analysis of pollen-expressed promoters. *In vitro Cell and Devl Biol. PLANT* **27**, (in press).

O'FARRELL, P. H., EDGAR, B. A., LAKICH, D. AND LEHNER, C. F. (1989). Directing cell division during development. *Science* **246**, 635–640.

PLASTOW, G. S. (1988). Molecular cloning and nucleotide sequence of the pectin methylesterase gene of *Erwinia chrysanthemi* B374. *Molec. Microbiol.* **2**, 247–254.

PRESSEY, R. AND REGER, B. J. (1989). Polygalacturonase in pollen from corn and other grasses. *Pl. Sci.* **59**, 57–62.

QIAN, S., HONGO, S. AND JACOBS-LORENA, M. (1988). Antisense ribosomal protein gene expression specifically disrupts oogenesis in *Drosophila melanogaster*. *Proc. natn. Acad. Sci. U.S.A.* **85**, 9601–9605.

RAO, K. S. AND KIRSTEN, U. (1990). The influence of the detergent Triton X-100 on the growth and ultrastructure of tobacco pollen tubes. *Can. J. Bot.* **68**, 1131–1137.

RAY, J., KNAPP, J., GRIERSON, D., BIRD, C. AND SCHUCH, W. (1988). Identification and sequence determination of a cDNA clone for tomato pectin esterase. *Eur. J. Biochem.* **174**, 119–124.

RICHARDSON, M. (1977). The proteinase inhibitors of plants and microorganisms. *Phytochem.* **16**, 159–169.

SAARILAHTI, H. T., HEINO, P., PAKKANEN, R., KALKKINEN, N. AND PALVA, E. T. (1990). Structural analysis of the *pehA* gene and characterisation of its protein product, endopolygalacturonase, of *Erwinia carotovora* sub species *carotovora*. *Molec. Microbiol.* **4**, 1037–1044.

SMITH, C. J. S., WATSON, C. F., RAY, J., BIRD, C. R., MORRIS, P. C., SCHUCH, W. AND GRIERSON, D. (1988). Antisense RNA inhibition of polygalacturonase gene expression in transgenic tomatoes. *Nature* **334**, 724–726.

SMITH, J. J., OLSON, J. R. AND KLAPPER, D. G. (1988). Monoclonal antibodies to denatured ragweed pollen allergen *Amb a* I: characterisation, specificity for the denatured allergen, and utilization for the isolation of immunogenic peptides of *Amb a* I. *Molec. Immun.* **25**, 355–365.

STAFF, I. A., TAYLOR, P. E., SMITH, P., SINGH, M. B. AND KNOX, R. B. (1990). Cellular localization of water soluble, allergenic proteins in rye-grass (Lolium perenne) pollen using monoclonal and specific IgE antibodies with immunogold probes. *Histochem. J.* **22**, 276–290.

STINSON, J. R., EISENBERG, A. J., WILLING, R. P., PE, M. E., HANSON, D. D. AND MASCARENHAS, J. P. (1987). Genes expressed in the male gametophyte of flowering plants and their isolation. *Pl. Physiol.* **83**, 442–447.

TANIAI, M., ANDO, S., USUI, M., KURIMORO, M., SAKAGUSHI, M., INOUYE, S. AND MATUHASI, T. (1988). N-terminal amino acid sequence of a major allergen of Japanese cedar pollen (*Cry j I*). *FEBS Lett.* **239**, 329–332.

TWELL, D., KLEIN, T. M., FROMM, M. E. AND MCCORMICK, S. (1989*b*). Transient expression of chimeric genes delivered into pollen by microprojectile bombardment. *Pl. Physiol.* **91**, 1270–1274.

TWELL, D., WING, R., YAMAGUCHI, J. AND MCCORMICK, S. (1989*a*). Isolation and expression of an anther-specific gene. *Molec. gen. Genet.* **217**, 240–245.

TWELL, D., YAMAGUCHI, J. AND MCCORMICK, S. (1990). Pollen-specific gene expression in transgenic plants: coordinate regulation of two different tomato gene promoters during microsporogenesis. *Development* **109**, 705–713.

TWELL, O., YAMAGUCHI, J., WING, R. A., USHIBA, J. AND MCCORMICK, S. (1991). Promoter analysis of genes that are coordinately expressed during pollen development reveals pollen-specific enhancer sequences and shared regulatory elements. *Genes Dev.* **5** (in press).

Ursin, V. M., Yamaguchi, J. and McCormick, S. (1989). Gametophytic and sporophytic expression of anther-specific genes in developing tomato anthers. *Plant Cell* **1**, 727–736.

Van der Krol, A. R., Lenting, P. E., Veenstra, J., Van der Meer, I. M., Koes, R. E., Gerats, A. G. M., Mol, J. N. M. and Stuitje, A. R. (1988). An anti-sense chalcone synthase gene in transgenic plants inhibits flower pigmentation. *Nature* **333**, 866–869.

Van Tunen, A. J., Mur, L. A., Brouns, G. S., Rienstra, J.-D., Koes, R. E. and Mol, J. N. M. (1990). Pollen- and anther-specific *chi* promoters from petunia: tandem promoter regulation of the *chiA* gene. *Plant Cell* **2**, 393–401.

Wagner, V. T., Cresti, M., Salvatici, P. and Tiezzi, A. (1990). Changes in volume, surface area, and frequency of nuclear pores of the vegetative nucleus of tobacco pollen in fresh, hydrated, and activated conditions. *Planta* **181**, 304–309.

Wing, R. A., Yamaguchi, J., Larabell, S. K., Ursin, V. M. and McCormick, S. (1989). Molecular and genetic characterisation of two pollen-expressed genes that have sequence similarity to pectate lyases of the plant pathogen *Erwinia*. *Pl. mol. Biol.* **14**, 17–28.

Yamamoto, M., Hara, S. and Ikenaka, T. (1983). Amino acid sequences of two trypsin inhibitors from winged bean seed (*Psophocarpus tetragonolobus* (L)DC.). *J. Biochem.* **94**, 848–861.

Printed in Great Britain © Society for Experimental Biology 1991

MOLECULAR AND EVOLUTIONARY ASPECTS OF SELF-INCOMPATIBILITY IN FLOWERING PLANTS

SHAIO-LIM MAU, MARILYN A. ANDERSON, MARCUS HEISLER, VOLKER HARING, BRUCE A. McCLURE and ADRIENNE E. CLARKE

Plant Cell Biology Research Centre, School of Botany, University of Melbourne, Parkville, Victoria, 3052, Australia

Summary

Self-incompatibility (SI) is widely distributed in flowering plants. In this review, early work on the biology, genetics and distribution of SI is summarized. Approaches to understanding the molecular genetics of SI have been made in two systems – Solanaceous species, for example *Nicotiana alata*, which have gametophytic systems of SI, and *Brassica* spp, which have sporophytic systems of SI. The information in both systems is derived from cDNAs that encode pistil glycoproteins (*S*-glycoproteins) that segregate with *S*-genotype. Comparison of the sequence data indicates that the gametophytic and sporophytic systems of SI probably arose independently during the evolution of angiosperms. The *S*-glycoproteins of a solanaceous plant *Nicotiana alata*, are ribonucleases (RNases). Whether the RNase activity is directly involved in the characteristic arrest of pollen tube growth during self-(incompatible) pollination, is not known. An alternative possibility is that the RNase was 'recruited' during evolution for a function in SI, without involvement of its catalytic function. The nature of the *S*-gene in pollen is not yet known for either the gametophytic or sporophytic SI systems. This is a key piece of information that will be required to progress our understanding of how the growth of a pollen tube bearing a particular *S*-allele is arrested within the style bearing an identical *S*-allele, but is not arrested within the style bearing other *S*-alleles.

Introduction

Self-incompatibility is one of the mechanisms that has evolved to encourage outbreeding in flowering plants. Many plants are hermaphroditic, that is they have the male and female sexual organs on the same flower, often in close proximity, so that self-pollination and inbreeding would be favoured in the absence of a mechanism to prevent it. Self-incompatibility has been defined as the 'inability of a fertile hermaphrodite seed-plant to produce zygotes after self-pollination' (de Nettancourt, 1977). It is genetically controlled, in many cases by a single locus, the *S*-locus, which has multiple alleles. The number of alleles can be quite large,

Key words: self-incompatibility, evolution, gametophytic, sporophytic, *Nicotiana alata*.

several hundred in some species (de Nettancourt 1977 and Ockendon, 1974), and the *S*-locus is thus extremely polymorphic.

The nature of self-incompatibility has been the subject of study by many plant biologists since Darwin (1877) first described the phenomenon and put foward the idea of its central significance during the evolution of flowering plants. It is of interest to modern biologists from two points of view; firstly, the intrinsic interest in the genetics of self-incompatibility and its role in plant breeding systems, and secondly, as an example of self-recognition in plants. It may be that by studying and understanding the mechanism of self-recognition (and/or non-self recognition) between pollen and style, principles that may be more generally applicable will be established.

Descriptions of self-incompatibility, its role in the evolution of flowering plants and its fundamental genetics and biology have been reviewed on many occasions. Much of the early literature is collected and discussed in the classic work 'Incompatibility in Angiosperms' by de Nettancourt (1977). Other early reviewers are Lewis (1949, 1979); Pandey (1979); Heslop-Harrison (1975, 1982, 1983); De Nettancourt (1984); and Harris *et al.* (1984). The more recent work on the molecular genetics of self-incompatibility has been reviewed by Clarke *et al.* (1987); Cornish *et al.* (1988); Ebert *et al.* (1989); Nasrallah and Nasrallah (1989); and Haring *et al.* (1990).

In this review we concentrate on two aspects of the topic: firstly, a discussion of the theories proposed to describe the evolution of self-incompatibility, and secondly, a description of the current understanding of the molecular genetics of self-incompatibility. We then consider the implications that molecular genetics has for considerations of the evolution of self-incompatibility. To provide background for these discussions, a brief outline of the biology of self-incompatibility is given. Further details can be obtained from the reviews cited above.

Biology of self-incompatibility

The female organ of flowering plants, the pistil, consists usually of the stigma, style and ovary. The male gametes are carried in pollen. The pollen is carried to the stigma by wind or insect vectors. In a successful (compatible) pollination, the pollen lands on the stigma surface, imbibes fluid from the surface, and germinates to produce a pollen tube. This tube carries the two sperm cells as well as the vegetative nucleus, which controls pollen tube growth. The pollen tube penetrates the stigma surface and grows through the style to the embryo sac. The sperm cells are then involved in the double fertilization which is characteristic of flowering plants. One sperm cell fertilizes the ovum and the other sperm cell fuses with the polar nuclei. The two sperm cells are characteristically present in the mature grain in some families (e.g. Brassicaceae) but in other families (e.g. Solanaceae), there is a single generative cell in the mature pollen grain that undergoes a mitotic division after germination. At least in some systems, the sperm cells are linked by evaginations of the plasma membrane that surrounds them, and one of the cells is physically connected to the vegetative nucleus to form the male germ unit (Dumas *et al.* 1984). Examples are the tricellular pollen of *Brassica* (McConchie *et al.* 1987), *Plumbago* (Russell and Cass, 1983; Russell, 1984) and *Spinacia* (Wilms, 1986) and the bicellular pollen of *Rhododendron* (Kaul *et al.* 1987; Taylor *et al.* 1989).

The two major systems of self-incompatibility: heteromorphic and homomorphic

Plants with heteromorphic self-incompatibility produce flowers with different morphologies, particularly of style length and anther level. This morphological difference presents a barrier to illegitimate pollination. Heterostyly has been reported in 24 families and more than 164 genera (Ganders, 1979). The classic example is *Primula* (Shivanna *et al.* 1981) and the system has been reviewed by Gibbs (1986). There has, to date, been no attempt to describe the system in molecular terms and it will not be discussed further here.

Homomorphic self-incompatibility is the more widespread type of self-incompatibility. This group is divided into two sub-groups, gametophytic (GSI) and sporophytic (SSI). In gametophytic systems, the self-incompatibility phenotype of the pollen is determined by its own (haploid) *S*-genotype. The sporophytic type of self-incompatibility is characterized by the fact that the behaviour of the pollen is determined by the (diploid) *S*-genotype of the pollen-producing plant. In most, but not all, gametophytic self-incompatibility systems, the self-incompatibility is controlled by a single *S*-gene with multiple alleles. The number of different alleles in natural populations of a species has only been examined for a few systems; for *Papaver rhoeas* (Campbell and Lawrence, 1981) and *Oenothera organensis* (Emerson, 1938), there are about 40 different alleles. Some systems are more complex, for example some grasses have self-incompatibility systems controlled by two loci, *S* and *Z* (Lundqvist, 1964), and *Ranunculus aeris* (Osterbye, 1975) and *Beta vulgaris* (Lundqvist *et al.* 1973; Larsen, 1977) are reported to have systems controlled by four different loci. When self-incompatibility is controlled by a single gene with multiple alleles, a pollen tube carrying a particular allele is inhibited if the same allele is present in the pistil. A successful fertilization can result only if the *S*-allele carried by the pollen differs from either of the two carried by the pistil.

The sporophytic system is apparently less widespread. It has been described in seven families (Charlesworth, 1985) and studied mainly in two families Brassicaceae (Cruciferae) and Asteraceae (Compositae). The genetics is based on a single *S*-gene with multiple alleles. A difference from the gametophytic system in which the *S*-alleles act independently in both pollen and pistil, is that the *S*-alleles in sporophytic self-incompatibility systems may express dominance relationships or independent action in the pollen or pistil (Thompson, 1957, 1972; Thompson and Taylor, 1966). There are other differences that are more-or-less typical of the two types. For example, in gametophytic systems, arrest of pollen tube growth after self-pollination, usually occurs within the style, but in sporophytic systems the arrest is characteristically at the stigma surface, or at a very early stage of penetration (Kanno and Hinata, 1969). The papillar cells of the stigma are the major barrier and if the pollen tubes successfully pass this barrier, they can continue growth into the ovary (Ockendon, 1974). Another difference is the pattern of callose deposition. Callose is a polysaccharide that contains a high proportion of $(1-3)$-β-glucan; it gives an intense fluorescence when stained with the aniline blue fluorochrome (Evans and Hoyne, 1982). This fluorochrome occurs as an impurity in preparations of aniline blue and is often referred to as the 'aniline blue fluorochrome'. It binds preferentially to $(1-3)$-β-glucans and some $(1-3),(1-4)$-β-D-glucans (Evans et al. 1984; Stone et al. 1984). The fluorochrome stains pollen grains and pollen tubes from a number of different species. Methylation analysis of pollen wall preparations of N. alata show that 3-linked glucose is a major component (Rae et al. 1985). Hydrolysates of pollen tube preparations from other species show glucose as a major component, and it is assumed to be in 1–3 linkage (for review see Harris et al. 1984). In N. alata, callose is present as an inner component of the wall of the pollen tube (Anderson et al. 1987). In a compatible pollination, cross walls or 'plugs' of callose are deposited within the tube at intervals after germination. This gives the pollen tube a 'ladder-like' appearance when viewed by fluorescence microscopy after staining of the pollinated style with the aniline-blue fluorochrome. During self-pollination of species with gametophytic self-incompatibility, the pattern of deposition of callose in the pollen tubes is initially similar to that seen in compatible pollination. At some stage during growth through the style, the pattern of deposition becomes irregular, and in general there is more intense staining in the pollen tube walls. Often the pollen tubes swell at the tip and they may burst. In contrast, in self-pollination of species with a sporophytic system of self-incompatibility, callose is often deposited at the site of contact of the pollen and stigma, both in the tip of the pollen tube and at the point of contact with the papillar cell of the stigma (Heslop-Harrison, 1975). There are exceptions to these features, for example the grass, Secale cereale has a gametophytic system of self-incompatibility, but the morphological features of pollen tube arrest resemble that of the sporophytic self-incompatibility system, in that the tube arrest occurs close to the stigma surface (Heslop-Harrison, 1982).

Nature of the S-gene

It is assumed that the products of the S-locus in pollen and pistil interact in some way to produce a response that results in inhibition of pollen tube growth within the pistil as described above.

Study of self-compatible mutants led Lewis (1949, 1960) to propose the tripartite hypothesis of S-locus organization. In this hypothesis, the S-locus consists of three closely linked genetic elements, a pollen activity part, a style activity part and an S-allele specificity part. This theory is based on the observation that mutations could be produced in self-incompatible plants, which affected either pollen behaviour or style behaviour. Thus the two parts are independently mutable.

Self-incompatibility is a feature expressed by mature and not immature flowers in both the gametophytic and sporophytic systems. This property is valuable as a breeding tool, as it is possible to achieve seed set from self-pollination by using mature pollen from one flower to pollinate a style from an immature (bud) flower on the same plant. This 'bud-selfing' can be used to generate plants homozygous for a particular *S*-allele.

Evolutionary aspects of self-incompatibility

There is some debate as to the importance of self-incompatibility in determining the evolutionary success of the angiosperms. For example, Whitehouse (1951) suggested that the success was due to the development of efficient outbreeding systems that promoted sexual reproduction and genetic recombination. Of the many outbreeding systems, such as dioecy, monoecy, di-allelic homomorphic and multiallelic homomorphic self-incompatibility, Whitehouse (1951) suggested that multiallelic self-incompatibility would be the most efficient because of availability of partners for cross-breeding. On the other hand, Anderson and Stebbins (1984) suggest that dioecy may be a more efficient outbreeding system. Self-incompatibility may have an additional role, for example, in the assessment of male quality by the female before fertilization (Zavada and Taylor, 1986).

Distribution of self-incompatibility in the angiosperms

SI is estimated to be present in more than half of all species of angiosperms (deNettancourt, 1977; Brewbaker, 1959). There is little information about a substantial number of taxa, largely because it is, in some cases, difficult to demonstrate unequivocally that a species is self-incompatible (Charlesworth, 1985). Self-incompatibility has been described in 81 of the 383 families of angiosperms described by Cronquist (1981) (Fig. 1). Of the 81 families, 15 have GSI, 39 have SI of an undefined type and 21 families may have SI (Charlesworth, 1985). SSI has been described in six families but some of these have heteromorphic flowers and may have genetic control that differs from that in plants with homomorphic flowers (Gibbs and Ferguson, 1987).

Origin of homomorphic self-incompatibility

GSI and SSI systems have been detected in diverse taxa (Fig. 1) suggesting that either the SI systems arose early in angiosperm evolution, before these taxa diverged (Whitehouse, 1951), or that they arose independently on more than one occasion (Bateman, 1952). Mather (1944) and Whitehouse (1951) considered a single ancestor more likely because at least three alleles are required before SI can operate; they considered that the chance of a single gene mutating to form three functional alleles on more than one occasion was remote.

The idea that SSI was derived from the more widely distributed GSI has been discussed by Brewbaker (1959); Pandey (1958, 1960, 1980); and Crowe (1964). Pandey suggested that the difference in behaviour between pollen from SSI and GSI plants was due to a shift in timing of gene action. That is, the *S*-genes in the

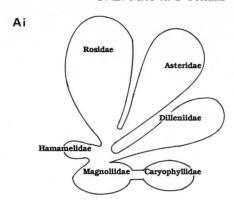

Ai

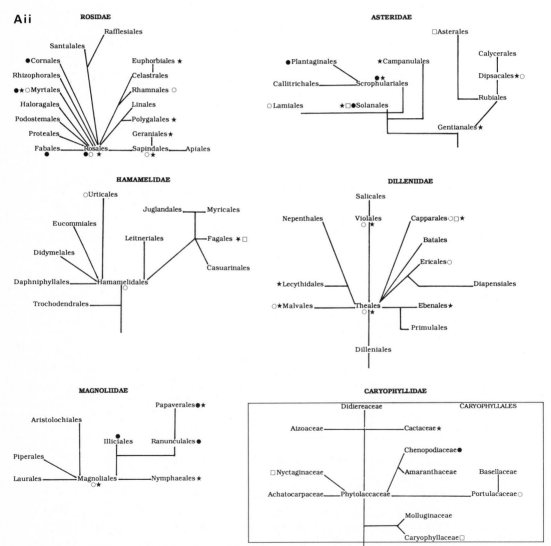

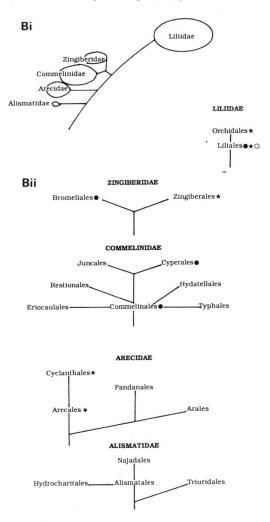

Fig. 1. The distribution of known self-incompatibility systems within sub-groups of the class MAGNOLIOPSIDA (A) and LILIOPSIDA (B). The classification is based on Cronquist's (1981) classification of flowering plants. The subclasses are shown in (i) and the orders within the subclasses are shown in (ii). The information concerning the presence of self-incompatibility types within orders was taken from Charlesworth (1985). Self-incompatibility types are indicated by the symbols: ●, gametophytic self-incompatibility; □, sporophytic self-incompatibility; *, self-incompatibility type unknown; ○, self-incompatibility possible.

SSI system were activated before meiosis, whereas those in GSI plants became activated after meiosis (Pandey, 1958).

An alternative view is that SSI arose earlier than GSI (Zavanda, 1984). This proposal was based on the observation that the presence of a perforate or reticulate exine on pollen grains is correlated with SSI whereas imperforate and microperforate exines are found in pollen from plants with GSI. The earliest fossils of angiosperm pollen have reticulate exines and thus, by analogy, are likely to be derived from plants that had SSI. However, Gibbs and Ferguson (1987) point out the uncertainties of predicting the type of SI system from the morphology of pollen.

Self-compatible taxa are also widespread and often closely related to taxa with SI. It is likely that they evolved from SI plants by mutation of the *S*-genes or genes involved in their regulation (Whitehouse, 1951; Jain, 1976; Mayo and Leach, 1987). The success of self-fertile plants depends on the strength of inbreeding depression within that plant population (Charlesworth and Charlesworth, 1987; Levin, 1989).

Our current knowledge of the molecular biology of SI gives a further insight into its possible origin. Genes encoding style glycoproteins that segregate with particular *S*-alleles, have been identified and cloned from plants with both gametophytic and sporophytic SI systems. The genes have no sequence homology, suggesting that the two systems arose independently, and that the generation of three original alleles may not be as rare an event as envisaged by Mather (1944). The possibility that the two systems have genes in common, which have not yet been identified, cannot be entirely discounted. Lewis and colleagues have recently produced evidence for the presence of gametophytic control as well as the well characterized sporophytic control in the Brassicaceae (Lewis *et al.* 1988).

Within the gametophytic SI system, gene sequences are now available for several members of the Solanaceae (Fig. 2). Within this group, some of the *Petunia S*-alleles are more closely related to some *N. alata S*-alleles than to other *S*-alleles from *N. alata* (Ai *et al.* 1990). Collectively, these data indicate that SI in the Solanaceae arose in an ancestor before speciation and that some alleles have remained relatively stable since that time.

Molecular genetics of self-incompatibility

The molecular genetics of self-incompatibility was built on the key observations that extracts of pistils of some species contained components that segregated absolutely with the *S*-genotype.

The pioneering work to detect components of pollen and pistils that corresponded to particular *S*-alleles was done by Lewis (1952). Some of the other early work was based on immunological techniques using antisera raised to extracts of pollen or pistils (Linskens, 1960; Nasrallah and Wallace, 1967). A major step forward was made by Nishio and Hinata (1979) and Hinata *et al.* (1982), who identified a glycoprotein in extracts of stigmas of *Brassica campestris* (genotype S_7), and showed that this component segregated absolutely with the S_7-function. They concluded that this component was either the product of the S_7-allele or of a gene closely related to, and segregating with, the S_7-allele.

A summary of the literature describing pistil components that correspond to particular *S*-alleles is given in Table 1. The only report on detection of *S*-allele specific components in pollen, is an early immunological study on pollen *S*-genotypes of *Oenothera organensis* (Onagraceae) (Lewis, 1952). There are no detailed reports of attempts to detect *S*-related components in pollen from other species.

The identification of style components that correspond to *S*-genotype, lead to the next stage of the work, which was to clone the genes encoding these components. Two systems have been studied in detail. *Nicotiana alata* (Solanaceae) was the first system having gametophytic self-incompatibility for which a cDNA encoding an *S*-allele-specific style component was described. The second example is *Brassica oleracea* (Brassicaceae), which was the first system having sporophytic self-incompatibility for which a cDNA encoding part of an *S*-locus-specific glycoprotein was described (Nasrallah *et al.* 1985). These two examples will now be described in more detail.

Nicotiana alata: An example of a solanaceous plant with a gametophytic system of self-incompatibility

Cloning of cDNAs encoding S-*glycoproteins of* N. alata

The cloning of a cDNA encoding the S_2-glycoprotein was approached by preparing a cDNA library from poly(A)$^+$ RNA of mature styles. A differential screening procedure was used to identify clones homologous to RNA species that were abundant in mature styles but not in immature styles or ovaries. The probe used for the second screening was derived by preparing a series of 14 bp oligonucleotides corresponding to part of the N-terminal sequence of the S_2-glycoprotein. The oligonucleotides were used to prime cDNA synthesis from mature style mRNA. A specific 100 bp product was obtained that provided sequence information for the preparation of a unique 30 bp oligonucleotide. This was used to screen the clones selected by differential screening. A full length cDNA encoding a putative signal sequence (22 amino acids in length) and an N-terminal hydrophobic region (15 amino acids in length) was obtained (Anderson *et al.* 1986). Northern and Southern analysis indicated wide divergence of the different *S*-allele sequences, as S_2-cDNA hybridized poorly to RNA and DNA isolated from plants with different *S*-genotypes. This was confirmed after cloning and sequencing two more *S*-alleles (S_3 and S_6) from *N. alata* (Anderson *et al.* 1989). The cloning of the cDNAs for these two alleles was done by taking advantage of their lack of homology. cDNA libraries were prepared from poly(A)$^+$ RNA isolated from mature styles of the S_3S_3- or S_6S_6-genotype and screened with cDNA prepared from S_3S_3- or S_6S_6-RNA from mature styles. Most plaques hybridized well with both probes. The few plaques in the S_3S_3-library that hybridized well with the S_3S_3-probe, but weakly with S_6S_6-probe, were the potential S_3-cDNAs. Similarly the few plaques in the S_6S_6-library that hybridized well with the S_6S_6-probe, but weakly with the S_3S_3-probe, were the potential S_6-cDNAs. The identity of the S_3- and S_6-cDNAs was confirmed by hybridization to the S_2-cDNA clone and by northern analysis. Comparison of the predicted amino acid sequence encoded by the S_2-, S_3- and S_6-cDNA clones shows only 63–70 % amino acid identity between the various pairs of alleles with only 51 % of the amino acids being conserved between all three alleles. The sequences were aligned and divided into a series of four 'hypervariable' regions that are interspersed with 'conserved' regions. The conserved features include all ten cysteine residues except that the S_2-sequence lacks one cysteine residue.

As cysteine residues are important for maintenance of the structural features of proteins, it is possible that these residues provide the essential structural framework. It is believed that the other conserved regions encode either essential structural or functional domains of the S-glycoproteins, although the 'hypervariable' regions may contain the information encoding S-allelic specificity (Anderson et al. 1989).

Glycosylation of S-allele glycoproteins of N. alata

All of the isolated S-glycoproteins contain N-glycosyl chains. The number of N-linked chains varies from 1 to 4 (Woodward et al. 1989). Comparison of the predicted amino acid sequence shows four potential N-glycosylation sites conserved between the S_2-, S_3- and S_6-alleles; the S_3-allele has one additional site. In each case at least one of the potential sites is left unsubstituted. The conservation of glycosylation sites emphasizes the possibility that the chains may play a structural role. At present, there is no detailed structural analysis of the individual carbohydrate chains.

Structure of the S-locus of N. alata

Genomic DNA from plants homozygous for the S_1-, S_2-, S_3,-S_6- or S_7-alleles was digested with various restriction enzymes and probed with the cDNAs corresponding to the S_2-, S_3- and S_6-alleles. In most cases single fragments that hybridized strongly with the probes were produced. The fragments generated were of a size such that only one or a few copies of the S-coding sequences could be accommodated. The data are consistent with this being a single locus, single-copy gene system (Bernatzky et al. 1988). The Southern analysis also showed an abundance of RFLPs and indicated that variability occurred in the flanking regions as well as the coding regions (Bernatzky et al. 1988). The level of variation was remarkably high, for example it was higher than that observed between alleles of random leaf cDNAs from Lycopersicon and Capsicum (Bernatzky and Tanksley, 1986).

A most important result from the Southern analysis is that the three cDNAs are products of the same gene. The evidence is that each S-allele has a characteristic set of HindIII restriction fragments that hybridize strongly to the cDNA encoded by that allele and weakly with cDNA encoded by the other alleles. If the cDNAs for the S_2-, S_3- and S_6-cDNAs were from different loci, the Southern analysis would be expected to show bands corresponding to each S-genotype in each DNA sample, unless there was deletion of genes corresponding to all but the expressed allele (Anderson et al. 1989). A 56 bp sequence, which is 499 bp upstream from the coding sequence of the transcription start site has high homology to a mitochondrial DNA with only a 3 bp substitution (Bernatzky et al. 1989). This sequence shows no open reading frames, and contains a short 8 bp direct repeat that immediately flanks the 5' region of homology. The presence of direct repeats is consistent with the features of transposable element excision (Nevers *et al.* 1986). This section of DNA may be the result of a transfer process between mitochondrial DNA and nuclear DNA. The significance of this sequence is not known.

Localization of S-*gene expression and* S-*gene products in* N. alata

The techniques of molecular cytology, *in situ* hybridization and immunocyto-chemistry, allow localization of S-gene expression and the S-gene product within the pistils. Immunocytochemical localization was based on antiserum raised to a synthetic 22 amino acid peptide starting from position 47 of the S_2-glycoprotein. This peptide corresponded to the major variable region of the protein. This region is hydrophilic and was chosen on the basis that it was specific to the S_2-glycoprotein and predicted to be located on the outer surface of the glycoprotein. The peptide was coupled to a carrier protein (human thyroglobulin) and injected into sheep. The antibody was affinity purified on an S_2-peptide-Sepharose column, and showed absolute specificity to the S_2-glycoprotein when tested on a western blot of style extracts from a range of S-glycoproteins. This antibody was used in both immunofluorescence and immunocytochemical studies of the pistil of *N. alata*, genotype S_2S_2. The antibody bound primarily to the extracellular material in the transmitting tract and the stigma. The epidermis of the placenta also bound the antibody. In the transmitting tract, there was very little binding to the cell walls. In contrast, the cell walls of the placental epidermis were the main site of antibody binding in the ovary (Anderson *et al.* 1989). This localization of the S-glycoprotein follows exactly the pattern of gene expression established by *in situ* hybridization, which was restricted to the transmitting tissue cells of the style and the inner epidermis of the placenta (Cornish *et al.* 1987). Thus, the S_2-glycoprotein is present in the pathway followed by the pollen tube as it grows through the style to the ovary.

S-*gene products in pollen*

Careful analysis of extracts both of pollen and pollen germinated *in vitro* has failed to reveal any protein that corresponded to S-genotype (S-L. Mau, unpublished results). Northern analysis using pollen RNA and the style glycoprotein cDNAs as probes did not show any strongly hybridizing species (B. McClure, M. Anderson, unpublished results). These observations do not rule out the possibilities that a homologous species is present in low abundance, or that there is only limited homology between the pollen and style S-allele products. It may be that the S-allele coding for the pollen product is distinct from that coding for the style product, although the two must be closely linked, as the two functions have never been separated by conventional breeding (see de Nettancourt, 1977). In this case, chromosome walking from the style S-gene might reveal regions that could represent the pollen S-gene.

Interaction of S-*glycoproteins with pollen tubes grown* in vitro

One approach to understanding how the product of the style *S*-gene acts to arrest growth of incompatible pollen tubes, is to study the interaction between pollen tubes growing *in vitro* and isolated style components, particularly the *S*-glycoproteins. Before methods for isolation of the style *S*-glycoproteins were established, the approach of studying the effect of unfractionated stigma or style extracts on *in vitro* grown pollen tubes was adopted. Generally, this approach showed that such extracts inhibited the growth of incompatible pollen more than that of compatible pollen. This approach has been used for *Brassica* (Ferrari and Wallace, 1977; Roberts *et al.* 1983), *Petunia hybrida* (Sharma and Shivanna, 1982), *Nicotiana alata* (Sharma and Shivanna, 1986), *Lilium longiflorum* (Dickinson *et al.* 1982), *Primula vulgaris* (Shivanna *et al.* 1981), and *Papaver rhoeas* (Franklin-Tong *et al.* 1988).

The first studies using defined pistil components were reported by Williams *et al.* (1982). In this case a glycoprotein preparation from styles of *Prunus avium* containing two components that probably corresponded to the S_3- and S_4-alleles, caused inhibition of tube growth of pollen from a tree bearing the S_3S_4 genotype. The genotype specificity of the interaction was not tested, but other style components of *Prunus avium* or glycoproteins of animal origin, had no effect. This study was important in that it was the first demonstration that an isolated style *S*-glycoprotein was an effective inhibitor of pollen tube growth. In an attempt to bypass the time-consuming process of measuring pollen tube length, Harris *et al.* (1987) attempted to quantify pollen growth using an ELISA assay based on the binding of a monoclonal antibody to the α-L-arabinosyl residues in the outer pollen tube wall. However, the relationship between length of tube and antibody bound was logarithmic rather than linear; that is, proportionally more antibody bound to short tubes than to long tubes. Thus the utility of the assay was limited. A further limitation to the assay became apparent during a subsequent study (Harris *et al.* 1989) in which the isolated S_2-glycoprotein from *N. alata* was shown to inhibit growth of pollen tubes bearing the S_2-allele, but not growth of tubes bearing other *S*-alleles. In this case, more antibody bound to the S_2-(inhibited) pollen tubes than the non-inhibited tubes, presumably due to an increased concentration of α-L-arabinosyl residues in the outer wall of the inhibited tubes.

The development of a method for rapid isolation of *S*-glycoproteins corresponding to a number of different *S*-alleles of *N. alata* (Jahnen *et al.* 1989a), allowed more stringent testing of the allelic specificity of the *in vitro* interaction. Using three *S*-glycoproteins, (Jahnen *et al.* 1989b) showed that pollen germination was not affected by the presence of the *S*-glycoproteins, but that pollen tube growth of all genotypes was inhibited to a certain degree. S_2-pollen was preferentially inhibited by the S_2-glycoprotein and S_3-pollen by the S_3-glycoprotein. The S_6-glycoprotein preferentially inhibited growth of both S_2- and S_6- pollen over S_3-pollen. Thus the *in vitro* assay using isolated style *S*-glycoproteins only partially reflects the *in vitro* events, and has limitations as a model system for studying self-incompatibility.

In contrast to this experience, Franklin-Tong *et al.* (1989) have developed an *in vitro* system for poppy, *Papaver rhoeas* in which the isolated *S*-glycoprotein gives allele-specific inhibition of pollen tube growth *in vitro*. This opens exciting possibilities for studying the pollen response to contact with the *S*-glycoproteins.

Ribonuclease function of the S-*glycoproteins*

A major advance in the understanding of self-incompatibility was made in 1989 when it was discovered that the *S*-glycoproteins of *N. alata* are ribonucleases (RNases). This discovery was made from the observation that there is some homology between the amino acid sequences of the S_2-, S_3- and S_6-glycoproteins from *N. alata* and the secreted fungal RNases from *Aspergillus oryzae* and *Rhizopus niveus* (McClure *et al.* 1989). The two longest regions of homology, (residues 67–72 and 132–136 of the S_2-glycoprotein) include the two histidine residues of the fungal RNases that are implicated in the catalytic sites (Fig. 2). Of the 122 amino acids conserved among the three *N. alata* *S*-glycoproteins, thirty are aligned with identical amino acids in the fungal RNases. Another 22 are aligned with closely related amino acids. Seven of the 9 cysteine residues are conserved between the *S*-glycoproteins and the RNases. This observation of sequence homology led to testing of the isolated *S*-glycoproteins for RNase activity. McClure *et al.* (1989) found that RNase activity co-eluted with individual *S*-glycoproteins in the standard cation exchange chromatographic procedure devised for isolation of the *S*-glycoproteins (Jahnen *et al.* 1989*a*). The specific activity of the purified fungal ribonuclease T_2 was 3900 A_{260} units min^{-1} mg^{-1}; under the same conditions, the specific RNase activity of the isolated *S*-glycoproteins was between 170 (S_1-glycoprotein) and 2200 units (S_2-glycoprotein). This RNase activity of the *S*-glycoproteins suggests that they could function by moving into the pollen tube and degrading a target RNA within the cytoplasm. This idea is compatible with the observation that the related self-compatible species have a relatively low level of RNase activity within the style mucilage, and with the phenomenon of unilateral self-incompatibility. In this phenomenon, crosses between *N. alata* and the self-compatible species *N. tabacum* are only successful if *N. tabacum* is used as the female partner. When *N. alata* is used as the female partner, it may reject *N. tabacum* pollen by virtue of the high specific activity RNase in the style. In the reverse situation, *N. tabacum* may allow growth of *N. alata* pollen tubes as it has only low levels of RNase associated with the style. A similar observation has been made for self-compatible and self-incompatible accessions of *Lycopersicon peruvianum* (McClure *et al.* 1989) but whether this correlation is more general is not known. The observation of the RNase activity of the isolated *S*-glycoproteins raises the question of whether the pollen *S*-gene products also have this activity. In the cases that have been tested, pollen extracts have a very low RNase activity (McClure *et al.* 1989). It seems then that if the pollen *S*-gene product is an RNase, which would be the result if it were identical with the style *S*-gene product, it must be expressed at a very low level. Alternatively, as discussed earlier, it may not be identical with the style *S*-gene

```
              1          20          40          60          80         100         120         140
                                                                  .           .           .           .

S1  N.alata     NFEYMQLVLTWPTAFCNVMN--C----------ERT-PTNFTIHGLWPDNVST----ELNYCD--RQKKFKLFE-DKKKQNLDDRDWPLTLDRDDCKKNGGQFWSYEYKKHGTC----
S2  N.alata     AFEYMQLVLTWPITFCRIKH--C----------ERT-PTNFTIHGLWPDNHT----MLNYCD--RSKPYNMFT-DGKKKNDLDERWPDLKTKFDSLLDKQAFWKDEYVKHGTC----
S3  N.alata     AFEYMQLVLQWPAAFCHTTPSPC----------KRI-PSNFTIHGLWPDNVST----MLNYCS--GEDEYEKLD-DKKKKNDLDRWPDLTIARADCIEHQVFWKHEYNKHGTC----
S6  N.alata     AFEYMQLVLQWPTAFCHTT--PC----------KNI-PSNFTIHGLWPDNVST----TLNFCG--KEDDYNIIM-DGPEKNGLYVRWPDLIREKADCMKTQNFWRREYIKHGTC----
cons. N.alata   FEYMQLVL WP aFC        C          P NFTIHGLWPDN aT       LNFC      f    D   L RWPDL   h D    Q FW   EY KHGTC

Sf11 N.alata    DFEYLQLVLTWPASFCY-ANHC----------ERIAPNNFTIHGLWPDNVKT----RLHNCK--PKPTYSFT--GKMLNDLDKHWMQLKFEQDYGRTEQPSWKYQYIKHGSC----
S2  N.alata     DFDYMQLVLTWPASFCY-PKNFC---------SRIAPKNFTIHGLWPDKVRG----RLQFCT--SE-KYVNFAQDSPILDDLDHHWMELKYHRDFGLENQFLWRGQYQKHGTC----
cons.           FEYMQLVLtWPASfcy PKN C          KR   a NFTIHGLWPE   F    RLEFC       GD KY    F    d  ild LERHW1QMhFDE  fA  hQPLW  EY hHGiC

S1  P.inflata   NFEYLQLVLTWPASFCQYCRPKNIC--------KRP-AKNFTIHGLWPEITGF--RLEFCT--GDPKYETFK-DNNIVDYLERHWVQMKFDENYAKYHQPLWSYEYRKHGMC----
S2  P.inflata   NFDYFQLVLTWPASFCY-PKNFC---------KRK-SNNFTIHGLWPENKHF----RLEFCT--GD-KYSRFK-EDNIINVLERHWIQMREDEKYASTKQPLWEHEYNRHGIC----
S3  P.inflata   NFDYIQLVLTWPASFCYRPKNIC---------KRI-PNNFTIHGLWPEKEHF----RLEFCD--GD-KFVSFSLKDRIVNDLERHWVQMKFDEKFAKIKQPLWTHEYNKHGIC----
cons. P.inflata NFdY QLVLTWPASFCf PKN C          KR a NFTIHGLWPE F    RLEFC      GD KY  F  d  lid LERHW1QMhFDE  fA  hQPLW  EY hHGiC

S1  S.tuberosum DFEELELVSTWPATFCY-AYGCS---------KRPIPNNFTIHGLWPDNKST----VLNFCNLAHEDEYIPIT-DHKILTELDKRWPQLRYDYLYGIRKQYLWKNEFIKHGSC----

cons. S-proteins  Fd dLV WP aFC          C            a NFTIHGLWPd           L C         f          L hW d1        Q W  df hHG C

Rh  R.niveus    MKAVLALATLIGSTLASSCSSTALSCCSNSAN--SDTCCSPEYGLVVLNMQ-WA--PGYGFDNAFTLHGLWPDKCSGAYAPSGGCDSNRASSIAVIKSKDSS-LYN--SMLTYWPSNQGNNNVFSHEWSKHGTCVSTY
T2  A.oryzae    --EFPSCPKDIPFSCQNSTAV-ADSCCPNSPGGALLQTQFWDTNPPSGPSDSWTIHGLWPDNDGSYGQF-CDKSREYSNTAILQEQGRTELLS--YMKKYWPNYEGGDEEFWEHEWNKHGTCINTI
cons. fungal RNases   aS    a SC NSa   aDaCC   G iLd Q W  Pa GP daffiHGLWPD C GaYa      CD R S Iaaii    L  M YWP dGddd FW HEW KHGTCi T

cons. overall     d      aa C                d         ft HGLWPd                C                 L                    d      df hHG C


            160          180         200         220         240         260

S1  N.alata     --CLPSYNQEQ----YFDLAMALKDKFDLLKSFRNHGIIPTKS---YTVQKYNNTVKAITKGF-PNLTCN---K-QMELQEIGICFDQKVKNV----IDCPRPKTCKA-TRNGITFP-----
S2  N.alata     --CSDKFDREQ----YFDLAMTLRDKFDLLSSLRNHGISRGFS---YTVQNLNNTIKAITGGF-PNLTCS---R-LRELKEIGICFDEIVKNV----IDCPNPKTCKP-TNKGVMFP----
S3  N.alata     --CSKSYNLTQ----YFDLAMALKDKFDLLTSLRKHGIIPGNS---YTVQKINSTIKAITQGY-PNLSCT---KRQMELLEIGICFDSKVKNV----IDCPHPKTCKPMGNRGIKFP----
S6  N.alata     --CSEIYNQVQ----YFRLAMALKDKFDLLISLKNHGIIRGYK---YTVQKINNTIKTVTKGY-PNLSCT---KGQ-ELWEVGICFDSTAKNV----IDCPNPKTCKTASNQGIMFP----
cons. N.alata     C   fd Q    YF LAMaLhDKFDLL S h HGI A    YTVQ N TIKaIT GFPNLaC       h  EL ElGICFD  KNV      IDCP PKTCK  a  Gl FP

Sf11 N.alata    --CQKRYNONT----YFGLALRLRDKFDLLRTLQTHRIIPGSS---YFPDIFDAIKTVSQE-NPDIKCAEVTKGTPELYEIGICTPNADSM----FRCPQSDTCDKTA--KVLFR-R----
S2  N.alata     --CIPRYNQMQ----YFLLAMRLKDKFDLLATLRTHGITPGTK---HTFNETRDAIKTVTNQVDPDLKCVEHIKGVRELYEIGICFPTADSF----FQCPHSNTCDETGITKILFR-R----
cons.            1YdQ a     YFLLA R1KdK DLLTTLRThGITPGTK  HTFGdIQKAIKTVTNd DPDLKCVE IKGV EL ElGICF PAaD F    ChHS TCDETDS  LFR

S1  P.inflata   --CSKIYNQKA----YFLLATRLKEKFDLLTTLRLHGITPGTK---HTFGDIQKAIKTVTNQVDPDLKCVEHIKGVQELNEIGICFNPAADNF----YPCHHSYTCDETDSKMILFR----
S2  P.inflata   --CKNLYDQEA----YFLLAIRLKDKFDLLTTLRTHGITPGTK---HTFGEIQKAIKTVTNNKDPDLKCVENIKGVKELNEIGICFNAAADSF----HDCRHSKTCDETDSTQTLFR-R----
S3  P.inflata   --CSSNLYDQRA---YFLLAMRVKDKFDLLTTLRLHGITPGTK---HTFGEIQKAIKTVTNNKDPDLKCVEHIKGVKELKEVGICFTPAADSF----HDCRHSNTCDETDSTKILFR----
cons. P.inflata   1YdQ a     YFLLA R1KdK DLLTTLRThGITPGTK  HTFGdIQKAIKTVTNd DPDLKCVE IKGV EL ElGICF PAaD F    ChHS TCDETDS  LFR

S1  S.tuberosum --SINRYKQPA----YFDLAMKIKDKFDLLGTLRNHGINPGST---YELDDIERAIMTVSIEV-PSLKCIQKPLGNVELNEIGICLDPEAKYM-----VPCPRTGSCHMG-HKIKFR----

cons. S-proteins  YF LA 1hdK DLL a  H I a       al aIa  P1 C     EL ElGIC          C  aaC         F

Rh  R.niveus    DPDCYDNYEEGEDIVDYFQKAMDLRSQYNVYKAFSNGITPGG-TYTATEMQSAIESYF-GAKAKIDCSSGT----LSDVALYFYVRGR--DTYVITDALSTGSCS-GDVEYPTK---
T2  A.oryzae    EPSCYKDYSPQKEVGDYLQKTVDLFKGLDSYKALAKAGIVPDSSKTYKRSEIESALAAIHDGKPKYISCEDGA---LNEIWYFNIKGNAITGEYQPIDTLTSPGCSTSGIKYLPKKSEN

cons. fungal RNases  dP CY dY   di DY QK 1DL   d YKA a  GI P    TY  aE1dSAl a   G Ka 1C Ga      Y  DaLaaaC  a l Y aK

cons. overall     f    Y      1   d    a    1         al a   a1 C            L dl     ff lhG        L d1     DaLaaaC   a l Y aK
```

Fig. 2. Alignment of deduced amino acid sequences of *S*-glycoproteins from three solanaceous plants with those of the fungal ribonucleases Rh and T_2. The amino acid sequence is numbered beginning with 1 for the first amino acid of the mature protein. The putative signal sequences are not shown. Potential N-glycosylation sites are underlined. A consensus sequence for the four *N. alata* sequences S_1, S_2, S_3 and S_6 is shown beneath the S_6 *N. alata* sequence. The perfectly conserved amino acids are shown by their single letter code; similar residues within the groups PAGST, EDNQ, RKH, MILV and FWY are shown in lower case as a, d, i, and f respectively. A consensus sequence for the three *P. inflata* *S*-glycoproteins and two *N. alata* sequences S_{f11} and S_z is shown in the same way. An overall consensus sequence for all the *S*-glycoproteins shown is given on the line beneath the sequence of the *S. tuberosum* S_1-glycoprotein. The bottom line gives residues that are also conserved in the T_2 and Rh fungal ribonucleases. The alignment is modified from Haring *et al.* (1990).

product. This observation of enzyme activity of the *S*-glycoproteins is a major new development in our understanding of self-incompatibility, and perhaps may have implications for other types of cell–cell communication. The first question raised by this finding is whether the RNase activity is directly involved in the arrest of pollen tube growth. One possibility is that it is simply an example of 'recruitment' of an existing protein (RNase) for a particular function (self-incompatibility), without involvement of its catalytic function. The next question is that of the nature of its substrate, if indeed its activity does involve the enzymic function. Is there a secreted RNA from the pollen? Is *S*-glycoprotein of the style taken into the pollen tube to act on a cytoplasmic RNA substrate? Where does the specificity of the interaction lie? Is it in the specificity of the enzyme or perhaps in the uptake mechanism? These are all questions that will be addressed in the future. The answers may well shed light on the role of ribonucleases that have been implicated in many aspects of plant growth and development. A phenomenon that is possibly related is that of the ribosome inactivating function of certain lectins such as ricin and abrin (Endo *et al.* 1988).

Self-incompatibility glycoproteins from other solanaceous plants

Cloning of the S_2-cDNA from *N. alata* styles (Anderson *et al.* 1986) and the subsequent work to establish the sequences of the S_3- and S_6-alleles, (Anderson *et al.* 1989) led to the understanding of which sequences within the cDNAs were conserved between alleles and which were variable. On the basis of that information, other groups have been able to obtain cDNAs encoding other *S*-glycoproteins from *N. alata* (Kheyr-Pour *et al.* 1990) and from four other self-incompatible solanaceous species, *Petunia inflata* (Ai *et al.* 1990), *Petunia hybrida* (Clark *et al.* 1990), *Solanum chacoense* (Xu *et al.* 1990*b*) and *Solanum tuberosum* (R. Thompson-personal communication). An alignment of ten sequences from three different species has been prepared by Haring *et al.* (1990) and is reproduced in Fig. 2. The *N. alata* sequences fall into two groups, firstly S_1-, S_2-, S_3- and S_6-sequences that have an identity between sequences of 51.5%. The second group includes S_Z and S_{F11} which have 61.2% of the amino acids in common with each other but only 33 % with the other group. This group seems more related to the *P. inflata* S-glycoproteins with which they share 46.0 % (S_{F11}) and 55.2 % (S_Z) identity. The sequences of the *P. inflata* S-glycoproteins show 69.5 % identity. Comparison of sequences across all 10 cDNAs shows that only 46 amino acids (22.4 %) are totally conserved. The conserved residues include 7 of the 8–10 cysteine residues of each glycoprotein, as well as the two histidine residues that are aligned with the active site histidine residues of the fungal RNases (Haring *et al.* 1990). The *P. hybrida*, *P. inflata* and *S. tuberosum* S-glycoproteins are also reported to have ribonuclease activity (Clark *et al.* 1990; T. Kao and R. Thompson personal communication).

Self-incompatibility in *Brassica oleracea* and *Brassica campestris*: Examples of cruciferous plants having a sporophytic system of self-incompatibility

The sporophytic system of self-incompatibility has been best studied in *Brassica oleracea* and *Brassica campestris* although it was first observed in radish *Raphanus sativus* (Stout, 1920). The interactions between the alleles of sporophytic self-incompatible plants are more complex than those observed in plants with gametophytic self-incompatibility systems. Some alleles are dominant over others, and in some combinations certain alleles may be weakened (Thompson and Taylor, 1966, 1971). The molecular biology of the system has been studied mainly by two groups, Nasrallah and co-workers at Cornell University and Hinata, Isogai and co-workers in Japan. A detailed review of the molecular genetics of self-incompatibility in *Brassica* was presented in 1989 (Nasrallah and Nasrallah, 1989).

S-*locus-specific glycoproteins*

S-locus-specific glycoproteins of stigma extracts were first detected immunologically (Nasrallah and Wallace, 1967; Nasrallah *et al.* 1970). These glycoproteins are known as SLSGs and they appear on SDS–PAGE as a group of bands in the M_r range 57×10^3 to 65×10^3 (Nasrallah *et al.* 1983; Nasrallah and Nasrallah, 1984). cDNAs encoding the SLSG from *Brassica oleracea* were obtained by differential screening of a stigma cDNA library with labelled cDNA from mature stigmas and whole seedlings. This cDNA was shown to encode at least part of the S_6-SLSG by western blot analysis of a β-galactosidase:S-gene product fusion protein produced in *E. coli*. Southern analysis showed restriction fragment length polymorphisms that segregated with particular genotypes. Subsequently cDNAs representing the S_2, S_5, S_6, S_7, S_{13}, S_{14}, S_{22} and S_{29} alleles have been obtained and sequenced (Table 1). SLSGs from *B. campestris* have been identified and sequenced by direct amino acid sequencing of peptide fragments (Takayama *et al.* 1987; Isogai *et al.* 1987).

The nucleotide sequence of the S_6-SLSG cDNA corresponds to a 436-amino acid protein, including a 31-amino acid hydrophobic signal peptide at the N terminus of the polypeptide. This is consistent with the finding that the SLSG is localized in the wall of the papillar cells (Kandasamy *et al.* 1989).

The cDNAs encoding four *S*-alleles of *B. oleracea* have 90% sequence similarity at the DNA level (Nasrallah and Nasrallah, 1989). At the amino acid level the sequences are considered in three regions; firstly, a relatively conserved amino-terminal region (residues 1–181, 80% conservation), secondly, a variable region (residues 182–268, 52% conservation) and finally, a relatively conserved carboxy-terminal region (78% conservation). This region contains eleven invariant cysteine residues. Some detailed information on the carbohydrate moieties of these glycoproteins is available (Takayama *et al.* 1986). The potential N-linked carbohydrate attachment sites are conserved between different alleles. In *B. campestris*, direct amino acid analysis indicated that seven of nine potential sites are glycosylated (Takayama *et al.* 1987). Five of these sites are conserved between the *B. oleracea* alleles.

The chains exist as N-linked chains of the 'complex' type (Takayama *et al.* 1986) There seem to be two forms with the (Man_3) $(GlcNAc_2)$ core structure; one form has Xyl and Fuc residues attached to the core, the other contains an additional GlcNAc residue. As the structures are similar for the carbohydrate chains of different SLSGs, it seems unlikely that they play a major note in determining specificity. However, experiments involving the application of the glycosylation inhibitor tunicamycin on excised styles implicates N-linked oligosaccharides in the specificity (Sarker *et al.* 1988) so that an involvement in determining specificity cannot be definitively discounted.

Structure of the sporophytic S-*locus*

Although the genetics of the system dictates a single gene system that determines specificity, Southern analysis of *Brassica* genomic DNA with certain SLSG-cDNAs shows multiple bands. These bands reflect the presence of a family of at least 11 sequences related to the SLSG structural gene (Nasrallah *et al.* 1985; Nasrallah *et al.* 1987). Two SLSG-related genes (SLRs) have been isolated from *B. oleracea* (Lalonde *et al.* 1989; Trick and Flavell, 1989). Protein encoded by an SLR gene from *Brassica campestris* has also been isolated and sequenced (Isogai *et al.* 1988). Both SLSG and SLR genes are expressed in the stigma in a developmentally controlled manner. The SLR1 sequences have been isolated from plants homozygous for different *S*-alleles. In contrast to the SLSG sequences, which have considerable sequence variability, the SLR1 sequences are highly conserved. As the SLR1 and SLSG polymorphisms segregate independently, it seems that the SLR1 gene is not involved in determination of allelic specificity. This view is reinforced by the finding that the SLR1 sequences are expressed even in self-compatible strains of *Brassica* (Nasrallah and Nasrallah, 1989). Two closely related (94%) SLG sequences, have been described for the S_2-allele of *B. oleracea*. In this case, both genes are genetically linked leading to the possibility that both genes may function in self-incompatibility (Chen and Nasrallah, 1990).

Table 1. *Summary of characteristics of pistil components identified as corresponding to particular S-alleles in different self-incompatible plants*

Species	Allele	M_r ($\times10^{-3}$)	pI	Glycosylation	cDNA	Reference
Gametophytic self-incompatibility						
Solanaceae						
Nicotiana alata	S_1*	–	–	–	+	Bredemeijer and Blass 1981
	S_2	32	>8.5	3 chains	+	Anderson et al. 1986; 1989
	S_3	34	>8.5	4 chains	+	Jahnen et al. 1989a
	S_6	32	>8.5	3 chains	+	Woodward et al. 1989
	S_7	32	>8.5	3 chains	–	Kheyr-Pour et al. 1990
	S_z	30	9.0	ND	+	
	S_{F11}	27	9.5	1 chain	+	
	S_a	–	–	–	+	
Lycopersicon peruvianum	S_1	28	7.5	ND	–	Mau et al. 1986
	S_2	27	9.5	ND	–	
	S_3	28	9.5	ND	–	
Petunia hybrida	S_1	27	8.7	+	+	Kamboj and Jackson 1986
	S_2	33	8.9	+	+	Broothaerts et al. 1989
	S_3	30	9.3	+	+	Clark et al. 1990
	S_4	31	8.6	+	–	
Petunia inflata	S_1	24	~9.3	+	+	Ai et al. 1990
	S_2	25	~9.3	+	+	
	S_3	25	~9.3	+	+	
Solanum chacoense	S_1	29	~8.6	ND	–	Xu et al. 1990a
	S_2	30	~8.6	ND	+	Xu et al. 1990b
	S_3	31	~8.6	ND	+	
Solanum tuberosum	S_1	27	8.3	+		Kirch et al. 1989
	S_2	24	~8.9	+		
	S_3	27	>9.1	+		
	S_4	23	~8.7	+		

Table 1. *cont.*

Species	Allele	M_r $(\times 10^{-3})$	pI	Glycosylation	cDNA	Reference
Rosaceae						
Prunus avium	S_3/S_4	37–39	8.8	ND	–	Mau *et al.* 1982
Papaveraceae						
Papaver rhoeas	S_1	22	7.4	+	–	Franklin-Tong *et al.* 1989
	S_2	22	7.6	+	–	
	S_3	22	8.6	+	–	
Sporophytic self-incompatibility						
Brassicaceae						
Brassica oleracea‡	S_6	63 and 65	>9.0	+	+	Nasrallah and Nasrallah 1984
(var. *acephala*)	S_7	57 and 59	>9.0	+	+	Nasrallah *et al.* 1985
	S_{13}	61 and 63	>9.0	+	+	Nasrallah *et al.* 1987
	S_{22}	–	–	–	+	Lalonde *et al.* 1989
(var. *capitata*)	S_{14}	62 and 64	6.2–7.9	+	+	Chen and Nasrallah 1990
(var. *alboglabra*)	S_2	–	–	–	+	Trick and Flavell 1989
	S_{29}					
	S_7	57	10.6	+		Nishio and Hinata 1982
	S_{22}	60 and 65	11.1	+		
	S_{29}	57	10.3	+		
(var. *gemmifera*)	S_5			+	+	Scutt *et al.* 1990
Brassica campestris	S_7	57	5.7	+		Isogai *et al.* 1987
	S_8	53	8.4	5–7 chains		Nishio and Hinata 1979
	S_9	51	6.9	5–7 chains		Takayama *et al.* 1986
	S_{12}	53	5.7	5–7 chains		Takayama *et al.* 1987

*This S_1-allele was described by Kheyr-Pour *et al.* (1990) and is distinct from the S_1-allele described by Woodward *et al.* (1989), which is equivalent to the S_{F11}-allele described by Kheyr-Pour *et al.* (1990).

+ Indicates carbohydrate detected by Schiff staining or lectin binding.

‡ var. *acephala* (inbred kale), var. *capitata* (cabbage), var. *alboglabra* (Chinese kale), var. *gemmifera* (brussels sprouts).

Conclusion

The style *S*-glycoproteins of solanaceous species displaying gametophytic self-incompatibility and the SLSGs of *Brassica* species displaying sporophytic self-incompatibility show no amino acid sequence homology. There have been no reports as to whether the isolated SLSGs have RNase activity. On the face of the limited data available, it seems unlikely that the two systems share a common ancestor. The data are more consistent with the idea that self-incompatibility arose separately on a number of occasions, as suggested by Bateman (1952). An interesting observation perhaps relevant to the question of expression of self-incompatibility genes was made by Moore and Nasrallah (1990). They transformed the self-incompatible species *N. tabacum* with the genomic clones representing the *B. oleracea* S_{13}- and S_{22}-alleles and found that the genes were expressed in the transmitting tract of the style; this is the same location as that of the *S*-genes of the gametophytic self-incompatibility system of *N. alata*. This implies similarities in the promoter regions of both genes. As knowledge is increased through studies of the systems, the questions of their evolutionary origin, the mechanism for generation of allelic diversity and the function of the *S*-genes will be addressed. A key piece of information that is still missing for both systems, is the nature of the *S*-gene in pollen. It may be that when this information is established, some common functional features of the two systems may be revealed.

References

Ai, Y., Singh, A., Coleman, C. E., Ioerger, T. R., Kheyr-Pour, A. and Kao, T-H. (1990). Self-incompatibility in *Petunia inflata*: Isolation and characterization of cDNAs encoding three *S*-allele-associated proteins. *Sex. Pl. Reprod.* **3**, 130–138.

Anderson, G. J. and Stebbins, G. L. (1984). Dioecy *versus* gametophytic self-incompatibility: a test. *Am. Nat.* **124**, 423–428.

Anderson, M. A., Cornish, E. C., Mau, S.-L., Williams, E. G., Hoggart, R., Atkinson, A., Bonig, I., Grego, B., Simpson, R., Roche, P. J., Haley, J. D., Penschow, J. D., Niall, H. D., Tregear, G. W., Coghlan, J. P., Crawford, R. J. and Clarke, A. E. (1986). Cloning of cDNA for a stylar glycoprotein associated with expression of self-incompatibility in *Nicotiana alata*. *Nature* **321**, 38–44.

Anderson, M. A., Harris, P. J., Bonig, I. and Clarke, A. E. (1987). Immuno-gold localization of α-L-arabinofuranosyl residues in pollen tubes of *Nicotiana alata* Link et Otto. *Planta* **171**, 438–442.

Anderson, M. A., McFadden, G. I., Bernatzky, R., Atkinson, A., Orpin, T., Dedman, H., Tregear, G., Fernley, R. and Clarke, A. E. (1989). Sequence variability of three alleles of the self-incompatibility gene of *Nicotiana alata*. *Pl. Cell* **1**, 483–491.

Bateman, A. J. (1952). Self-incompatibility systems in angiosperms. I. Theory. *Heredity* **6**, 285–310.

Bernatzky, R., Anderson, M. A. and Clarke, A. E. (1988). Molecular genetics of self-incompatibility in flowering plants. *Dev. Genet.* **9**, 1–12.

Bernatzky, R., Mau, S.-L. and Clarke, A. E. (1989). A nuclear sequence associated with self-incompatibility in *Nicotiana alata* has homology with mitochondrial DNA. *Theor. appl. Genet.* **77**, 320–324.

Bernatzky, R. and Tanksley, S. D. (1986). Majority of random cDNA clones correspond to single loci in the tomato genome. *Mol. gen. Genet.* **203**, 8–14.

Bredemeijer, G. M. M. and Blaas, J. (1981). *S*-specific proteins in styles of self-incompatible *Nicotiana alata*. *Theor. appl. Genet.* **59**, 185–190.

BREWBAKER, J. L. (1959). Biology of the angiosperm pollen grain. *Indian J. Genet. Pl. Breed.* **19**, 121–133.

BROOTHAERTS, W. J., VAN LAERE, A., WITTERS, R., PREAUX, G., DECOCK, B., VAN DAMME, J. AND VENDRIG, J. C. (1989). Purification and N-terminal sequencing of style glycoproteins associated with self-incompatibility in *Petunia hybrida*. *Plant molec. Biol.* **14**, 93–102.

CAMPBELL, J. M. AND LAWRENCE, M. J. (1981). The population genetics of the self-incompatibility polymorphism in *Papaver rhoeas*. I. The number and distribution of S-alleles in families from three localities. *Heredity* **46**, 69–79.

CHARLESWORTH, D. (1985). Distribution of dioecy and self-incompatibility in angiosperms. In *Evolution*: Essays in honour of John Maynard Smith (ed. P. J. Greenwood, P. H. Harvey, and M. Slatkin), pp. 237–268. Cambridge University Press.

CHARLESWORTH, D. AND CHARLESWORTH, B. (1987). Inbreeding depression and its evolutionary consequences. *Ann. Rev. Ecol. Syst.* **18**, 237–268.

CHEN, C-H. AND NASRALLAH, J. B. (1990). A new class of S-sequences defined by a pollen recessive self-incompatibility allele of *Brassica oleracea*. *Molec. gen. Genet.* **222**, 241–248.

CLARK, K., OKULEY, J. J., COLLINS, P. D. AND SIMS, T. L. (1990). Sequence variability and developmental expression of S-alleles in self-incompatible and pseudo-self-compatible *petunia*. *The Plant Cell* **2**, 815–826.

CLARKE, A. E., ANDERSON, M. A., BACIC, A., CORNISH, E. C., HARRIS, P. J., MAU, S. L. AND WOODWARD, J. R. (1987). Molecular aspects of self-incompatibility in flowering plants. In UCLA Symposia on Molecular and Cellular Biology New Series. *Plant Gene Systems and their Biology* **62** (ed. J. L. Key and L. McIntosh), pp. 53–64.

CLARKE, A. E., ANDERSON, M. A., BACIC, T., HARRIS, P. J. AND MAU, S.-L. (1985). Molecular basis of cell recognition during fertilization in higher plants. *J. Cell. Sci. Suppl.* **2**, 261–285.

CORNISH, E. C., ANDERSON, M. A. AND CLARKE, A. E. (1988). Molecular aspects of fertilization in flowering plants. *Ann. Rev. Cell Biol.* **4**, 209–228.

CORNISH, E. C., PETTITT, J. M., BONIG, I. AND CLARKE, A. E. (1987). Developmentally controlled expression of a gene associated with self-incompatibility in *Nicotiana alata*. *Nature* **326**, 99–102.

CRONQUIST, A. (1981). An integrated system of classification for flowering plants. Columbia University Press.

CROWE, L. K. (1964). The evolution of outbreeding in plants. I. The angiosperms. *Heredity* **19**, 435–457.

DARWIN, C. (1877). *The Different Forms of Flowers on Plants of the Same Species*. John Murray, London.

DE NETTANCOURT, D. (1977). *Incompatibility in Angiosperms*. Springer-Verlag: Berlin, Heidelberg, New York.

DE NETTANCOURT, D. (1984). Incompatibility. In *Cellular Interactions*. Encyclopedia of Plant Physiology, New series **17** (ed. H. F. Linskens and J. Heslop-Harrison), pp. 624–639. Springer-Verlag; Berlin, Heidelberg, New York, Tokyo.

DICKINSON, H. G., MORIARTY, J. F. AND LAWSON, J. R. (1982). Pollen–pistil interaction in *Lilium longiflorum*: the role of the pistil in controlling pollen tube growth following cross- and self-pollinations. *Proc. R. Soc. Lond. B.* **215**, 45–62.

DUMAS, C., KNOX, R. B., McCONCHIE, C. A. AND RUSSELL, S. D. (1984). Emerging physiological concepts in fertilization. *What's New In Pl. Physiol.* **15**, 17–20.

EBERT, P. R., ANDERSON, M. A., BERNATZKY, R., ALTSCHULER, M. AND CLARKE, A. E. (1989). Genetic polymorphism of self-incompatibility in flowering plants. *Cell* **56**, 255–262.

EMERSON, S. (1938). The genetics of self-incompatibility in *Oenothera organesis*. *Genetics* **23**, 190–202.

ENDO, Y., TSURUGI, K. AND LAMBERT, J. M. (1988). The site of action of six different ribosome-inactivating proteins from plants on eukaryotic ribosomes: The RNA N-glycosidase activity of the proteins. *Biochem. biophys. Res. Commun.* **150**, 1032–1036.

EVANS, N. A. AND HOYNE, P. A. (1982). A fluorochrome from aniline blue: structure, synthesis and fluorescence properties. *Aust. J. Chem.* **35**, 2571–2575.

EVANS, N. A., HOYNE, P. A. AND STONE, B. A. (1984). Characteristics and specificity of the interaction of a fluorochrome from aniline blue (Sirofluor) with polysaccharides. *Carbohydr. Polym.* **4**, 215–230.

FERRARI, T. E. AND WALLACE, D. H. (1977). A model for self-recognition and regulation of the incompatibility response of pollen. *Theor. appl. Genet.* **50**, 211–225.

FRANKLIN-TONG, V. E., LAWRENCE, M. J. AND FRANKLIN, F. C. H. (1988). An *in vitro* bioassay for the stigmatic product of the self-incompatibility gene in *Papaver rhoeas* L. *New Phytol.* **110**, 109–118.

FRANKLIN-TONG, V. E., RUUTH, E., MARMEY, P., LAWRENCE, M. J. AND FRANKLIN, F. C. H. (1989). Characterization of a stigmatic component from *Papaver rhoeas* L. which exhibits the specific activity of a self-incompatibility (*S*-) gene product. *New Phytol.* **112**, 307–315.

GANDERS, F. R. (1979). The biology of heterostyly. *N. Z. J. Bot.* **17**, 607–635.

GIBBS, P. E. (1986). Do homomorphic and heteromorphic self-incompatibility systems have the same sporophytic mechanisms? *Plant Syst. Evol.* **154**, 285–323.

GIBBS, P. E. AND FERGUSON, I. K. (1987). Correlations between pollen exine sculpturing and angiosperm self-incompatibility systems – a reply. *Pl. Syst. Evol.* **157**, 143–159.

HARING, V., GRAY, J. E., McCLURE, B. A., ANDERSON, M. A. AND CLARKE, A. E. (1990). Self-incompatibility: A self-recognition system in plants. *Science* **250**, 937–941.

HARRIS, P. J., ANDERSON, M. A., BACIC, A. AND CLARKE, A. E. (1984). Cell–cell recognition in plants with special reference to the pollen-stigma interaction. In *Oxford Surveys of Plant Molecular and Cell Biology* **1** (ed. B. J. Miflin), pp. 161–203. Oxford University Press, Oxford.

HARRIS, P. J., FREED, K., ANDERSON, M. A., WEINHANDL, J. A. AND CLARKE, A. E. (1987). An enzyme-linked immunosorbent assay (ELISA) for *in vitro* pollen growth based on binding of a monoclonal antibody to the pollen tube surface. *Pl. Physiol.* **84**, 851–855.

HARRIS, P. J., WEINHANDL, J. A. AND CLARKE, A. E. (1989). Effect on *in vitro* pollen growth of an isolated style glycoprotein associated with self-incompatibility in *Nicotiana alata*. *Pl. physiol.* **89**, 360–367.

HESLOP-HARRISON, J. (1975). Incompatibility and the pollen-stigma interaction. *A. Rev. Pl. Physiol.* **26**, 403–425.

HESLOP-HARRISON, J. (1982). Pollen-stigma interaction in the grasses: a brief review. *N. Z. J. Bot.* **17**, 537–546.

HESLOP-HARRISON, J. (1983). Self-incompatibility: phenomenology and physiology. *Proc. R. Soc. Lond. B* **218**, 371–395.

HINATA, K., NISHIO, T. AND KIMURA, J. (1982). Comparative studies on *S*-glycoproteins purified from different *S*-genotypes in self-incompatible *Brassica* species. II. Immunological specificities. *Genetics* **100**, 649–657.

ISOAGAI, A., TAKAYAMA, S., SHIOZAWA, H., TSUKAMOTO, C., KANBARA, T., HINATA, K., OKAZAKI, K. AND SUZUKI, A. (1988). Existence of a common glycoprotein homologous to *S*-glycoproteins in two self-incompatible homozygotes of *Brassica campestris*. *Pl. Cell Physiol.* **29**, 1331–1336.

ISOGAI, A., TAKAYAMA, S., TSUKAMOTO, C., UEDA, Y., SHIOZAWA, H., HINATA, K., OKAZAKI, K. AND SUZUKI, A. (1987). *S*-locus-specific glycoproteins associated with self-incompatibility in *Brassica campestris*. *Pl. Cell Physiol.* **28**, 1279–1291.

JAHNEN, W., BATTERHAM, M. P., CLARKE, A. E., MORITZ, R. L. AND SIMPSON, R. J. (1989*a*). Identification, isolation and N-terminal sequencing of style glycoproteins associated with self-incompatibility in *Nicotiana alata*. *Pl. Cell* **1**, 493–499.

JAHNEN, W., LUSH, W. M. AND CLARKE, A. E. (1989*b*). Inhibition of *in vitro* pollen tube growth by isolated *S*-glycoproteins of *Nicotiana alata*. *Pl. Cell* **1**, 501–510.

JAIN, S. K. (1976). The evolution of inbreeding in plants. *A. Rev. Ecol. Syst.* **10**, 469–495.

KAMBOJ, R. K. AND JACKSON, J. F. (1986). Self-incompatibility alleles control a low molecular weight, basic protein in pistils of *Petunia hybrida*. *Theor. appl. Genet.* **71**, 815–819.

KANDASAMY, M. K., PAOLILLO, D. J., NASRALLAH, J. B., FARADAY, C. D. AND NASRALLAH, M. E. (1989). The *S*-locus specific glycoproteins of *Brassica* accumulate in the cell wall of developing stigma papillae. *Devl Biol.* **134**, 462–472.

KANNO, T. AND HINATA, K. (1969). An electron microscopic study of the barrier against pollen-tube growth in self-incompatible *Cruciferae*. *Pl. Cell Physiol.* **10**, 213–216.

KAUL, V., THEUNIS, C. H., PALSER, B. F., KNOX, R. B. AND WILLIAMS, E. G. (1987). Association of the generative cell and vegetative nucleus in pollen tubes of *Rhododendron*. *Ann. Bot.* (London) **59**, 227–235.

KHEYR-POUR, A., BINTRIM, S. B., IOERGER, T. R., REMY, R., HAMMOND, S. A. AND KAO, T.-H. (1990). Sequence diversity of pistil *S*-proteins associated with gametophytic self-incompatibility in *Nicotiana alata. Sex. Pl. Reprod.* **3**, 88–97.

KIRCH, H. H., UHRIG, H., LOTTSPEICH, F., SALAMINI, F. AND THOMPSON, R. D. (1989). Characterization of proteins associated with self-incompatibility in *Solanum tuberosum. Theor. appl. Genet.* **78**, 581–588.

LALONDE, B., NASRALLAH, M. E., DWYER, K. D., CHEN, C. H., BARLOW, B. AND NASRALLAH, J. B. (1989). A highly conserved *Brassica* gene with homology to the *S*-locus specific glycoprotein structural gene. *Pl. Cell* **1**, 249–258.

LARSEN, K. (1977). Self-incompatibility in *Beta vulgaris* L. I. Four gametophytic, complementary *S*-loci in sugar beet. *Hereditas* **85**, 227–248.

LEVIN, D. A. (1989). Inbreeding depression in partially self-fertilizing *Phlox. Evolution* **43**, 1417–1423.

LEWIS, D. (1949). Incompatibility in flowering plants. *Biological Reviews* **24**, 472–496.

LEWIS, D. (1952). Serological reactions of pollen incompatibility substances. *Proc. R. Soc. Lond. B.* **140**, 127–135.

LEWIS, D. (1960). Genetic control of specificity and activity of the *S*-antigen in plants. *Proc. R. Soc. Lond. B.* **151**, 468–477.

LEWIS, D. (1979). *Sexual incompatibility in plants*. The Institute of Biology's Studies in Biology No. 10. Edward Arnold, London.

LEWIS, D., VERMA, S. C. AND ZUBERI, M. I. (1988). Gametophytic-sporophytic incompatibility in the Cruciferae *Raphanus sativus. Heredity* **61**, 355–366.

LINSKENS, H. F. (1960). Zur Frage der Entstehung der Abwehr-Körper bei der Inkompätibilitatsreaktion von *Petunia*. III. Mitteilung: Serologische Teste mit Leitgewebs- und Pollen-Extrakten. *Zeitschrift für Botanik* **48**, 126–135.

LUNDQVIST, A. (1964). The nature of the two-loci incompatibility system in grasses. IV. Interaction between the loci in relation to pseudo-compatibility in *Festuca pratensis* Huds. *Hereditas* **52**, 221–234.

LUNDQVIST, A., OSTERBYE, U., LARSEN, K. AND LINDE-LAURSEN, I. B. (1973). Complex self-incompatibility systems in *Ranunculus acris* L. and *Beta vulgaris* L. *Hereditas* **74**, 161–168.

MATHER, K. (1944). Genetical control of incompatibility in angiosperms and fungi. *Nature* **153**, 392–394.

MAU, S.-L., RAFF, J. AND CLARKE, A. E. (1982). Isolation and partial characterization of components of *Prunus avium* L. styles, including an antigenic glycoprotein associated with a self-incompatibility genotype. *Planta* **156**, 505–516.

MAU, S.-L., WILLIAMS, E. G., ATKINSON, A., ANDERSON, M. A., CORNISH, E. C., GREGO, B., SIMPSON, R. J., KHEYR-POUR, A. AND CLARKE, A. E. (1986). Style proteins of a wild tomato (*Lycopersicon peruvianum*) associated with expression of self-incompatibility. *Planta* **169**, 184–191.

MAYO, O. AND LEACH, C. R. (1987). Stability of self-incompatibility systems. *Theor. appl. Genet.* **74**, 789–792.

MCCLURE, B. A., HARING, V., EBERT, P. R., ANDERSON, M. A., SIMPSON, R. J., SAKIYAMA, F. AND CLARKE, A. E. (1989). Style self-incompatibility gene products of *Nicotiana alata* are ribonucleases. *Nature* **342**, 955–957.

MCCONCHIE, C. A., RUSSELL, S. D., DUMAS, C., TUOHY, M. AND KNOX, R. B. (1987). Quantitative cytology of the sperm cells of *Brassica campestris* and *B. oleracea. Planta* **170**, 446–452.

MOORE, H. M. AND NASRALLAH, J. B. (1990). A *Brassica* self-incompatibility gene is expressed in the stylar transmitting tissue of transgenic tobacco. *The Pl. Cell* **2**, 29–38.

NASRALLAH, J. B., DONEY, R. C. AND NASRALLAH, M. E. (1983). Two-dimensional electrophoretic analysis of *S*-glycoproteins in *Brassica. Incompatibility News Letter* **15**, 2–8.

NASRALLAH, J. B., KAO, T.-H., CHEN, C.-H., GOLDBERG, M. L. AND NASRALLAH, M. E. (1987). Amino-acid sequence of glycoproteins encoded by three alleles of the *S*-locus of *Brassica oleracea. Nature* **326**, 617–619.

NASRALLAH, J. B., KAO, T.-H., GOLDBERG, M. L. AND NASRALLAH, M. E. (1985). A cDNA clone encoding an *S*-locus-specific glycoprotein from *Brassica oleracea. Nature* **318**, 263–267.

NASRALLAH, J. B. AND NASRALLAH, M. E. (1984). Electrophoretic heterogeneity exhibited by the S-allele specific glycoproteins of Brassica. Experientia **40**, 279–281.

NASRALLAH, J. B. AND NASRALLAH, M. E. (1989). The molecular genetics of self-incompatibility in Brassica. A. Rev. Genet. **23**, 121–139.

NASRALLAH, M. E., BARBER, J. T. AND WALLACE, D. H. (1970). Self-incompatibility proteins in plants: detection, genetics and possible mode of action. Heredity **25**, 23–27.

NASRALLAH, M. E. AND WALLACE, D. H. (1967). Immunogenetics of self-incompatibility in Brassica oleracea L. Heredity **22**, 519–527.

NEVERS, P., SHEPHERD, N. S. AND SAEDLER, H. (1986). Plant transposable elements. Adv. Bot. Res. **12**, 103–203.

NISHIO, T. AND HINATA, K. (1979). Purification of an S-specific glycoprotein in self-incompatible Brassica campestris. L. Jap. J. Genet. **54**, 307–311.

NISHIO, T. AND HINATA, K. (1982). Comparative studies on S-glycoproteins purified from different S-genotypes in self-incompatible Brassica species. I. Purification and chemical properties. Genetics **100**, 641–647.

OCKENDON, D. J. (1974). Distribution of self-incompatibility alleles and breeding structure of open-pollinated cultivars of Brussels sprouts. Heredity **33**, 159–171.

OSTERBYE, U. (1975). Self-incompatibility in Ranunculus acris L. Genetic interpretation and evolutionary aspects. Hereditas **80**, 91–112.

PANDEY, K. K. (1958). Time of S-allele action. Nature **181**, 1220–1221.

PANDEY, K. K. (1960). Evolution of gametophytic and sporophytic systems of self-incompatibility in angiosperms. Evolution **14**, 98–115.

PANDEY, K. K. (1979). Overcoming incompatibility and promoting genetic recombination in flowering plants. N. Z. J. Bot. **17**, 645–663.

PANDEY, K. K. (1980). Evolution of incompatibility systems in plants: Origin of 'independent' and 'complementary' control of incompatibility in angiosperms. New Phytol. **84**, 381–400.

RAE, A. L., HARRIS, P. J., BACIC, A. AND CLARKE, A. E. (1985). Composition of the cell walls of Nicotiana alata Link et Otto pollen tubes. Planta **166**, 128–133.

ROBERTS, I. N., GAUDE, T. C., HARROD, G. AND DICKINSON, H. G. (1983). Pollen-stigma interactions in Brassica oleracea; a new pollen germination medium and its use in elucidating the mechanism of self incompatibility. Theor. appl. Genet. **65**, 231–238.

RUSSELL, S. D. (1984). Ultrastructure of the sperm of Plumbago zeylanica. II. Quantitative cytology and three-dimensional organization. Planta **162**, 385–391.

RUSSELL, S. D. AND CASS, D. D. (1983). Unequal distribution of plastids and mitochondria during sperm cell formation in Plumbago zeylanica. In Pollen: Biology and Implications for Plant Breeding (ed. D. L. Mulcahy, and E. Ottaviano), pp. 135–140. Elsevier; Amsterdam, New York.

SARKER, R. H., ELLEMAN, C. J. AND DICKINSON, H. G. (1988). Control of pollen hydration in Brassica requires continued protein synthesis, and glycosylation is necessary for intraspecific incompatibility. Proc. natn. Acad. Sci. U.S.A. **85**, 4340–4344.

SCUTT, C. P., GATES, P. J., GATEHOUSE, J. A., BOULTER, D. AND CROY, R. D. (1990). A cDNA encoding an S-locus specific glycoprotein from Brassica oleracea plants containing the S_5 self-incompatibility allele. Mol. Gen. Genet. **220**, 409–413.

SHARMA, N. AND SHIVANNA, K. R. (1982). Effects of pistil extracts on in vitro responses of compatible and incompatible pollen in Petunia hybrida. Vilm. Indian. J. exp. Biol. **20**, 255–256.

SHARMA, N. AND SHIVANNA, K. R. (1986). Self-incompatibility, recognition and inhibition in Nicotiana alata. In Biotechnology and Ecology of Pollen (ed. G. B. Mulcahy, E. Mulcahy and E. Ottaviano), pp. 179–184. Springer-Verlag, New York.

SHIVANNA, K. R., HESLOP-HARRISON, J. AND HESLOP-HARRISON, Y. (1981). Heterostyly in Primula. 2. Sites of pollen inhibition, and effects of pistil constituents on compatible and incompatible pollen tube growth. Protoplasma **107**, 319–337.

STONE, B. A., EVANS, N. A., BONIG, I. AND CLARKE, A. E. (1984). The application of Sirofluor, a chemically defined fluorochrome from aniline blue for the histochemical detection of callose. Protoplasma **122**, 191–195.

STOUT, A. B. (1920). Further experimental studies on self-incompatibility in hermaphroditic plants. J. Genet. **9**, 85–129.

TAKAYAMA, S., ISOGAI, A., TSUKAMOTO, C., UEDA, Y., HINATA, K., OKAZAKI, K., KOSEKI, K. AND SUZUKI, A. (1986). Structure of carbohydrate chains of *S*-glycoproteins in *Brassica campestris* associated with self-incompatibility. *Agric. biol. Chem.* **50**, 1673–1676.

TAKAYAMA, S., ISOGAI, A., TSUKAMOTO, C., UEDA, Y., HINATA, K., OKAZAKI, K. AND SUZUKI, A. (1987). Sequences of *S*-glycoproteins, products of the *Brassica campestris* self-incompatibility locus. *Nature* **326**, 102–105.

TAYLOR, P., KENRICK, J., LI, Y., GUNNING, B. E. S. AND KNOX, R. B. (1989). The male germ unit of *Rhododendron*: quantitatitive cytology, three-dimensional reconstruction, isolation and detection using fluorescent probes. *Sex Pl. Reprod.* **2**, 254–264.

THOMPSON, K. F. (1957). Self-incompatibility in marrow-stem kale, *Brassica oleraceae* var. *acephala*. I. Demonstration of a sporophytic system. *J. Genet.* **55**, 45–60.

THOMPSON, K. F. (1972). Competitive interaction between two *S* alleles in a sporophytically-controlled incompatibility system. *Heredity* **28**, 1–7.

THOMPSON, K. F. AND TAYLOR, J. P. (1966). Non-linear dominance relationships between *S* alleles. *Heredity* **21**, 345–362.

THOMPSON, K. F. AND TAYLOR, J. P. (1971). Self-incompatibility in Kale. *Heredity* **27**, 459–471.

TRICK, M. AND FLAVELL, R. B. (1989). A homozygous *S*-genotype of *Brassica oleracea* expresses two *S*-like genes. *Molec. gen. Genet.* **218**, 112–117.

WHITEHOUSE, H. L. K. (1951). Multiple-allelomorph incompatibility of pollen and style in the evolution of the angiosperms. *Ann. Botan.* New Series **14**, 198–216.

WILLIAMS, E. G., RAMM-ANDERSON, S., DUMAS, C., MAU, S.-L. AND CLARKE, A. E. (1982). The effect of isolated components of *Prunus avium*. L. styles on *in vitro* growth of pollen tubes. *Planta* **156**, 517–519.

WILMS, H. J. (1986). Dimorphic sperm cells in the pollen grain of *Spinacia*. In *Biology of reproduction and motility in plants and animals* (ed. M. Cresti, and R. Dallai), pp. 193–198. University of Sienna, Sienna.

WOODWARD, J. R., BACIC, A., JAHNEN, W. AND CLARKE, A. E. (1989). N-linked glycan chains on *S*-allele-associated glycoproteins from *Nicotiana alata*. *Pl. Cell* **1**, 511–514.

XU, B., GRUN, P., KHEYR-POUR, A. AND KAO, T.-H. (1990a). Identification of pistil-specific proteins associated with three self-incompatibility alleles in *Solanum chacoense*. *Sexual Pl. Reprod.* **3**, 54–60.

XU, B., MU, J., NEVINS, D. L., GRUN, P. AND KAO, T-H. (1990b). Cloning and sequencing of cDNAs encoding two self-incompatibility associated proteins in *Solanum chacoense*. *Molec. gen. Genet.* **224**, 341–346.

ZAVADA, M. S. (1984). The relation between pollen exine sculpturing and self-incompatibility mechanisms. *Pl. Syst. Evol.* **147**, 63–78.

ZAVADA, M. S. AND TAYLOR, T. N. (1986). The role of self-incompatibility and sexual selection in the gymnosperm-angiosperm transition: a hypothesis. *Am. Nat.* **128**, 538–550.

Printed in Great Britain © Society for Experimental Biology 1991

ENGINEERED MALE STERILITY IN PLANTS

C. MARIANI[1]*, R. B. GOLDBERG[2] and J. LEEMANS[1]

[1] Plant Genetic Systems NV, J. Plateaustraat 22, B-9000 Gent, Belgium
[2] Department of Biology, University of California, Los Angeles, California 90024-1606, U.S.A.

Summary

We constructed two chimaeric ribonuclease genes that are specifically expressed in anthers of tobacco and rapeseed plants. The expression of these genes affects the production of functional and viable pollen yielding plants that are male sterile. This dominant gene for nuclear male sterility should facilitate the production of hybrid seed in various crops.

Introduction

In flowering plants, male gamete formation is a highly regulated developmental process that occurs in the anthers. In this organ system haploid microspores are produced after meiotic division of the pollen mother cells. The microspores develop further in pollen grains that carry the sperm cells. One of the tissues of the anther, the tapetum, plays an important role in the development and maturation of pollen. Initially, the tapetal cells function as a nutritive tissue for the microspores and contribute the sporopollenin precursor for the deposition of the exine. Afterwards, the tapetum provides the enzymes that release the microspores from the callose wall (Mascarenhas, 1990; Tiwari and Gunning, 1986).

Over the past years, cytoplasmic and nuclear mutations have been identified that lead to the absence of pollen grains and, consequently, to male sterility. In several cases these mutations are associated with a defective tapetum, suggesting that this cell type is essential for pollen formation. In some crops cytoplasmic male sterility has proved to be very useful for the efficient production of hybrid seed.

Here we report the isolation of a tapetal-specific gene from tobacco (*Nicotiana tabacum*) and the identification of the region essential for tapetal-specific expression. We show that in transgenic plants this promoter can be used to direct the expression of chimaeric ribonuclease genes exclusively in the tapetal cells. The ribonuclease impairs normal functioning of the tapetum and leads to production of male sterile plants.

Molecular analysis of a tapetum specific tobacco gene

Two anther-specific cDNA clones designated as *TA29* and *TA13* were isolated

* Author for correspondence.

Key words: tapetum-specific genes, ribonuclease, male sterility.

from a tobacco anther cDNA library (Goldberg, 1988). These clones are 85 % similar at the nucleotide level, and are complementary to a 1.1 and 1.2 kb anther mRNA species. Hybridization with anther, pistil, petal, leaf, root and stem mRNAs showed that the *TA29* and *TA13* messengers are highly prevalent in anthers and are undetectable in other vegetative or floral organ systems (Koltunow *et al.* 1990). Furthermore, *TA29* and *TA13* mRNAs accumulate early during flower growth and disappear as the tapetum degenerates at later stages of anther development. *In situ* hybridization studies showed that these transcripts are localised specifically in the tapetal cells (Fig. 1C and Koltunow *et al.* 1990).

The *TA29* genomic clone was isolated by screening a tobacco genomic library with the *TA29* cDNA. DNA sequence studies and R-loop mapping proved that the anther-specific gene does not contain introns (Seurinck *et al.* 1990; Koltunow *et al.* 1990) and has a coding sequence of 963 nucleotides that may encode a glycine-rich protein. Primer extension and S1 mapping revealed that the *TA29* message initiates 50 nucleotides upstream from the proximal putative translation start codon (Seurinck *et al.* 1990) and that it carries two 3' end formation and polyadenylation sites. This implies that the *TA29* gene specifies two mRNA species which differ in their 3' end sequence (Koltunow *et al.* 1990)

We wished to identify the region required for temporal and spatial expression of *TA29*. Therefore, a DNA fragment containing 1510 nucleotides 5' of the translation start codon was fused to the coding sequence of the β-glucuronidase gene (*gus*; Jefferson *et al.* 1987) and introduced into tobacco by T-DNA transfer. Expression of the chimaeric gene was detected exclusively in anthers where the *gus* messenger was most abundant at the developmental stage at which the *TA29* mRNA reaches its peak level. *In situ* hybridization on anther sections showed that the *gus* message accumulates as the endogenous *TA29* gene solely in tapetal cells (Fig. 1D) and the activity of the *gus* gene product is strictly confined to the tapetum (Fig. 1B). These results show that the chimaeric *TA29–gus* gene is regulated in the same fashion as the endogenous *TA29* gene.

More recently, Koltunow *et al.* (1990) have demonstrated that a 122 bp fragment of the 5' region is sufficient and necessary to direct correctly spatial and temporal tapetal-specific expression of the chimaeric *TA29–gus* gene within the anthers.

We probed genomic DNA of several crop plants with the coding region of *TA29* to see whether the *TA29* gene is conserved. In all cases a discrete hybridization pattern was observed suggesting that this gene is a member of a small gene family which is well conserved in distantly related species (Mariani and Gossele, unpublished results). Based on these results, we introduced the chimaeric *TA29–gus* gene into oilseed rape (*Brassica napus*) to determine whether its anther-specific expression pattern is maintained in plants other than tobacco. Our results showed that also in oilseed rape the activity of the β-glucuronidase is specifically confined to the tapetal cell layer but it appears when meiosis of the pollen mother cells has already occurred and the pollen grains begin to form (data not shown). This indicates that the expression of the chimaeric *TA29–gus* gene in oilseed rape seems to occur at a later developmental stage than in tobacco.

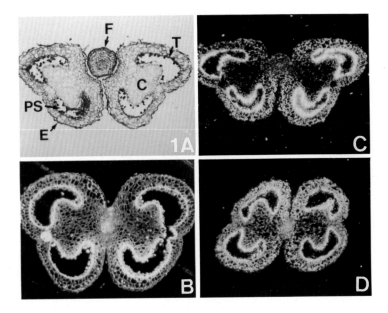

Fig. 1. Localization of *TA29* and *TA29–gus* gene products in anthers of transgenic plants. (A) Bright-field photograph of a transverse section of an anther at stage 2 of flower development (Goldberg, 1988). C, connective tissue; E, epidermis; F, filament; PS, pollen sac and T, tapetum. (B) *In situ* localization of GUS enzyme activity. Anthers of transgenic plants carrying the *TA29–gus* gene were cut and incubated in 1 mM X-gluc in 50 mM P-buffer, pH 7 at 37 °C for several hours (Jefferson, 1987). After the reaction took place, the anthers were fixed in glutaraldehyde, embedded in LR-white resin (Polysciences) and sectioned to a width of 1–2 μM with a glass knife. Pink areas represent the regions with enzyme activity. Photograph taken with dark-field microscopy. (C) *In situ* hybridization of a *TA29* anti-mRNA probe with sections of anthers at stage 2 of flower development from transgenic plants carrying the *TA29–gus* gene. White grains outlining the tapetum represent RNA–RNA hybrids. (D) *In situ* hybridization of a *gus* anti-mRNA probe with anther sections from the same flower bud used for the previous hybridization as described in (C). Dark-field photograph. For *in situ* localization of mRNAs, the apices of the anthers were sliced off to facilitate fixative penetration, and anthers were embedded in paraffin. Sections of 5 μM width were hybridized *in situ* with single stranded ^{35}S-labelled anti-mRNA and ^{35}S-labelled mRNA (control) probes as described by Cox and Goldberg (1988). All photographs were taken with the same magnification factor (×100). To construct the *TA29–gus* gene a promoter cassette was engineered introducing a *Nco*I site to the first ATG of the TA29 open reading frame. The *Cla*I–*Nco*I fragment of 1.5 kb which was obtained was fused to the β-glucuronidase coding sequence (*gus*: Jefferson, 1987) followed by the 3′*ocs*, a 706 bp *Pvu*II fragment containing the polyadenylation site of the octopine synthase gene (Gielen *et al.* 1984). This chimaeric gene was inserted between the T-DNA borders of the plant expression vector pTTM3, which also contains the selectable marker gene for bialophos resistance *bar* (DeBlock *et al.* 1987). pTTM3 was mobilized into the *Agrobacterium* recipient C58C1Rif^R (pGV2260) (Deblaere *et al.* 1987) and the recombinant plasmid was used to transform *N. tabacum* var. Petit Havana SR1 (Maliga *et al.* 1973) with the method of leaf disk infection (Horsch *et al.* 1984).

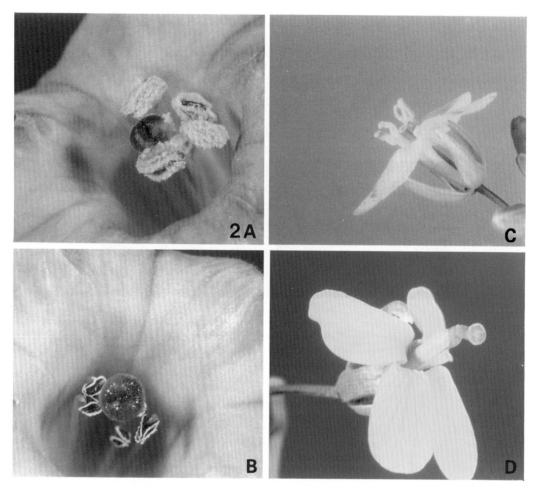

Fig. 2. Male sterile flowers of transgenic tobacco and oilseed rape. (A) Wild-type
N. tabacum Petit Havana SR1. (B) Flower of a transgenic male sterile plant. (C) Wild-
type *B. napus* var. Drakkar. (D) Flower of a transgenic male sterile plant carrying the
TA29–RNAase gene. The chimaeric *TA29–RNAase* genes were constructed by fusing
the *Cla*I-*Nco*I cassette with the coding sequence of the RNAase T1 gene (Quaas *et al.*
1988) and Barnase gene (Hartley, 1988). A 200 bp fragment containing the
polyadenylation site of the nopaline synthase gene (Depicker *et al.* 1982) was provided
at the 3′ end of the coding sequence. The chimaeric genes were cloned between the T-
DNA borders of the expression vectors pTTM6 (carrying RNAase T1) and pTTM8
(carrying Barnase). Both vectors contain also the bialaphos resistance gene (*bar*) as a
selectable marker. pTTM6 and pTTM8 were transferred separately to *N. tabacum* var.
Petit Havana SR1 and *Brassica napus* var. Drakkar using standard *Agrobacterium*
transformation procedures (Maliga *et al.* 1973; Horsch *et al.* 1984; Deblaere *et al.* 1987;
De Block *et al.* 1989). Transformants were selected on media containing $5 \, mg \, l^{-1}$ of
phosphinothricin and allowed to grow in a green house until flowering.

Taken together, these results show that the 1500 nucleotide 5′ region of the *TA29* gene contains the information for temporal and spatial regulation that is characteristic of *TA29* and that these regulatory elements are conserved in other crops.

Chimaeric *TA29–ribonuclease* genes induce male sterility

Ultrastructural and cytochemical studies showed that the morphology of the tapetum varies extensively in different taxa. Tapetal-cells either form a secretory epithelium or an invasive plasmodium around the microspores in the pollen sac. In both cases, the tapetum acquires specialized features while it is functioning as nutritive tissue for the microspores. During this period the ER and ribosomes become more abundant, the cytoplasm accumulates lipidic globules, and vesicles or vacuoles accumulate near the spores (Tiwari and Gunning, 1986).

The intensive interaction of this cell layer with the developing pollen indicates that a well functioning tapetum is essential for microspore development. The selective destruction of the tapetal-cells would demonstrate to what extent this tissue contributes to spore formation and to the differentiation and the function of other anther cell types. In addition, ablation of tapetal-cells would mimic naturally occurring male sterile mutants that exhibit a damaged tapetum.

To this end we introduced into tobacco and oilseed rape two chimaeric genes whose product should lead to the specific destruction of the tapetal-cells. We chose ribonucleases as cell-lethal enzymes because the activity of these proteins was expected to be detrimental for the cytoplasmic mRNA pool and thus for cell functioning. The lethal chimaeric genes were made by fusing the *TA29* promoter to the coding sequence of RNAase T1, a ribonuclease secreted by *Aspergillus oryzae*, or to Barnase, a ribonuclease secreted by *Bacillus amyloliquefaciens* (Quaas *et al.* 1988; Hartley, 1988). RNAase T1 and Barnase are two small monomeric enzymes of 104 and 110 amino acids respectively that are active at various ranges of pH and temperature and do not require cofactors. The fungal RNAase T1 is a guanosine-specific ribonuclease and contains two disulfide bonds, whereas the bacterial Barnase is an aspecific RNAase and lacks disulfide bonds in its chain.

The two chimaeric ribonuclease genes were individually introduced in plant vectors that also carry the bialphos resistance gene (De Block *et al.* 1987) and transferred into tobacco and oilseed rape by *Agrobacterium* mediated transformation. The tobacco and oilseed rape transformants grew like wild-type plants and developed completely normal leaves, stems, corollas and pistils. However, 2 out of 20 tobacco plants and 17 out of 24 rapeseed transformants carrying the *TA29–RNAase* T1 gene shed no pollen grains. Amongst the transformants carrying the *TA29–barnase* gene we obtained 106 male sterile tobacco plants out of a total of 115 plants and 10 male sterile rapeseed plants out of a total of 13 plants. Fig. 2 shows the difference in flower morphology between the wild-type and the male steriles in both species. The tobacco male sterile plants develop normal

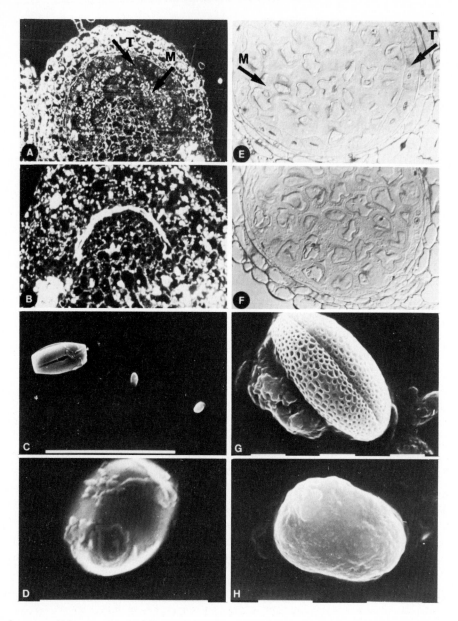

anthers until late stages of flower development, but at the moment of dehiscence they do not release pollen. In wild-type flowers of oilseed rape the stamens protrude from the corolla and stretch out above the pistil. By contrast, male sterile oilseed rape has smaller anthers and shorter filaments. It is noteworthy that male sterile flowers of the engineered oilseed rape are identical to the flowers of some naturally occurring male sterile *Brassica napus*.

We sectioned the anthers of the male sterile tobacco plants and compared them with anthers of the wild-type plants. Fig. 3A shows a transverse section of a tobacco anther in which the pollen sac is surrounded by a well developed tapetum

Fig. 3. Abnormalities in male sterile tobacco and oilseed rape anthers. (A) Dark-field photograph of anther section of untransformed tobacco plant. M, microspores enveloped in the callose wall; T, active tapetum. (B) Transverse section of anther from a male sterile tobacco plant carrying the *TA29–RNAase* gene. Dark-field photograph. (C) Scanning electron micrograph of pollen grains produced by wild-type (left) and male sterile (right) tobacco plants. White bar, 100 µm. (D) Higher magnification of male sterile tobacco pollen grain shown in (C). White bar 10 µm. (E) Bright-field photograph of anther section of untransformed oilseed rape. M, microspores; T, tapetum. Photograph taken with Nomarski interference microscopy. (F) Anther section of a male sterile transgenic oilseed rape plant carrying the *TA29–RNAase* gene. (G) Oilseed rape pollen grain from a wild-type plant. White bar 10 µm. (H) Scanning electron micrograph of pollen-like body from a male sterile oilseed rape plant. White bar 10 µm. Stage 1 tobacco anthers were fixed in glutaraldehyde, embedded in LR-white and sliced in 1–2 µm transverse sections with a glass knife. Oilseed rape anthers were harvested from 3 mm flower buds and processed like the tobacco anthers. Pollen grains were collected from dehisced anthers in open flowers.

and the tetrads are enveloped in a thick callose wall. By contrast, in a transverse section of anthers of a male sterile tobacco plant at the same stage of development, the pollen sac has caved in, the tapetal layer is not distinguishable and microspores are not detectable (Fig. 3B).

One of the tobacco plants carrying the chimaeric ribonuclease T1 gene shed very few pollen grains, which failed to germinate. Scanning electron microscopy showed three major differences with wild-type pollen. Whereas pollen of wild-type tobacco has an oblong shape and a regular surface with formation of the sulci and folding along them, the grains of the male sterile plant have a round shape without visible sulci, and the deposition of the exine on the surface is very irregular (Fig. 3C,D).

The male sterile tobacco plants failed to set seeds by selfing, even when several anthers were used for manual pollination. By contrast, when they were pollinated with pollen of a fertile control plant, seed set was normal, indicating that female fertility of the transformed plants was not affected.

A similar analysis was carried out to examine the morphology of the anthers of the sterile rapeseed plants in flower buds of 2.5 to 3 mm length. In accordance with the expression of the *TA29–gus* gene, it seems that in oilseed rape the expression of the chimaeric *TA29–RNAase* genes affects the tapetum at later stages of its development (Fig. 3E,F). After meiosis of the pollen mother cells the tapetum is still present as remnant tissue, with the cells devoid of cytoplasm. In wild-type plants, the tapetum is at this stage still a very active tissue. The locules of the pollen sac of the male sterile plants contain pollen grains that will degenerate at later stages of development (Mariani *et al.* 1990). It is not clear yet whether the disappearance of pollen is a consequence of tapetum death or if the ribonuclease is secreted in the pollen sac, enters the pollen grain and destroys the cell. At the moment of dehiscence, however, only a few pollen-like bodies are released. These grains fail to germinate and to pollinate the flower. E.M. studies show that these pollen-like structures lack the shape and the characteristic wall that is typical of this plant species (Fig. 3G,H).

To ascertain whether the male sterile phenotype is due to the expression of the *TA29–ribonuclease* genes, we examined the segregation of the male sterility trait in 2200 F_1 tobacco plants in the greenhouse and the offspring of oilseed rape plants in the field. Our results show that the sterile phenotype correlates with the presence of the chimaeric *TA29–ribonuclease* genes, and that the engineered male sterility trait is also stable under variable climatic conditions in the field (De Greef *et al.* unpublished data).

Taken together, these results demonstrate that the expression of the RNAase genes is confined to the tapetum and that this enzyme causes cell death. The consequence of the destruction of the tapetum is that the anthers become incapable of producing viable pollen grains. The chimaeric *TA29–ribonuclease* gene can therefore be successfully used to introduce male sterility into tobacco and oilseed rape and possibly also into other crops. To address this question, we transferred the *TA29–ribonucleases* genes into lettuce and chicory. Preliminary results confirm that transgenic plants carrying these chimaeric genes are male sterile, implying that male sterility can be engineered and extended to various crops.

Male sterility in the production of hybrid seeds

Hybridization of plants is recognized as an important process to produce offspring that combine the desirable traits of the parent plants. The resulting hybrid offspring often perform better than the parents with regard to different traits such as yield, adaptability to environmental changes, and disease resistance. This improved performance is called heterosis or hybrid vigor. For this purpose the process of hybridization has been used extensively to improve major crops such as corn, sugarbeet, sunflower and oilseed rape. However, the controlled cross-pollination of plants to produce a harvest of hybrid seeds, has been difficult to achieve for many crops.

Today's hybrid seed production employs, in many cases, natural male sterile mutants that have been isolated and identified in several plant species. As the male sterile plants do not produce pollen, they cannot sustain self-pollination and can only produce seeds when they receive pollen from neighbouring fertile plants. The seeds obtained by these obligatory crosses will give rise to a homogeneous hybrid F_1. The availability of naturally occurring male sterile lines bypasses the need for mechanical and hand emasculation. In some instances though, the mutations that lead to male sterility have side effects on the plants. The Texas male-sterile cytoplasm (cms-T) of maize, for example, is associated with the susceptibility to a fungal disease (Laughnan and Gabay-Laughnan, 1983; Levings III and Brown, 1989) which renders these male sterile plants worthless to the breeders. In other examples, the male sterility trait is not stable under variable climatological conditions. Furthermore, the transfer of mutations leading to male sterility is often troubled by the lack of an associated morphological marker, which would allow the trait to be followed more easily (Kaul, 1988).

We succeeded in overcoming the latter disadvantage by linking the

TA29–RNAase chimaeric gene to the *bar* gene under the control of a constitutive promoter (De Block *et al.* 1987). This gene confers the resistance to phosphinothricin (ppt), a selective agent contained in the herbicide Basta. Because the two genes are physically linked on the same T-DNA, they will segregate as a single locus. Therefore, the male sterile line can be maintained through a back-cross with fertile wild-type plants followed by the application of the herbicide. In this way half of the progeny survives, yielding a uniform population of sterile plants.

If the herbicide is not applied, the progeny obtained from the back-cross with a different line will be entirely hybrid and will be composed of 50 % male sterile and 50 % male fertile plants. This system for hybrid production is applicable to those crops in which the harvested product is the vegetative part of the plants, for example carrot, lettuce and cabbages. In these cases the offspring will provide a 100 % hybrid harvest, irrespective to whether plants are sterile or fertile. However, if the harvested product is the fruit or the seeds and the plants are self-pollinators, only 50 % of the offspring (the fertile plants) will be able to set seeds. The 50 % male sterile plants from this cross will not be able to produce fruit and/or seeds. This implies that in those crop plants where the harvested product is the fruit or the seeds it will be necessary to envisage a way to restore full male fertility in the hybrid offspring.

Discussion

The isolation of the tapetum specific *TA29* gene of tobacco permitted us to demonstrate the importance of tapetum cells for spore development. As the expression of the *TA29* gene is controlled at the transcriptional level (Koltunow *et al.* 1990), we could use the *TA29* promoter to direct the synthesis of a lethal gene product in the tapetal cells and evaluate the effect on spore development. We chose ribonuclease genes as cell lethal genes because we expected that their activity would be more detrimental to the cells than any other enzymatic activity. RNAases irreversibly cleave RNA in the cytoplasm, and do not require transport to other cell compartments to impair cell function. Interestingly, McClure *et al.* (1989) have reported that the style self-incompatibility gene products of *N. alata* have a ribonuclease activity and share sequence homology with RNAase T2 of *Aspergillus oryzae*. This cytotoxic agent would be taken up by the pollen carrying the same S-alleles of the style and would inhibit the growth of the pollen tube.

We introduced the chimaeric ribonuclease genes in tobacco, oilseed rape, lettuce and chicory. Transgenic plants are as viable as wild-type, but the anthers invariably fail to produce viable pollen. We could prove by genetic analysis that in both tobacco and oilseed rape the sterility was due to the presence of the chimaeric gene we had introduced.

The normal appearance of the male sterile plants indicates that transcriptional control of the *TA29* promoter is tightly regulated in the anthers and that there is no side effect of the ribonuclease gene in other parts of the plants. As the control of the *TA29* promoter is maintained upon reintroduction in heterologous systems such as oilseed rape, lettuce and chicory, the function of the *TA29* gene seems to be conserved amongst different plant species.

The engineered male sterile plants also have an agronomical value, as they can be used in crossing programs to produce hybrid seeds, particularly when natural male sterile lines are not available.

References

COX, K. AND GOLDBERG, R. (1988). Analysis of plant gene expression. In *Plant Molecular Biology: A Practical Approach*. (C. Shaw, ed.). Oxford: IRL Press, pp. 1–35.

DEBLAERE, R., REYNAERTS, A., HOFTE, H., HERNALSTEENS, L., LEEMANS, J. AND VAN MONTAGU, M. (1987). Vectors for cloning in plant cells. *Methods in Enzymology* **153**, 272–291.

DE BLOCK, M., BOTTERMAN, J., VANDEWIELE, M., DOCKX, J., THOEN, C., GOSSELE, V., MOVVA, N., THOMPSON, C., VAN MONTAGU, M. AND LEEMANS, J. (1987). Engineering herbicide resistance in plants by expression of a detoxifying enzyme. *EMBO J.* **6**, 2513–2518.

DE BLOCK, M., DE BROUWER, D. AND TENNING, P. (1989). Transformation of *Brassica napus* and *Brassica oleracea* using *Agrobacterium tumefaciens* and the expression of the *Bar* and *Neo* genes in the transgenic plants. *Pl. Physiol.* **91**, 694–701.

DEPICKER, A., STACHEL, S., DHAESE, P., ZAMBRYSKI, P. AND GOODMAN, H. (1982). Nopaline synthase transcript mapping and DNA sequence. *J. molec. appl. Genet.* **1**, 561–573.

GIELEN, J., DE BEUCKELEER, M., SEURINCK, J., DE GREVE, H., LEMMERS, M., VAN MONTAGU, M. AND SCHELL, J. (1984). The complete nucleotide sequence of the TL DNA region of the *Agrobacterium tumefaciens* plasmid pTIACH-5. *EMBO J.* **3**, 835–845.

GOLDBERG, R. B. (1988). Plants – novel developmental processes. *Science* **240**, 1460–1467.

HARTLEY, R. W. (1988). Barnase and Barstar – expression of its cloned inhibitor permits expression of a cloned ribonuclease. *J. molec. Biol.* **202**, 913–915.

HORSCH, R., FRALEY, R., ROGERS, S., SANDERS, P., LLOYD, A. AND HOFFMANN, N. (1984). Inheritance of functional foreign genes in plants. *Science* **223**, 496–498.

JEFFERSON, R. (1987). Assaying chimeric genes in plants: the GUS gene fusion system. *Pl. molec. Biol. Report.* **5**, 387–405.

KAUL, M. L. H. (1988). *Monograph on Theor. and Appl. Genet. Vol. 10*. Male Sterility in Higher Plants (Springer, Berlin.).

KOLTUNOW, A. M., TRUETTNER, J., COX, K. H., WALLROTH, M. AND GOLDBERG, R. B. (1990). Different temporal and spatial gene expression patterns occur during anther development. *Pl. Cell* **2**, 1201–1224.

LAUGHNAN, J. R. AND GABAY-LAUGHNAN, S. (1983). Cytoplasmic male sterility in maize. *A. Rev. Genet.* **17**, 27–48.

LEVINGS III, C. S. AND BROWN, G. G. (1989). Molecular biology of plant mitochondria. *Cell* **56**, 171–179.

MALIGA, P., BREZNOWITZ, A. AND MARTON, A. (1973). Streptomycin-resistant plants from callus culture of haploid tobacco. *Nature* **244**, 29–30.

MARIANI, C., DE BEUCKELEER, M., TRUETTNER, J., LEEMANS, J. AND GOLDBERG, R. B. (1990). Induction of male sterility in plants by a chimeric ribonuclease gene. *Nature* **347**, 737–741.

MASCARENHAS, J. (1990). Gene activity during pollen development. *A. Rev. Pl. Physiol. Pl. molec. Biol.* **41**, 317–338.

McCLURE, B. A., HARING, V., EBERT, P. R., ANDERSON, M. A., SIMPSON, R. J., SAKIYAMA, F. AND CLARKE, A. E. (1989). Style self-incompatibility gene products of *Nicotiana alata* are ribonucleases. *Nature* **342**, 955–957.

QUAAS, R., McKEOWN, Y., STANSSENS, P., FRANK, R., BLOCKER, H. AND HAHN, U. (1988). Expression of the chemically synthesized gene for ribonuclease-T1 in *Escherichia coli* using a secretion cloning vector. *Eur. J. Biochem.* **173**, 617–622.

SEURINCK, J., TRUETTNER, J. AND GOLDBERG, R. B. (1990). The nucleotide sequence of an anther-specific gene. *Nucl. Acids Res.* **18**, 3403.

TIWARI, S. AND GUNNING, B. (1986). An ultrastructural, cytochemical and immunofluorescence study of postmeiotic development of plasmodial tapetum in *Tradescantia virginiana* L. and its relevance to the pathway of sporopollenin secretion. *Protoplasma* **133**, 100–114.

INDEX OF SUBJECTS

Abscisic acid (ABA)
  autonomy of mutants, 14
  deficient mutants, in *Arabidopsis thaliana*, 8
  increase *in vivo*, guard cells, 188
  signal transduction pathways, 47–48
  stomatal closure, 182
Agrobacterium
  binary vector, resistance to kanamycin, 66
  overproduction of plant hormone genes, 240–241
  T-cytokinin gene, 28
  T-DNA, moss transformation system, 38–40
  T-DNA gene
    iaaH, 70
    insertional mutagenesis, 6, 111
    mapping onto *Arabidopsis thaliana* RFLP map, 63–75
Angiosperms *see* Flowering plants
Antibiotic-resistance, transformation system, 36–37
Antirrhinum
  advantages for genetic studies, 4–5
  GBF-like activity, 211
  haploid genome and genetic data, 4
Antirrhinum centroradialis mutation, 93
Antirrhinum deficiens, 98, 103–104
Antirrhinum floricaula, 93
Antirrhinum globosa, 98
Antirrhinum ovulata, 97
Antirrhinum pleniflora, 99
Antirrhinum sepaloidea, 98
Antirrhinum squamata, 93
Antirrhinum squamosa, 93
Antirrhinum sterilis mutation, 93
Antisense RNA
  and gene function, 125
  technology, tomato fruit softening, 117–127
  vectors, polygalacturonase (PG) cDNA fragments, 120
Arabidopsis thaliana
  abscisic acid deficient mutants, 8
  abscisic acid-insensitive genes
    abi-1, *abi-2*, and *abi-3*, 58
    mutants, 53
  Ac and *Ds*, two element transposon system, 68–70
  benzyladenine, 24
  colony hybridization, 58

cytokinin-insensitive mutants, greening phenotypes, 27
cytokinins, role during greening, 21
deleted *Ac*, germinal excision frequency, 68
*det*1 mutants, 21–29
development of system for efficient chromosome walking, 57–62
developmental and physiological processes affected by different classes of mutants, 11
double mutants
  abi mutants, 7
  hy mutants, 7
ecotypes Columbia and Landsberg *erecta*, 58–61
embryonic morphogenesis, 11–12
ethylene responses, 149–158
  locus *ETR*, LC5 gene, 154–155
  mutants, isolation, 152–154
etioplasts, 25
evocation of flowering, 12–13
evolutionary considerations, 110–111
excision markers, 66–67
flower development, 13, 89–115
  classes of mutations, 91–100
    apetala2 (*AP2*) locus, 94
    clavata loci, 93
    flower primordia and transition from inflorescence meristem, 92
    homeotic flower mutants, 95
    organ primordia and their subsequent specification, 93–100
    summary of phenotypes, 94
  gene transcription factors, 103
    agamous, DNA-binding proteins, 103–104
  genetic interactions between genes, 100–103
  models of floral morphogenesis, 106–110
  mosaic organs, 109
  specification of organ identity and organ primordia formation, 104–106
  structure, 90–92
G Box, 203
G-box-binding protein (GBF), 211–218
  similarity to wheat leucine zipper protein (HBP-1), 211–218
GBF-1
  leucine zipper protein, 212
  and other plant leucine zipper proteins, 215

sequence homology with HBF-1,
    212–216
gbf-1, single or low-copy sequence, 217
genome mapping, 45–56
    cloning of hormone biosynthetic and
        response loci, 53–55
    fingerprinting strategy, 48–50
    GA-defective mutants, 53
    physical mapping, 46–52
    and RFLP mapping, 52–53
gibberellin deficient mutants, 8
greening phenotypes, wild-type and *det1*
    plants, 23
GUS fusion, 70
haploid genome and genetic data, 4
homology to maize ABP cDNA, 129
in vivo footprinting, 211
2-isopentenyl adenine, effect on hypocotyl
    length, 24
kinetin, 24
λZAP cDNA expression library, 212, 213
Landsberg *erecta* transformants
    2'-SPT fusion, 66–67
    root transformation experiments, 67
LC5 gene, *ETR* locus, 154–155
light-dark expression of *cab* and *chs*
    promoters, seedlings treated with
    cytokinins, 25–27
maize autonomous element *Ac*, 64
monitoring *Ac* activity, 64–66
mutant *hls1*, 13
mutant induction, advantages for genetic
    studies, 6
NAM-resistant, 70
other excision markers, 67–68
phytochrome mutants *hy1*, *hy2*, *hy3* and
    hy6, 13
rbcS-1A promoter, 211
RFLP linkage maps, 58
root transformation procedure, 66
root-hair (*rhd*) mutants, 12
streptomycin fusion, 66
streptomycin-resistant seedlings, 65
targeted *vs* non-targeted tagging
    experiments, 70–72
transposon tagging system, 63–75
trichome formation, 12, 77–87
    development of trichomes on
        normal/mutant plants, 78–80
    gl1 mutants, 84–85
    GL1 as *myb*-related gene, 83
    gl2 and *gl3* mutants, 80–81
    isolation of *gl1*, 82–83
    myb-related genes, 85–86
    ttg and *gl1*, 81–82
vegetative growth, 90

yeast artificial chromosome (YAC) library,
    51, 57, 58–61
    cloning of high relative molecular mass
        DNA, 59
    YAC vector pYAC4, 58
Aspergillus
    pectate lyases, 230–231
    RNAase T1, 273
Aspergillus nidulans, *nim* gene,
    Physcomitrella patens, 41
Auxin-binding protein, 129–148
    epitopes, conservation, between maize and
        tobacco, 146
    homologous genes in monocot and dicot
        plants, 140–142
    monoclonal antibodies (mAbs) detect
        auxin-induced co-formational change,
        131–132
    role as auxin receptor, 143
    single locus, strawberry, 142
Auxin-binding sites, first identification, 130
Auxin(s)
    development of *P. patens* gametophyte, 35
    insensitive *diageotropica* mutant, 10
    and phospholipase A2 activity, 169
    receptors, characterisation, 130
Auxotrophs, mutants, lethal, 10

Bacillus amyloliquefaciens, Barnase, 273
Barley
    eceriferum mutants, 6
    haploid genome and genetic data, 4
Benzoic acid analogues, effect on mAbs,
    131–132
Benzyladenine, *Arabidopsis thaliana*, 24
Brassica napus (oilseed rape)
    chimaeric *TA29-ribonuclease* genes,
        induction of male sterility, 273–276
    tapetum-specific gene, molecular analysis,
        272–273
Brassica oleracea and *B. capestris*,
    sporophytic self-incompatibility (SSI),
    260–261
Brassicaceae, pistil components, 263

Ca^{2+}
    animal cells, 186
    identity of intracellular stores, 185–186
    release from internal stores, 184
    transients, and guard cell closure
        response, 186–187
$[Ca^{2+}]_i$
    evidence for fluxes across plasma
        membrane, 183–184
    Fluo-3, 177
    fluorescence intensity, two wavelengths,
        178

fluorescence ratio changes, 180
linkage to $[Ca^{2+}]_e$, 184
role in guard cells of *Commelina communis*, 177–190
signal-response coupling, 178
transients, visualisation and measurement, 179–183
Callose, structure and fluorescence, 248
Carrot
advantages for genetic studies, 5
inositol phospholipids
distribution, 161
and vanadate-sensitive ATPase activity, 167
phospholipase A2 activity, 169
protoplasts, antisense RNA experiments, 119
Cauliflower mosaic virus (CaMV)
promoter-driven neomycin phosphotransferase, 195–197
transgenic tomato *LAT52* and *LAT59*, 236
Caulonemal apical cell, moss, development, 32–34
cDNA subtractive hybridisation techniques, 40
Chalcone synthase (CHS)
activity, 191
β-glucuronidase
expression data, 201
reporter gene, 198
unit 1 for light-dependent expression, 202
Box II
cis-acting elements, control by *trans*-acting factors, 206–208
G Box family of *cis*-acting elements, 203
replacement by Box III, 204
sequence elements identical to functional core, 207
single base substitutions, 203–204
chimaeric CHS gene promoter constructions, 195–197
encoding by one gene, in *Petroselinum crispum*, 191–210
functional redundancy in parsley CHS gene promoter, 200–202
functional representation, 204
genomic footprinting to parsley CHS gene promoter, 197–198
in vivo footprinting, 197
induction of CHS gene expression by light, 193
light-dependent CHS gene promoter activity, four boxes, 198–200
in *Petunia*, 121
Chimeras
construction

grafting, 14
irradiation of heterozygous genotypes, 14
reciprocal crosses, mutants and wild-type, 14–15
loss on main stem, 6
Chloronemata, moss, development, 32–34
Chloroplasts, development, 21
Chromosome walking, 226–227
Arabidopsis thaliana, 57
Cinnamic acid 4-hydroxylase (C4H), flavonoid biosynthetic pathway, 192
Commelina communis, guard cells, role of $[Ca^{2+}]_i$, 177–190
Confocal laser scanning microscopy (CLSM), 178
Ca^{2+} release from endoplasmic reticulum/vacuole, 185
guard cells, 186
techniques, 179
Contigs
computer image, 50
multiplex hybridization, 52
see also Arabidopsis thaliana, genome mapping
4-Coumarate:CoA ligase (4CL), flavonoid biosynthetic pathway, 192
Crop plants
marker-based analysis of complex traits, 223–224
molecular markers and trait analysis, 221–226
RFLP analysis of complex traits, 219–228
see also Flowering plants
Cruciferae
pistil components, 263
pollen, 246
S-locus specific glycoproteins, 260–261
Cytokinins
benzylaminopurine, polymerase chain reaction (PCR), 40
chloroplast development, absence of light, 25
development of *P. patens* gametophyte, 35

Daucus carota, see Carrot
Developmental mutants, examples, 9–11
DNA polymorphism
systems for analysis, 220–221
technology for RFLP analysis, 220–221
DNASTAR, PROSCAN program, 155
Double (digenic) mutants, 7
Driselase, wall degrading enzymes, 166–168
Drosophila
antisense inhibition of genes, 236

sternopleural chaetal number, 222
transposon tagging, 63

Echinocloa crusgalli, epitope mapping, 132
Endosperm, *LAT52-GUS* (β-glucuronidase)
    fusions, 235
Epitope mapping
    Echinocloa crusgalli, 132
    kit, 129
    maize auxin-binding protein, 132
Erwinia, pectate lyases, similarity to *LAT56*
    and *LAT59* proteins, tomato pollen,
    230–231
Escherichia coli
    β-glucuronidase (GUS) reporter gene, 26
        in parsley protoplast system, 195
    LAT52, *LAT56* and *LAT59* proteins,
        production of polyclonal antisera, 235
Ethylene
    agonists and antagonists, 151
    biosynthetic pathway, 150, 153
    cloning of *ETR* locus
        analysis of cDNA clones, 154–155
        RFLP mapping, 154
        using chromosome walking, 154
    ethylene response mutants, *Arabidopsis
        thaliana*, 152–154
    functions, 149
    high affinity binding sites, 151
    responses, molecular mechanisms, 150–151
    synthesis, and ethylene-mediated
        responses, 150, 153
Ethylmethane sulphonate (EMS), 6
Etioplasts, *Arabidopsis thaliana*, 25

Flowering plants
    breeding programmes, marker-based
        selection, 225
    development, *Arabidopsis thaliana*, 89–115
    developmental mutants
        cell autonomous, 2
        genetic analysis, 6–9
        isolation, 1–19
        molecular techniques, 2–3
        mutagenesis, 5–6
        non-cell autonomous, 2
    double fertilization, 230, 246
    hybrid seed production, 276–277
    male sterility, engineered, 271–279
    plant lines
        dihaploid plants, 224
        'substitution mapping', 224
        successful dissection of complex traits,
            225–226
    plant lipid phosphorylation, phosphatidic
        acid, 162–163
    self-incompatibility

defined, 245–246
distribution, 249
evolutionary aspects, 249–252
heteromorphic *vs* homomorphic,
    247–248
molecular genetics, 252–261
S-locus in pollen, 248–249
sporophytic (SSI) *vs* gametophytic
    (GSI), 249–252
see also Crop plants
Fluorescence ratio photometry, in
    measurement of [Ca^{2+}]$_i$, 179–183
Fusicoccin, reversibility of guard cell turgor,
    187

Gametophores, mosses, 36
Gametophyte development, gene regulation,
    229–242
Gametophytic self-incompatibility (GSI),
    249–252, 253–260
λgbf recombinant phage, 213
GBF-1 and other plant leucine zipper
    proteins, 215
Gene cloning, 111–112
Genetic analysis of mutants, missing
    mutants, 8–9
Genetic model systems, characteristics, 3–5
Genome mapping
    'fine structure mapping' approach, 53–54
    'fingerprinting' strategy, 47
    physical map
        advantages for genetic studies, 46
        Arabidopsis thaliana, 46–52
    restriction fragment length polymorphism
        (RFLP), 47, 52–53
Gibberellin (GA)
    deficient mutants, in *Arabidopsis thaliana*,
        8
    maternal effect, 14
    signal transduction pathways, 47–48
β-Glucuronidase
    chalcone synthase (CHS), 198, 201, 202
    LAT52, *LAT56* and *LAT59*, fusion,
        229–242
β-Glucuronidase (GUS) gene, *Escherichia
    coli*, 26, 195
β-Glucuronidase-fusion plasmids, 061/H and
    061/I, 204
Glycine max, *see* Soybean
Glycoproteins, *S*-locus specific glycoproteins,
    257–259, 260–261
Grafting, in Solanaceae species, 14
Guard cells
    Commelina communis, role of [Ca^{2+}]$_i$,
        177–190
    loss of turgor, 187

signal transduction, 188
  pathways, 178
GUS *see* β-Glucuronidase

HBP-1, *see* Wheat leucine zipper protein
Hemicellulase, wall degradation, 166–168
Herbicide resistance, phosphinothricin, 277
Hormones, signal transduction, 10

Inositol phospholipids
  $[Ca^{2+}]_i$ role, 178
  metabolism in plants, 160–164
  phospholipase A2, activation, 168–169
  'PtdIns shuttle', 164–165
    cell wall loosening, 166
    light exposure, 165
    osmotic stress, 167
    vanadate-sensitive ATPase activity,
      166–167
  role in animal cells, 159–160
  as second messengers, 159–175
  stimulus-response pathways, 170–171
Irradiation, construction of chimeras, 14
2-Isopentenyl adenine
  effect on hypocotyl length, *Arabidopsis
    thaliana*, 24
  mimicking, 27

Japanese cedar, pollen allergens, similarity
  to *LAT56* and *LAT59* proteins, 229, 233

Kanamycin, monitoring *Ac* activity, 64
Kinetin, *Arabidopsis thaliana*, 24
Kunitz trypsin inhibitors
  KTi1 and *KTi2* gene transfer, 234
  similarity to *LAT52* gene, 234
  soybean, 229, 234–235

LAT52, *LAT56* and *LAT59*
  pollen-expressed genes, tomato, 229–242
    antibodies, 235–236
  sporophyte level, antisense *vs* wild-type
    controls, 236–287
LAT52, and *LAT59* coding regions,
  antisense constructs, 236
LAT-GUS chimeric genes, 241
Lemna gibba, cytokinin effects, 27
Light-regulated developmental pathways, 22
Light-regulated genes, expression in
  darkness, presence of cytokinins, 25–27
Liliopsida, self-incompatibility systems, 251
Lycopersicon chmielewskii, RFLP maps, 225
Lycopersicon esculentum, see Tomato
Lysolipids, plant cells, 161

Magnoliopsida, self-incompatibility systems,
  251

Maize
  Aux311, promoter, 139
  Aux381, length of intron-1, 139
  auxin-binding protein, 129–148
    cDNA probes, Southern blot
      hybridisations, 146
    DNA sequences compared, 138
    epitope mapping, 132
    impermeant auxins, 133
    multiple genes, 133–134
  C1 gene, 83
    myb homologue, 41
  D1 mutation, GA insensitivity, 14
  endogenous transposons, 63
  haploid genome and genetic data, 4
  pollen-expressed genes, 230
  sequences of genes
    from cDNA clones, 134–136
    from genomic clones, 136–137
  trait dissection, 225
  transposon Ac, in moss *Physcomitrella
    patens*, 38–40
Male sterility, engineered, angiosperms,
  271–279
Man, transcription factors, MCM1 and SRF,
  103
Mapped genes, sources, 10
Marker-trait linkages, construction of
  defined genotypes, 226
Maxam-Gilbert piperidine reaction, 197, 199
Microseris, hairy achenes, 110
Microsporogenesis *see* Pollen-expressed
  genes
Molecular markers, and trait analysis, crop
  plants, 221–226
Mosses, developmental genetic studies,
  31–43
Mutants
  cell and tissue autonomy, 13–14
  mutagenic procedure, tissue culture, 5–6
  physiological mutants, examples, 9–11
myb genes, *Arabidopsis thaliana*, *GL1* as
  myb-related gene, 83
myb oncogenes, identification of
  developmentally-relevant genes, using
  heterologous probes, 41

Naphthaleneacetic acid (NAA) derivatives,
  protein conjugates, 144
Neomycin phosphotransferase II (*npt*II),
  moss transformation, 37
Neurospora crassa, inositol phospholipids,
  162
Nicotiana alata, gametophytic self-
  incompatibility (GSI), 253–260

Nicotiana plumbaginifolia, G-box-like
   sequence, 211
Nicotiana tabacum see Tobacco
*npt*II, *see* Neomycin phosphotransferase II

Oenothera
   pollen-expressed genes, 230–231, 235
   S-alleles, 252–253
Oilseed rape *see Brassica napus* (oilseed
   rape)
Oryza sativa, see Rice

Parsley, *see Petroselinum crispum*
Pea
   advantages for genetic studies, 4–5
   haploid genome and genetic data, 4
Petroselinum crispum
   chalcone synthase gene, 191–210
   protoplasts, transient assay system for
      inducible gene expression, 194–197
Petunia
   advantages for genetic studies, 4–5
   antisense RNA experiments, 119
   chalcone synthase (CHS), 121
   GBF-like activity, 211
   haploid genome and genetic data, 4
   rbcS gene, removal of introns, 145
   S-alleles, 252, 259
Phenoxyacetic acid (POA) analogues, effect
   on mAbs, 131–132
Phenylalanine ammonia-lyase (PAL),
   flavonoid biosynthetic pathway, 192
Phosphatidic acid (PA), plant lipid
   phosphorylation, 162–163
Phosphinothricin, herbicide resistance, 277
Phospholipase A2
   inositol phospholipids, activation, 168–169
   plasma membrane, 168–169
Physcomitrella patens (moss)
   behaviour of maize transposon Ac, 38–40
   cytokinin-regulated sequences, 41
   developmental genetic studies, 31–43
   gametophyte development, 32–34
   identification of developmentally-regulated
      genes, 40–41
      using heterologous probes, 41
   transformation system, 36–37
Phytochrome
   primary mode of action, 27
   transcription of specific genes, 9–10
Piperidine reaction, Maxam-Gilbert, 197,
   199
Pisum sativum, see Pea
pKU plasmid family, moss transformation,
   38–40

Plasma membrane
   phospholipase A2, 168–169
   stimulus-response pathways, 170–171
Pollen
   allergens, ryegrass, exine localization, 236
   callose deposition, 248
   generative nucleus, transcriptional activity,
      241
   mutagenic treatment, 6
   promoters, direction of tissue-specific
      expression, 240–242
   reporter gene fusions, 241–242
   S-gene, tripartite hypothesis, 248–249
   self-incompatibility, *S*-locus in pollen,
      248–249
Pollen tubes, corkscrew growth pattern, 237
Pollen-expressed genes, tomato
   gene regulation, 229–242
   sequence motifs important for pollen,
      238–239
Polyethyleneglycol, treatment of protoplasts,
   195
Polygalacturonase (PG)
   activity in transgenic tomato, 120–122
   antisense inhibition of genes, 236
   biochemical role, 124
   biological role, 125
   cDNA fragments, antisense RNA vectors,
      120
   function in fruit ripening, 118–120
   pg antisense gene CaMV, 121
   reduction in levels, *pg* antisense gene,
      122–123
   RNA analysis, 123–124
Promoter-driven neomycin
   phosphotransferase (NPTII), cauliflower
   mosaic virus, 195–197
PROSCAN program (DNASTAR), 155
Protonemata, moss
   development, 32–34
      effect of light and gravity, 35–36
Prunus avium, pistil components, *S*-
   glycoproteins, 256, 263
Pseudomonas, polygalactonuronase gene,
   231

Quantitative trait locus (QTL)
   likelihood maps, 223
   linkage of markers, 224–225

Ragweed, allergens (tryptic peptides),
   similarity to *LAT56* and *LAT59*, 229, 233
RFLP analysis
   of complex traits, crop plants, 219–228
   differentiation of homo- and
      heterozygotes, 226
   multi-locus hybridisation probes, 221

random oligonucleotides/polymerase chain reaction (PCR), 221
simple-sequence variation and dinucleotide repeats, 221
single-locus hybridisation probes, 220
Ribonuclease genes, anther-specific expression, construction in tobacco and rapeseed, 271–272
Ribulose-bisphosphate carboxylase genes, promoters, L-, I- and G-boxes, 211
Rice
  advantages for genetic studies, 4–5
  haploid genome and genetic data, 4
Rosaceae, pistil components, 256, 263
Ryegrass, pollen allergens, exine localization, 236

S-locus specific glycoproteins, 260–261
  ribonuclease function, 257–259
Secale cereale, GSI, 248
Second messengers, inositol phospholipids, 159–175
Self-incompatibility
  gametophytic (GSI), *Nicotiana alata*, 253–260
  homomorphic
    homomorphic *vs* heteromorphic, 247–248
    origin, 249–252
  mechanisms, 245–249
  sporophytic (SSI)
    Brassica oleracea and *B. capestris*, 260–261
    vs gametophytic (GSI), 249–252
Signal transduction
  guard cells, 188
  pathways, key pathways, 169
SLSG, *see S*-locus specific glycoproteins
Solanaceae
  grafting, chimeras, 14
  pistil components, summary, 262
  pollen, 246, 252, 253–260
  S-glycoproteins, 257–259, 259–260
Soybean
  advantages for genetic studies, 5
  Kunitz trypsin inhibitors, 229, 234–235
Spinach, homology to maize ABP cDNA, 129
Sporophytic self-incompatibility (SSI), 249–252, 260–261
Stigma, style, *S*-glycoproteins, 260–261
Stimulus-response pathways, 170–171
  'cross-talk', 170–171
Stomatal pores, *see* Guard cells
Strawberry
  ABP gene
    single locus, 142

structure, 143
  cloning, advantages, 146
  homology to maize ABP cDNA, 129
Sunflower, inositol phospholipids and vanadate-sensitive ATPase activity, 167

T7 system, *LAT52*, *LAT56* and *LAT59* transcripts, 235
Tapetum, morphology, 273
Tapetum-specific gene, tobacco, molecular analysis, 271–272
Tissue culture, mutagenic procedure, 5–6
Tobacco
  anther-specific expression of cDNA clones, 271–272
  antisense RNA experiments, 119
  chimaeric *TA29-ribonuclease* genes, induction of male sterility, 273–276
  conservation of ABP epitopes, 146
  GBF-like activity, 211
  Kunitz trypsin inhibitors *KTi1* and *KTi2* gene transfer, 234
  male sterility in production of hybrid seeds, 276–277
  tapetum-specific gene, molecular analysis, 271–272
Tomato
  advantages for genetic studies, 4–5
  diageotropica, mutant, 151
  fruit ripening system, 118–119
  fruit softening, antisense RNA technology, 117–127
  G box, 203
  GBF, 211
  gene regulation
    LAT promoters, *cis* and *trans*-acting, 239–240
    sequence motifs, 238–239
  genetic mapping, 225
  grafting, 14
  haploid genome and genetic data, 4
  LAT52, *LAT56* and *LAT59*, pollen-expressed genes, gene regulation, 229–242
  mutants, 10
  orange fruit, RNA analysis, 123
  pJR16A transformants, analysis of selfed progeny, 122
  polygalacturonase (PG)
    activity in transgenic plants, 120–122
    antisense effect, specificity, 123
    biochemical role, 124
    biological role, 125
  polygalacturonase (PG) antisense gene
    copy number effect, 122
    and reduction in PG level, 122
    RNA experiments, 119

vectors, 120
pTOM13, 125
RFLP maps, 225
RNA analysis, 123–124
Transcription factor, yeast and humans, MCM1 and SRF, 103
Transposon tagging system, in *Arabidopsis thaliana*, 63–75
Triticum aestivum, *see* Wheat
L-tryptophan, effect on mAbs, 131–132

UV-absorbing substances, plants, 191–193

Wall degrading enzymes
Driselase, 166–168

hemicellulase, 166–168
see also Polygalacturonase (PG)
Wheat
advantages for genetic studies, 4–5
Em gene, ABA-responsive promoter, 207
leucine zipper protein (HBP-1), 211
proteins, EmBP-1, and HBP-1, 211, 212–216

Yeast, transcription factors, MCM1 and SRF, 103
Yeast artificial chromosome (YAC)
hybridization probes, 51
library, 51, 57

Zea mays, *see* Maize

The Journal of
Experimental
Biology

Review volumes

A series of volumes covering important areas in modern biological research. They are provided free to subscribers to the Journal.

Order no.

E81 **Cellular oscillators** (1979)
Edited by M. J. Berridge, P. E. Rapp and J. E. Treherne
306 pp   Price £15.00 (US $30.00)   ISBN 0 521 22948 0

E89 **Neurotransmission, neurotransmitters and neuromodulators** (1980)
Edited by E. A. Kravitz and J. E. Treherne
286 pp   Price £15.00 (US $30.00)   ISBN 0 521 23651 7 (sold out)

E95 **Glial–neurone interactions** (1981)
Edited by J. E. Treherne
240 pp   Price £15.00 (US $30.00)   ISBN 0 521 24556 7

E100 **Control and co-ordination of respiration and circulation** (1982)
Edited by P. J. Butler
319 pp   Price £25.00 (US $60.00)   ISBN 0 521 25348 9

E106 **Epithelial and cellular mechanisms in osmoregulation** (1983)
Edited by J. Phillips and S. Lewis
299 pp   Price £25.00 (US $60.00)   ISBN 0 9508709 0 0

E112 **Mechanisms of integration in the nervous system** (1984)
Edited by M. Burrows
357 pp   Price £25.00 (US $60.00)   ISBN 0 9508709 3 5

E115 **Design and performance of muscular systems** (1985)
Edited by C. R. Taylor, E. Weibel and L. Bolis
412 pp   Price £25.00 (US $60.00)   ISBN 0 9508709 5 1 (sold out)

E124 **Ion channels and receptors** (1986)
Edited by P. D. Evans and I. B. Levitan
392 pp   Price £20.00 (US $50.00)   ISBN 0 948601 02 7 (3rd impression)

E132 **Neural repair** (1987)
Edited by J. G. Nicholls and J. E. Treherne
300 pp   Price £35.00 (US $60.00)   ISBN 0 948601 07 8

E139 **The secretory event** (1988)
Edited by W. T. Mason and D. B. Sattelle
345 pp   Price £40.00 (US $75.00)   ISBN 0 948601 18 3

E146 **Principles of sensory coding and processing** (1989)
Edited by S. B. Laughlin
350 pp   Price £40.00 (US $75.00)   ISBN 0 948601 21 3

E153 **Synapse formation** (1990)
Edited by A. J. Aguayo and E. A. Howes
303 pp   Price £38.00 (US $65.00)   ISBN 0 948601 28 0

Volumes may be purchased from:
Portland Press Ltd, PO Box 32, Commerce Way, Colchester CO2 8HP, UK.

24-hour ordering by FAX: Colchester (0206) 549331/International +44 206 549331

590020